Vegetation Ecology of Socotra

PLANT AND VEGETATION

Volume 7

Series Editor: M.J.A. Werger

For further volumes:
http://www.springer.com/series/7549

Gary Brown • Bruno A. Mies

Vegetation Ecology of Socotra

Springer

Gary Brown
Department of Biology
College of Science
Sultan Qaboos University
Muscat, Oman

Bruno A. Mies
Department of Botany
Universität Duisburg-Essen
Gebäude S05 V05 E32
Essen, Germany

ISSN 1875-1318
ISSN 1875-1326 (electronic)
ISBN 978-94-007-4140-9
ISBN 978-94-007-4141-6 (eBook)
DOI 10.1007/978-94-007-4141-6
Springer Dordrecht Heidelberg New York London

Library of Congress Control Number: 2012937039

Printed on acid-free paper

Springer is part of Springer Science+Business Media (www.springer.com)

Preface and Acknowledgements

The impetus for this book came from the initial detailed studies carried out by Bruno A. Mies, which commenced with his first field trip to Socotra in the mid-1990s, nearly 20 years ago. Both authors have accumulated a considerable amount of field experience in the Arabian Peninsula, and Gary Brown has lived and carried out botanical research in three of its countries (Kuwait, UAE, Oman) for over 12 years since 1995, apart from travelling widely throughout the region.

As highlighted in the introduction, the Socotra Archipelago is a major biodiversity hotspot in the region on account of its remarkable endemic flora and fauna. Apart from dealing briefly with certain aspects of the flora itself, this book covers a variety of topics, including island biogeography, ecology, evolutionary biology, vegetation and conservation. Overviews of the lichen flora, based mainly on our own work, partly together with co-workers, and of the bryophytes are also given. There has been comparatively little vegetation work conducted throughout the entire region, even less so in an ecological context. Many of the studies that have been undertaken, some of which are of a very high standard, have been published in relatively obscure journals or volumes that are often difficult to access. Apart from summarising the results of our own work and more widely available publications, an important task of this book therefore was to trawl through lesser known information, filter out what appeared to us to be relevant, and present a comprehensive overview of the existing state of knowledge.

We hope that this book, in spite of its various shortcomings, will give a useful overview of the vegetation ecology of the remarkable island of Socotra. Furthermore, we hope that it will highlight the many gaps in our knowledge and therefore inspire more detailed studies. A final aim of this book, which we suspect might be viewed as being wildly optimistic given the reality on the ground, is that it will serve to promote conservation of the vegetation, and therefore a large part of the natural heritage of the archipelago as a whole, which is at serious risk of calamitous destruction and degradation.

Given the problems of transliterating Arabic/Socotri place names into English, we have striven at least to maintain consistency throughout the book, rather than to adhere to any of the "accepted" spellings, which are invariably a source of controversy.

We have even used the spelling "Socotra", rather than the more correct transliteration, "Soqotra", as most non-Arabic speakers would have difficulty in correctly pronouncing the Arabic "q".

In carrying out the studies for this book and right up until its completion, a large number of people provided assistance in many different ways. Bruno A. Mies would like to thank the various persons who accompanied him in the field, in particular Gaby Beyer, Friedrich E. Beyhl, Peter Hein, Gary J. James, John J. Lavranos, Thomas A. McCoy, Christian Printzen, Matthias Schultz and Hans and Helga Zimmer. He also acknowledges financial support provided by the Heinrich and Erna Walter-Stiftung (Germany), the Cactus and Succulent Society of America, the Deutsche Kakteen-Gesellschaft and the German BMBF Program Biolog E14. Gary Brown thanks Mohammed Al-Khamis (assistance with literature acquisition), Ahmed Al-Saaed (accompaniment on field trips), his wife Ina, Richard Porter and Torsten Weber. Sultan Qaboos University, Muscat, Oman granted him research leave to enable completion of the book.

Several persons kindly allowed us to use their photographs, including AbdulRahman Al-Sirhan, Frank Boltz, Abdulkadir Elshafie, John J. Lavranos and Wolfgang Wranik. These persons are mentioned in the relevant figure captions.

John J. Lavranos provided various comments on the text, and identified a number of plant species for us. Bernhard Pracejus reviewed the initial draft of the geology chapter, and Fiona Sewell edited the final manuscript. Many thanks to all at Springer for their patience and support.

We are especially grateful for the substantial support of Professor Marinus J.A. Werger, who provided detailed comments on the draft manuscript and was a great source of encouragement. All remaining errors and omissions are claimed by the authors, and we would appreciate any feedback or corrections that improve the contents.

Sultan Qaboos University, Muscat, Sultanate of Oman — Gary Brown
University of Duisburg-Essen, Duisburg, Germany — Bruno A. Mies

Contents

Chapter 1
Introduction

Abstract Due to its remarkable flora and fauna, the Socotra Archipelago has been the focus of a number of ecological studies dating back to the beginning of the nineteenth century. Its unique biodiversity and – until very recently at least – relatively intact ecosystems have earned it the status of a World Heritage Site, and the islands are frequently referred to as the "Galápagos of the Indian Ocean". Despite its small size, the archipelago is one of the major island biodiversity hotspots in the world with respect to its endemic flora, with roughly 37% of plant species and 15 genera unknown from elsewhere. The objective of this book is to summarise the existing state of knowledge on the vegetation in an ecological context. Apart from providing descriptive accounts of the various vegetation units, the book also deals therefore with the factors, predominantly abiotic, affecting the distribution and abundance of species. As with other isolated islands or archipelagos, Socotra is an ideal model system, relatively unaffected by many complicating factors, to test fundamental ecological theories from a wide variety of biological disciplines (including biogeography, evolutionary biology and colonisation).

Due to its remarkable flora and fauna, the Socotra Archipelago, which lies in the north-western part of the Indian Ocean north-east of the Horn of Africa, has been the focus of a number of ecological studies dating back to the beginning of the nineteenth century. Indeed, its unique biodiversity and – until very recently at least – relatively intact ecosystems based on age-old traditional practices (Cronk 1986) have earned it the status of a World Heritage Site (UNESCO 2008), and the islands are frequently referred to as the "Galápagos of the Indian Ocean". Despite its small size, the archipelago is one of the major island biodiversity hotspots in the world with respect to its endemic flora, with roughly 37% of species and 15 genera unknown from elsewhere (see Chap. 4). Although Socotra itself, by far the largest of the islands and the main focus of this book, is undoubtedly the most important in terms of biodiversity, the smaller islands also harbour remarkable endemic species, such as the extremely local *Euphorbia abdelkuri* on Abdalkuri and *Begonia samhaensis* on Samhah. Even the rocky islet of Saboniya can boast a minor claim to

G. Brown and B.A. Mies, *Vegetation Ecology of Socotra*, Plant and Vegetation 7,
DOI 10.1007/978-94-007-4141-6_1,

fame, albeit from an ornithological perspective, hosting the only known breeding colony of the globally vulnerable Socotra cormorant in the archipelago, although it breeds elsewhere very locally along the coast of mainland Arabia (IUCN 2010).

The objective of this book is to summarise the existing state of knowledge on the vegetation in an ecological context. Apart from providing descriptive accounts of the various vegetation units, the book also deals therefore with the factors, predominantly abiotic, affecting the distribution and abundance of species. Very little information is forthcoming on biotic interactions concerning the plants of Socotra, and so this aspect, equally crucial for comprehending an underlying order to vegetation, has been rather neglected. Regarding species nomenclature, this book closely follows the *Ethnoflora of the Soqotra Archipelago* (Miller and Morris 2004), a truly excellent effort that will remain the reference work for the foreseeable future. A few newly recorded species for the archipelago have been added, and taxonomy at the family level has been updated somewhat in accordance with the most recent edition of the *Flowering Plant Families of the World* (Heywood et al. 2007). Some colleagues may be disappointed that we have not used the opportunity to be more adventurous in our approach to the taxonomy of the vascular plants, as the *Ethnoflora* is quite conservative. However, although we readily agree that a number of species in the *Ethnoflora* could indeed be further segregated, we felt that this would be beyond the scope and remit of this book. In addition, it is also clear that any further splitting of taxa would be unfairly biased towards species that are reasonably well studied on account of their high profile, in particular succulents.

The account of the vegetation in Chap. 6 is purely descriptive and highly subjective. Little attempt has been made to force the described vegetation units into the strait-jacket of the Braun-Blanquet floristic association system, which enjoys a wide following in Central Europe for reasons explained in that chapter. In the same vein, quantitative approaches to vegetation analysis and description involving multivariate procedures have not been utilized because suitable data are completely lacking. Given the potential power of quantitative approaches, carefully designed studies could, however, help in the development of reliable models to shed light on important plant community processes in the archipelago. In attempting to identify underlying patterns in the vegetation of Socotra, but also to uncover peculiarities that may elucidate interesting ecological processes, a certain emphasis has been placed on comparisons with other related ecosystems in the wider region, most prominently the Canary Islands and Dhofar in southern Oman.

As with other isolated islands or archipelagos, Socotra is an ideal model system, relatively unaffected by many complicating factors, to test fundamental ecological theories from a wide variety of biological disciplines (including biogeography, evolutionary biology, colonisation), as underlined by Whittaker and Fernández-Palacios (2007). Due to its relative geographic remoteness and the problems that formerly (and again more recently) existed for researchers in reaching the islands, much basic work remains to be undertaken, and this book occasionally emphasises the substantial gaps in our knowledge. A large part of this work has to be based on fieldwork – unpalatable as that may appear to many funding agencies in the current research climate, dominated almost exclusively by biotechnological and molecular approaches. Although such "modern" techniques are of great benefit, the significance of Socotra as a "natural laboratory" for carrying out field research, which also gives rise to the important basic questions for molecular studies, should not be underestimated. Meticulously documented field observations are generally of enormous scientific value, and we hope this book in particular has profited from them. They are at the heart of fundamental conceptual models relevant to various aspects of ecology, including efforts to restore seriously damaged ecosystems.

Basic ecological fieldwork is therefore essential, if – from an idealistic point of view – further information can help to protect the archipelago from the "anthropogenic tsunami" (van Damme and Banfield 2011) that is currently sweeping over it, or – from a more pessimistic, but perhaps realistic perspective – to document the natural history of the archipelago before it is irrevocably destroyed. Apart from the fact that very few exhaustive studies on the vegetation of the islands have been carried out to date, the rapid deterioration of many of the more accessible ecosystems means that we shall be forever ignorant of their true nature. In this context, extreme caution is required in the interpretation of results from ecological studies, as the current status quo may not represent what is presumed to be the intact situation. On the relevant time-scale, which is not necessarily always that of a human being, natural vegetation processes are highly dynamic. However, the rapid degradation of the landscape in recent years and the associated reduction of perennial plant cover and simplification of the vegetation structure are ominous indicators of the massive detrimental anthropogenic influences afflicting the archipelago, as has been comprehensively documented for the whole North-African–Arabian region by Le Houérou (1996). In this context, it should be emphasised that solutions to these problems will not be easy to achieve, because combating "desertification" is also a major socio-economic challenge (Agnew and Warren 1996).

Chapter 2
Topography, Climate and Soils

Abstract The Socotra Archipelago is located in the north-western part of the Indian Ocean, close to the Horn of Africa. Socotra itself (ca. 3,600 km^2) can be divided into three main topographical regions: (1) the granitic Hagghier mountains; (2) limestone plateaus, which occur between 300 and 700 m and occupy by far the largest part of the island; and (3) coastal plains. The arid tropical climate of Socotra is characterised by pronounced seasonal, altitudinal, spatial and inter-annual variability, with the seasonally reversing monsoons exerting a major influence on weather patterns. The overall arid macroclimate is also greatly modified by the diverse topography of the island and the extent of cloud cover. Rainfall is the chief form of available water at the lower elevations, either directly or indirectly through surface and subsurface redistribution. However, at higher altitudes, where forest is developed, fog and mist provide an important input of moisture, substantially augmenting the amount of plant-available water. Due to the arid nature of the climate, soils are poorly developed over much of the island. Pedogenesis involving chemical processes and the synthesis of organic matter is mainly restricted to the upper mesic montane zone where there is dense vegetation cover and the climate is more humid.

2.1 Introduction

The Socotra Archipelago is located in the north-western part of the Indian Ocean, close to the Horn of Africa and to the south of the Arabian mainland (Fig. 2.1). It consists of the main island of Socotra itself (ca. 3,600 km^2), the much smaller islands of Samhah (ca. 45 km^2) and Darsah (ca. 17 km^2), known together as "The Brothers", as well as Abdalkuri (ca. 150 km^2), the western-most island of the archipelago. Also included are the two islets of Ka'al Firawn and Saboniya. In total, the archipelago covers a surface area of about 3,800 km^2. A cartographical survey was conducted by the British military in the 1960s, leading to the production of the first detailed topographical map (Royal Geographical Society 1978). The project was subsequently continued by the Russian military (Russian Military Map 1978).

G. Brown and B.A. Mies, *Vegetation Ecology of Socotra*, Plant and Vegetation 7,
DOI 10.1007/978-94-007-4141-6_2,

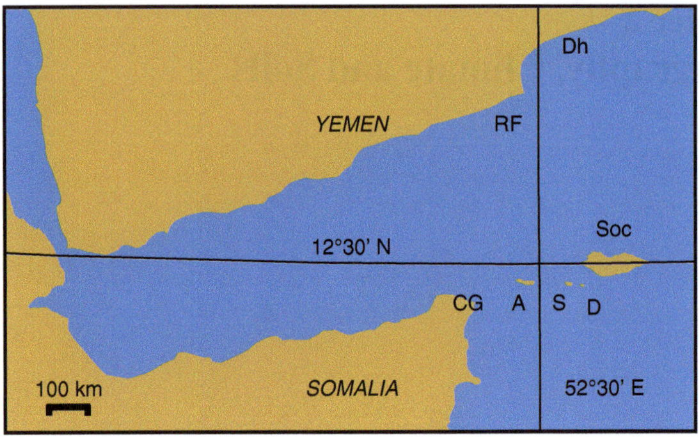

Fig. 2.1 Location of Socotra off the Horn of Africa. Abbreviations: *A* Abdalkuri, *CG* Cape Guardafui, *D* Darsah, *Dh* Dhofar, *RF* Ra's Fartaq, *S* Samhah, *Soc* Socotra

Socotra, by far the largest island, is about 133 km from west to east, has a maximum north–south breadth of 43 km, and accounts for ca. 94% of the surface area of the archipelago (Fig. 2.2). It is located between latitude 12°42′35″ in the north and 12°17′50″ in the south, roughly equidistant from the Tropic of Cancer in the north and the Equator in the south, and longitude 53°18′16″ in the west and 54°32′03″ in the east. The distance from Socotra to the nearest point on the African mainland, Cape Guardafui at the north-eastern tip of Somalia (Horn of Africa), is 232 km. The far west of Abdalkuri lies only about 95 km from Cape Guardafui, and is therefore a few kilometres closer to the African mainland than to Socotra. Although the archipelago belongs territorially to Yemen, the closest point on Socotra to the Arabian mainland is Ra's Fartaq in southern Yemen, some 351 km away. As noted in Chap. 4, a number of Socotran plant taxa have close relatives in Dhofar, southern Oman, possibly in part due to the geographical proximity of the two areas prior to the Gulf of Aden rifting (see Sect. 3.2), and some of these taxa also extend westwards into the adjacent Hawf region of southern Yemen. The Dhofar border with Yemen is about 440 km away from the nearest point on Socotra.

2.2 Topography

The present structure and morphology of Socotra are a result of post-Lower Miocene uplift with arching, block-faulting and tilting (Beydoun and Bichan 1970), although some authors (see Sect. 3.9) have suggested that the granite core of the Hagghier mountains was possibly subject to Cretaceous uplift. In accordance with Popov (1957), Socotra can be divided into three main topographical regions, as indicated in Fig. 2.2.

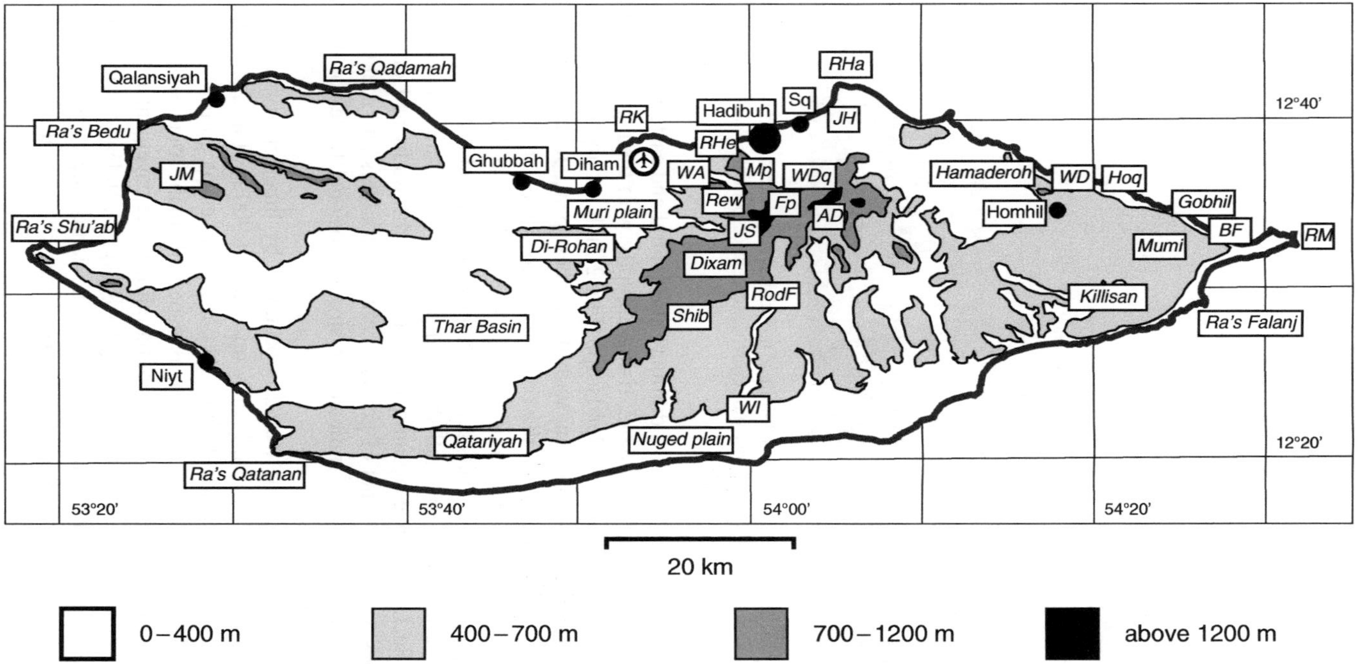

Fig. 2.2 Place names and locations mentioned in the text. Abbreviations: *AD* Adho Dimele, *BF* Bandar Fikhah, *Fp* Fieri peaks, *JH* Jebel Hauwari, *JM* Jebel Ma'alah, *JS* Jebel Skent, *Mp* Muqadrihon pass, *Rew* Rewged (Rughid), *RHa* Ra's Haulaf, *RHe* Ra's Hebak, *RK* Ra's Kharmah, *RM* Ra's Mumi, *RodF* Rokeb di Firmihin, *Shib* Shibehon plateau, *Sq* Suq (Shiq), *WA* Wadi Ayhaft, *WD* Wadi Dizyaf (Wadi Shi'faar), *WDq* Wadi Dihzafaq (Wadi Denegen), *WI* Wadi Ireh

Fig. 2.3 The granitic Hagghier mountains dominate the central and eastern parts of the island. They appear mainly white due to the dense growth of crustose lichens (especially *Pertusaria* spp.)

1. The granitic Hagghier mountains dominate the central and eastern parts, and are an impressive feature of the island, rising abruptly with their jagged pinnacles and peaks from the surrounding landscape (Fig. 2.3). It is only here on the island that the altitude exceeds 1,000 m, with the highest peak, Jebel Skent, at some 1,550 m. The area is strongly dissected by deep wadis, sheltered gullies and cliffs, but gentler slopes also occur. Boulder-strewn slopes at the base of the massif can be quite extensive. Soils are generally very thin and barely developed in the steeper parts, except in small sheltered pockets, but fertile red soils (cambisols – see Sect. 2.4) have accumulated locally in the wadis and on the more gentle slopes. Large amounts of gravels, cobbles, boulders and rocks that have been deposited in the wadis indicate the strong erosional forces of occasional heavy rain and flooding.
2. By far the largest part of the island is occupied by limestone plateaus, which occur between 300 and 700 m, locally up to about 900 m (Fig. 2.4). The individual plateaus are often undulating and cut by numerous gullies, ravines and cliffs, which makes walking in many areas, especially to the south of the Hagghier massif, very arduous. Extensive areas of bare limestone pavement are evident on many of the plateaus (e.g. Dixam, Qatariyah, Rewged), and sink holes and caves are scattered throughout. Karstification of the landscape is therefore a pronounced feature. In some places, the surface has been worn smooth, but it is

Fig. 2.4 Limestone plateaus occupy large parts of the island

usually rough and fissured, pockmarked by erosion. Locally, the limestone is interrupted by small areas of sandstone. In the east of the island, the limestone plateau extends to the coast, and impressive, steep cliffs are developed that plunge up to 400 m to the sea. To the south, the plateau drops abruptly up to several hundred metres onto the Nuged plain, forming an extensive escarpment. Soils are at best poorly developed on the limestone plateau, mainly in rock fissures, cracks and other such sheltered sites, but in some larger depressions, thin deposits of soil (rendzina) may accumulate. In general, however, the surface topography is characterised by the eroded limestone bedrock, and any soils that do form are usually seriously deficient in organic material.

3. Coastal plains are developed where the mountains and limestone plateaus do not reach the coast. The most extensive of these, the Nuged plain, runs parallel to the south coast and is about 70 km in length and up to 5 km wide (Fig. 2.5). Smaller plains, separated from each other where the limestone plateau juts out to the immediate coastline, are found along the northern and western coasts. They are typically quite narrow, usually less than a few kilometres in width, and in the north are dissected by numerous run-off channels, gullies and shallow wadis. The plains are covered by Pleistocene to Holocene sediments consisting of sands, gravels and coarser materials, depending on the location.

Fig. 2.5 Nuged plain on the south coast. December 2008

2.3 Climate and Weather Patterns

In-depth information on the climate of Socotra is largely lacking, and most published studies give rather general accounts, with more specific details restricted to localised areas. These studies have enabled broad statements to be made concerning variables such as wind and temperature, but contradictions and gaps remain, particularly in respect of precipitation. It is also possible to infer a certain amount of information on the climatic situation of specific ecosystems from the prevailing vegetation (at least where reasonably intact), especially when compared to analogous vegetation units from elsewhere in the world. However, detailed information on the climate and its variability, essential for an understanding of ecosystem processes and also for conservation purposes, requires a network of meteorological stations providing reliable measurements over a longer period of time from various parts of the island.

Useful overviews of the general climate have been provided by Kopp (1999), Mies (1999a, 2001), Miller and Morris (2004) and, in particular, Cheung and DeVantier (2006). An assessment of monsoonal influences on precipitation and vegetation, in part summarising the work of other authors, but also including recent data obtained from various weather stations over an altitudinal range from 3 to ca. 800 m asl from the year 2002 to 2006, has been presented by Scholte and de Geest (2010). Fleitmann et al. (2004) and Shakun et al. (2007) give details on the Quaternary climate history of Socotra.

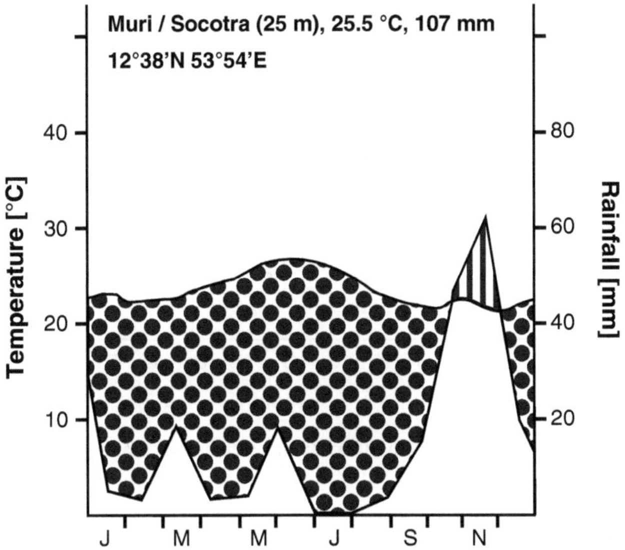

Fig. 2.6 Climate diagram based on data from various literature sources. Measurements were taken on the Muri plain in the north of the island

The arid tropical climate of Socotra is characterised by pronounced seasonal, altitudinal, spatial and inter-annual variability (Mies 1999a), with the seasonally reversing monsoons exerting a major influence on weather patterns (see Gadgil 2003). The overall macroclimate is also greatly modified by the diverse topography of the island and the extent of cloud cover, as reiterated by Mies (1999a) and by Scholte and de Geest (2010). The distinctly arid nature of the climate of large parts of the island is underlined by the fact that potential evapotranspiration greatly exceeds precipitation during most of the year (Fig. 2.6). Mies (2001) estimated that ca. 80% of Socotra receives a mean annual rainfall of less than 200 mm. More recent data reported by Scholte and de Geest (2010) show that mean annual rainfall from 2002 to 2006, measured at a network of stations below 800 m, was about 216 mm. However, there was considerable geographical variability, ranging from just 67 mm (Qalansiyah) to about 400 mm at Mathre on the northern coastal plain. Interestingly, the amount of rainfall does not appear to be correlated with altitude, as the station that registered the largest amount (Mathre) was located just 20 m asl, whereas a station at 760 m asl received only 168 mm.

Rainfall is generally essential for terrestrial plant growth, but due to its overriding importance in arid desert ecosystems, it has been described as the "master input" by Noy-Meier (1973), and three aspects deserve special mention: (1) arid deserts are systems with a discontinuous input; (2) arid deserts are systems with a stochastic input; (3) edaphic factors in arid ecosystems greatly modify the water regime, including infiltration and storage capacity, horizontal and vertical redistribution,

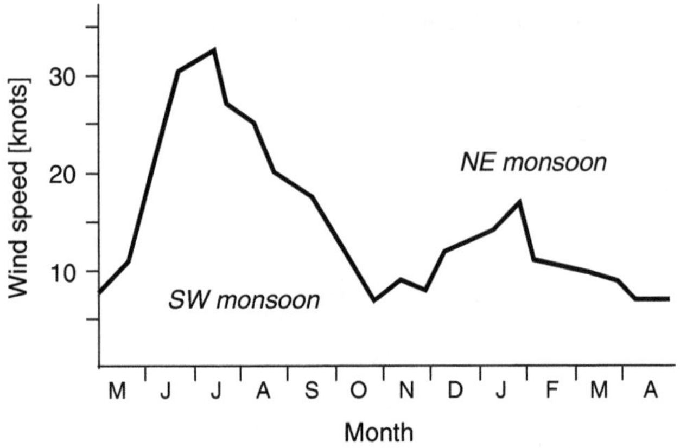

Fig. 2.7 Monsoon wind speeds in the northern Indian Ocean over the year (Based on Haake et al. 1993)

and the potential use of rainwater for biomass production. This last point is essential for understanding ecosystem processes, and also for vegetation restoration attempts (see Sect. 7.5.1).

As indicated above, the annual weather pattern of Socotra is governed to an overwhelming extent by the seasonal shift of the inter-tropical convergence zone (ITCZ), which gives rise to the predominance of the SW monsoon during the summer months, and the NE monsoon in the winter. The winds associated with the SW monsoon are overall much stronger than those of the NE monsoon, with a relatively calm period occurring during the spring and autumn transitional periods, i.e. from February to May as well as during September and October (Fig. 2.7). In the following, the complex weather patterns on Socotra are summarised in very general terms. Information on temperatures and rainfall applies mainly to the areas below 800 m.

In winter, moderate north-easterly winds are a consequence of the atmospheric pressure gradient that develops between the high-pressure cell over the Eurasian continent and the low-pressure cell over the southern Indian Ocean, at its furthest south situated at roughly the same latitude as northern Madagascar. This is the season of the north-east winter monsoon on Socotra, coinciding with the coolest months of the year, December to February. Mean monthly temperatures are usually below 28°C in most locations. At the highest elevations, nocturnal temperatures can drop to below 10°C on some days, although frosts appear to be unknown. Wind speeds rarely exceed 35 km h^{-1} during the winter. Due to the moist, cool air arriving from the north-east, the northern side of the Hagghier mountains is shrouded in rain clouds and mist (Fig. 2.8), whereas on the leeward side, only patchy cloud formation usually occurs, because the mountains act as an effective barrier to the clouds. Rainfall is therefore concentrated in the northern part of the island during the winter. As already indicated by Mies and Beyhl (1998), the main rainy period in the north is between November and February, and more specific data evaluated by Scholte and de Geest (2010) for the years 2002–2006 show that most rainfall was

Fig. 2.8 Dense cloud cover over the north of Socotra. December 2008

received in November. Typical rainfall amounts during the winter at lower to medium elevations range from about 45 to 350 mm.

In the spring transitional period between the two main monsoon seasons, and as the ITCZ has begun its seasonal shift northwards, moderate south-westerly winds predominate as the northern tropical and subtropical landmasses gradually warm up. North-westerly winds are also occasionally recorded. Late spring (late April and May) is the warmest period of the year, with mean daily temperatures reaching ca. 31°C during May. Relative humidity is generally high, with values sometimes exceeding 95%. A distinct rainy period occurs during the late spring, although the amounts received are much less than in the early winter.

In summer, the ITCZ has migrated northwards, and assumes its most northerly position over southern Pakistan by July. The summer monsoon has typically commenced by early May. The wind direction is from the south-west, and increases noticeably in speed as the month progresses. When the monsoon is at its most intense, wind speeds are on average more than 50 km h^{-1}, with substantially stronger gusts (up to hurricane-force) interspersed. Own measurements in the northern wadis of Socotra at the beginning of July 2002 recorded gale-force winds exceeding 160 km h^{-1}. These extreme windy conditions meant that in the past, Socotra was often cut off during the summer, and life in the northern part of the island still becomes exceedingly uncomfortable. Activities such as fishing virtually grind to a halt. The winds originate from Africa, and are mainly hot and dry as they pass over Socotra. As the summer progresses, the southern half of the island becomes increasingly affected by cloud cover, which often envelops large parts of the south and the

Hagghier mountains (see Scholte and de Geest 2010), although this is not accompanied by any significant amounts of rainfall, at the lower altitudes at least.

With the onset of the autumn in late September, the ITCZ retreats southwards, and this is marked by a decline in the intensity of the monsoon. During the autumn transitional period (mainly late September and throughout most of October), a second distinctly warm spell usually occurs, with mean temperatures around 29°C at the lower altitudes.

Rainfall is the chief form of available water at the lower elevations, either directly or indirectly through surface and subsurface redistribution. However, at higher altitudes, where forest is developed, fog and mist provide an important input of moisture, substantially augmenting the amount of plant-available water. No specific data are available to quantify this additional source, but Mies (2001) estimates it to be in the region of at least twice the amount of annual rainfall in the forested mesic montane zone (and so in excess of 800 mm, probably even equivalent to as much as 1,000 mm total precipitation). Beyhl and Mies (1996), Mies and Beyhl (1998) and Mies (2001) have emphasised the crucial role of fog for vegetation development at the higher altitudes, and also for the abundance of epiphytic and even epiphyllic lichens and bryophytes.

Quantifying the amount of fog precipitation is problematic for various methodological reasons, as is the precise assessment of its ecological significance in some cases, because of the close interrelationship with other factors. On the northern slopes of several of the Canary Islands, which receive a regular input of moisture from the trade winds, various types of evergreen laurel forests are developed. Mean annual rainfall on these slopes ranges from about 350 to 900 mm, and Kämmer (1974) estimated the amount of additional fog precipitation intercepted by the trees to be in the region of 300 mm. Miller and Morris (1988) underlined the importance of trees and vegetation in condensing moisture on the escarpment forest of Dhofar, a unique deciduous cloud-forest formation characterised by the endemic *Anogeissus dhofarica*, and located in an overall arid macroclimatic area. Rainfall is between 200 and 500 mm, and the coolest monthly mean temperatures (ca. 27–28°C) are recorded in January and February. This forest is undoubtedly developed below the lowest extreme in terms of rainfall requirements, as also shown by Hildebrandt et al. (2007), and therefore relies heavily on the monsoon mists in the summer for its survival. It is probably for this reason that it is a deciduous forest type, coming into leaf during the moist late-summer months, in contrast to the mainly evergreen types of Socotra and the Canary Islands. Equally important for the development of forest on Socotra and the other above-mentioned areas is regular high cloud cover, which dramatically reduces radiation loads, and thus temperatures and evapotranspiration, as made abundantly clear in the case of the escarpment forest in Dhofar, which is dormant during the cooler, but dry winter period. This has also been demonstrated in Dhofar by detailed ecohydrological studies (e.g. Hildebrandt et al. 2007).

Figure 2.9 shows a 24-h cycle of measurements (end of March 1996) of relevant climatic parameters from the Hagghier mountains at 1,050 m asl (taken from Mies and Beyhl 1998), which the authors believe to be representative of the situation for a large portion of the year, as supported by cloud cover data provided by Scholte and

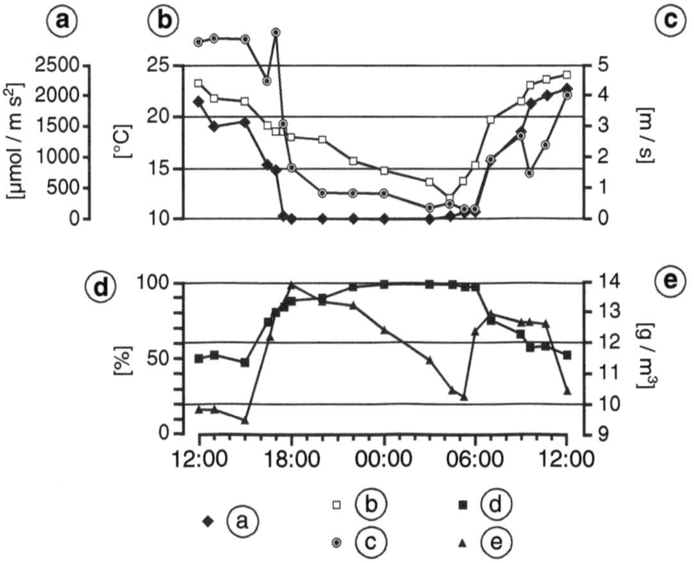

Fig. 2.9 Time course of important meteorological parameters over a 24-period in the Hagghier mountains at 1,050 m asl. Abbreviations: *a* global radiation, *b* air temperature, *c* wind speed, *d* relative humidity, *e* absolute humidity

de Geest (2010). It is interesting to note the sudden drop in temperature and wind speed after 18:00. This is accompanied by an equally steep rise in relative humidity, which is at, or close to, 100% for most of the period of darkness. The nocturnal longwave radiation loss to the atmosphere is so great that moisture will readily condense on any structures such as vegetation or the soil surface in the form of dewfall. One of the consequences of this additional input of moisture is that substrate-water availability is substantially higher in the mountains.

In their discussion of the cloud forest of the Canary Islands, Walter and Breckle (1999) emphasised an aspect that is probably also highly relevant to the cloud forest of Socotra, namely that the frequency of moisture input is just as important as the overall amounts of precipitation received, if not more so. As underlined by Mies (1999a), also important for the overall hydrological situation on Socotra is that the large amounts of precipitation received at higher altitudes are retained on the island due to absorption by the soil and vegetation, and at lower altitudes due to percolation into the karstic limestone.

Cloud cover also plays an important role at lower altitudes, i.e. on the limestone plateau (Fig. 2.10), as indicated by fog measurements taken during the SW monsoon by Scholte and de Geest (2010). Depending on aspect and location, these authors collected between 0.18 and 10.14 l m^{-2} per day, reaching up to the equivalent of 10 mm of rainfall per day. This moisture input, and in particular its regular occurrence, probably play a key role in the occurrence of a relative abundance of epiphytic lichens on trees on the plateau. However, as noted by Popov (1957), it is possible that the strong desiccating winds may prevent development of phanerogams on exposed limestone plateaus, which are often conspicuously poorly vegetated.

Fig. 2.10 Cloud cover and mist are a regular feature on many parts of the limestone plateau. December 2008

2.4 Soils

Due to the overall arid nature of the climate, soils are poorly developed over much of the island (Fig. 2.11). Pedogenesis involving chemical processes and the synthesis of organic matter is mainly restricted to the upper mesic montane zone (above ca. 800–1,000 m) where there is dense vegetation cover and the climate is more moist. At medium and lower altitudes, the parent bedrock is very much in evidence over extensive areas, but limited soil formation has taken place over large tracts on the plains and in basins. Small pockets of humus and "terra rossa"-like substrates (see below) accumulate in rock crevices and other such favourable microsites, and these are of enormous ecological significance for the establishment of many plant species. It is important to realise that large-scale soil formation involving chemical and biological processes once took place during more favourable climatic periods (pluvials), especially during the pre-Holocene. The last extended period of high rainfall in the region probably occurred between about 6,000 and 10,500 years ago (Fleitmann et al. 2003), but since the general aridification of the climate, soil degradation has occurred, exacerbated substantially by recent human activities (see Sect. 7.3.2), as has been the case in other arid and semi-arid parts of the world. Recent soil formation processes over much of the island are therefore restricted primarily to mechanical weathering, i.e. the disintegration of the parent rock into smaller particles with the same properties, rather than chemical or biological processes.

Fig. 2.11 Only very limited soil formation has taken place over large areas, so that the parent material is very much in evidence

However, an important physico-chemical process associated with the aridification of the climate in the Arabian region during the Quaternary is the secondary accumulation of calcium carbonate, which can be viewed as an indicator of soil degradation, along with loss of soil structure. Pietsch (2006) assumes that all the calcium-rich soils investigated by her on Socotra are of recent origin. Calcretes are near-surface accumulations of predominantly $CaCO_3$ in unconsolidated sediments, sedimentary rocks and soils (Goudie 1973), and various types are extensively distributed at lower altitudes on Socotra over carbonate-rich parent material. Rainfall chemistry appears to be a particularly important control on carbonate distribution in the regolith, and carbonates are generally thought to precipitate in areas where annual rainfall is less than about 400–500 mm. The regular alternation of short periods of excess water followed by drought is thought to be particularly beneficial for the formation of calcretes (Eren et al. 2008).

In accordance with the classification system of the recently published World Reference Base for Soil Resources (IUSS Working Group WRB 2006), a number of main reference soil groups (RSG) play an important role on Socotra, namely cambisols, calcisols, fluvisols, arenosols, leptosols and regosols. Detailed soil taxonomic information on several specific sites on Socotra is available from Pietsch (2006), Pietsch and Kühn (2009) and Pietsch and Morris (2010).

Cambisols, relatively young soils derived from various rock types, combine a wide range of different soil types, but all are characterised by at least the beginnings of horizon differentiation in the subsoil. Brownish discoloration, increasing clay fraction and carbonate removal are other general features of this group. These soils are the most widespread group on Socotra at lower altitudes, but also high in the mountains. At lower elevations, cambisols are often deep red in colour ("red soils", in part equivalent to the relictual "terra rossa" soils that are common in the Mediterranean region – see Pietsch 2006). This is due to the enhanced formation of haematite, and occurs when limestone-rich soils weather to liberate large amounts of iron oxide that was closely bound to clay minerals (Durn 2003). It involves the soil formation process known as "rubification", which is associated with cool, wet winters alternating with warm, dry summers, i.e. occurring in a typical Mediterranean-type climate. During the moist winter period, carbonate dissolution in the upper soil layers takes place, which in turn leads to the oxidation of iron (accounting for the orange to red colour), and in the summer, high evaporation rates lead to the accumulation of calcium carbonate and the precipitation and irreversible crystallisation of iron oxide (haematite).

The development of cambisols is usually associated with moist climatic conditions, which would have prevailed on Socotra during the last pluvial. Since this time, i.e. over the past ca. 6,000 years, soil evolution processes that can be interpreted as regressive have prevailed, including the accumulation of calcium carbonate and loss of soil structure, leading in part to the formation of calcisols.

Calcisols (which mainly belong to the calcids in the US soil taxonomy) are widespread in arid and semi-arid regions that have calcium carbonate-rich parent materials. Calcisols are developed where capillary action facilitates the movement of calcium carbonate to the soil surface, and are usually the result of more recent soil formation processes. On Socotra, they are fairly widely distributed, and have been described in detail by Pietsch (2006) and Pietsch and Morris (2010) at Homhil. They are generally silty, but also partly clayic.

Fluvisols are young, azonal soils typically formed in alluvial deposits. On Socotra, fluvisols are restricted to the wadis and coastal plains, and the parent material consists predominantly of recent fluvial or marine deposits. The sediments vary in size substantially, and this feature can be used to characterise the different types of fluvisols further.

Arenosols (sandy soils) are present locally, especially on the plains of the north and south coasts, and along the coastline itself. On Socotra, arenosols are mainly derived from calcareous parent material, but pockets of extremely weathered siliceous rock probably also occur. Sand dunes are typically characterised by arenosols. One of the most important characteristics is their coarse texture, which accounts for the generally high permeability and low water and nutrient storage capacity of these soils.

Leptosols are very shallow soils over continuous rock, and soils that are extremely gravelly and/or stony. Leptosols include lithosols in some soil classification systems. They strongly restrict the ability of plants to root, resulting in a very patchy vegetation cover. Leptosols are highly characteristic of strongly eroding landscapes, often in very hot (or very cold) climates, at medium to high altitudes. On Socotra, they are widely distributed on the limestone plateau, and also on the granite of the Hagghier massif. In some national systems, leptosols correspond with rendzinas over calcareous rocks and rankers over acidic ones (IUSS Working Group WRB 2006). The B-horizon in such soils is therefore absent or very poorly developed.

Regosols encompass a range of weakly developed mineral soils in unconsolidated material that are not rich in gravels (i.e. leptosols), sand (arenosols) or fluvic material (fluvisols). Regosols can be regarded as a soil-taxonomic remnant group and a convenient category in which to place many soils showing only incipient formation processes that are otherwise difficult to accommodate. Regosols are widespread, for instance, on the limestone plateau.

Over large tracts of the limestone plateau, continuous rock occurs at the surface and this is generally considered as non-soil in many soil classification systems.

As clearly outlined in Sect. 7.2, land degradation in the form of soil deterioration is a key element of "desertification", and in fact soil degradation and erosion are used as the primary indicators of this phenomenon in the *World Atlas of Desertification* (Middleton and Thomas 1997). Specific examples of the impacts and consequences of soil degradation on biodiversity are given in Sects. 7.3.2 and 7.5.1. Pietsch and Morris (2010) highlight the importance of soil protection for the preservation of biodiversity on Socotra, and give an account of traditional and modern techniques aimed at conserving soil resources.

over calcareous rocks and rankers over acidic ones (IUSS Working Group WRB 2006). [illegible]

[illegible]

Chapter 3
Geology

Abstract Socotra is a typical example of an ancient continental island, but in biological evolutionary terms, it is more akin to an oceanic island due its relatively long duration of isolation. Post-Lower Miocene uplift with arching, block-faulting and tilting has played a major role in the present structure and morphology of the island. Most parts of Socotra, with the possible exception of the central Hagghier mountains, were intermittently submerged until the Miocene. As a consequence, the Precambrian basement rocks are overlain unconformably by Cretaceous and Tertiary limestones, but are exposed in three main uplift areas, most prominently in the Hagghier mountains. Prior to the Gulf of Aden rifting in the Oligocene, Socotra was probably located adjacent to southern Arabia. The precise time at which the Socotra Archipelago became detached from the African mainland is unclear, but could be of considerable biogeographical significance. The climate history of the Horn of Africa, with Socotra, differed from that of the adjacent regions, because since about the Late Cretaceous, and extending into the Middle Eocene, the evidence suggests that arid conditions persisted there, at least intermittently. During the glacial phases of the Pleistocene, it seems that as the rainforests contracted, "arid corridors" existed between South Africa and the north-east of the continent, including the Horn of Africa, and that repeated plant migrations were taking place in both directions. Such corridors could help explain the present-day disjunct distributions of various taxa.

3.1 Introduction

In accordance with the classification of islands of Alfred Russel Wallace (1911), Socotra is a typical example of a continental fragment or ancient continental island. Tectonic drift caused the separation of fragments from the mainland tens of millions of years ago, and Socotra is one such fragment originating from Gondwana. Deep waters now isolate Socotra from the mainland landmasses of Arabia and Africa. Despite being a continental fragment, in biological evolutionary terms, Socotra is

G. Brown and B.A. Mies, *Vegetation Ecology of Socotra*, Plant and Vegetation 7,
DOI 10.1007/978-94-007-4141-6_3, © Springer Science+Business Media Dordrecht 2012

more akin to an oceanic island on account of its relatively long duration of isolation (at least 15 My, possibly substantially longer).

The geology of Socotra was described in detail by Beydoun and Bichan (1970), and much of the present account is based on this paper. More recent studies have been undertaken by a number of authors, including Birse et al. (1997) and Samuel et al. (1997), and these results have been incorporated here.

3.2 General Geology and Structure

The present structure and morphology of Socotra are a result of post-Lower Miocene uplift with arching, block-faulting and tilting (Beydoun and Bichan 1970), although some authors (see below) have suggested that the granite core of the Hagghier mountains, which dominate the central and eastern part of the island, was possibly subject to uplift during the Cretaceous. Most parts of Socotra and the ancillary islands were intermittently submerged until the Miocene (<20 Mya). As a consequence, basement rocks are overlain unconformably by Cretaceous and Tertiary limestones throughout much of the area, but are exposed in three main uplift areas on Socotra. This can be most prominently observed in the Hagghier mountains (Figs. 3.1 and 3.2). The two other crystalline massifs crop out in the west at Ra's Shu'ab and in the vicinity of Qalansiyah (Fig. 3.1).

The coastal ranges and extensive plateaus on Socotra consist mainly of thick layers of white to light-grey Cretaceous and Tertiary limestone, as well as reef complexes from the Tertiary, and are underlain by igneous intrusive basement. The limestone strata, especially those from the Tertiary, diminish markedly in thickness towards the Hagghier mountains (see Beydoun and Bichan 1970).

Plate tectonic reconstructions indicate that prior to the Gulf of Aden rifting in the Oligocene, Socotra was located adjacent to the Dhofar region of southern Oman and south-eastern Yemen, which explains the stratigraphic similarity between the two areas.

3.3 Basement Rocks

Beydoun and Bichan (1970) and Kopp (1999) assumed that the basement of Socotra dates back to the Precambrian (i.e. more than 500 Mya). Recent geochronological measurements from crystalline rocks of the Arabian–Nubian Shield on the flanks of the Red Sea using the Rb–Sr isotope ratios have shown the anorogenic peralkaline granite series to be at least 476 ± 2 My old (Abdel-Rahman and Doig 1987). A similar age can therefore be assumed for the basement rocks of Socotra. This granite series represents the last manifestation of igneous activity in the region.

The Hagghier mountains are composed of a plutonic nucleus consisting largely of amphibolite facies meta-sediments and meta-igneous rocks that were intruded by syn-kinematic granites, and late-kinematic gabbros. Post-kinematic igneous activity has resulted in the formation of a sequence of peralkaline granites,

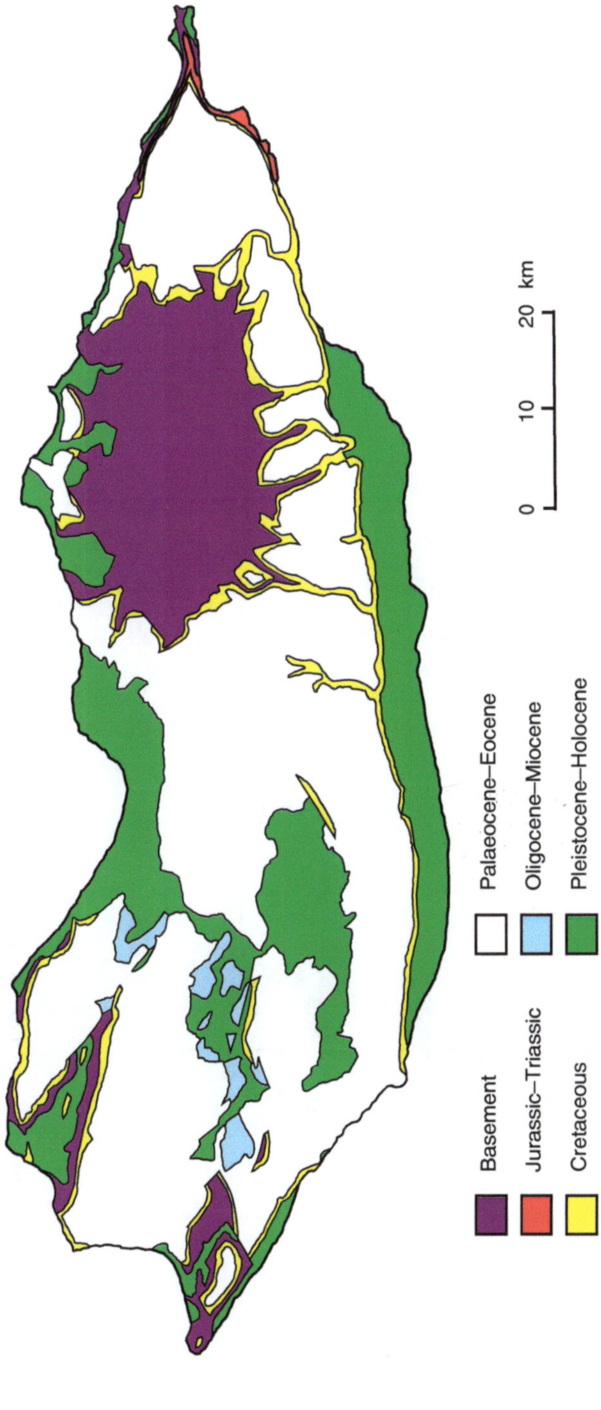

Fig. 3.1 Geological map of Socotra (Adapted from Beydoun and Bichan 1970)

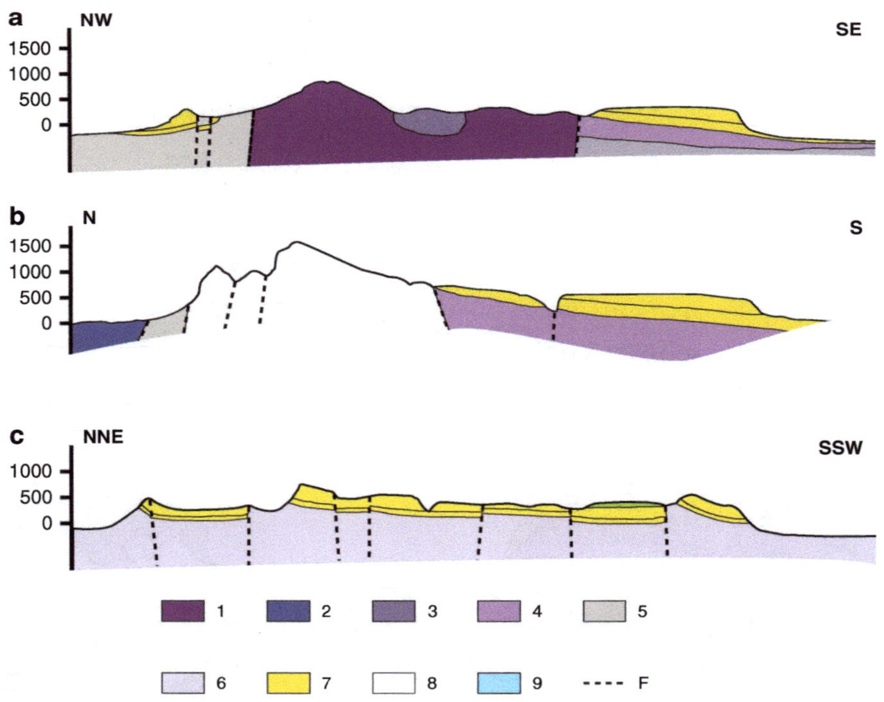

Fig. 3.2 Geological cross-sections of different parts of Socotra. (**a**) Eastern Hagghier mountains from Jebel Hauwari to the Nuged plain; (**b**) Hagghier mountains from SE of Hadibuh to the Nuged plain; (**c**) Western Socotra from Ra's Qadamah to Niyt. Legend: *1* peralkaline granite, *2* Hadibuh series, *3* gabbro, *4* effusive association, *5* biotite granite, *6* biotite schist, amphibolite granite and granite gneiss, *7* Cretaceous limestones, *8* Palaeocene to Eocene limestones, *9* Oligocene to Miocene sediments, *F* Fault. The shallow Quaternary cover sediments on the northern and southern plains are not shown (Data from Beydoun and Bichan 1970, Naumkin 1993 and own observations)

hornblende, biotite granites and gabbros. These account for the bulk of rocks making up the Hagghier mountains. It was from these exposed rocks on Socotra that the mineral riebeckite was first described, named in honour of the nineteenth-century German explorer Emil Riebeck. Riebeckite also occurs in the Precambrian crystalline rocks along the flanks of the Red Sea (Arabian–Nubian Shield) (Abdel-Rahman and Doig 1987). The metamorphic association largely comprises muscovite-biotite-schists and -gneisses, with outcrops on the eastern flanks of the Hagghier range as well as near Ra's Shu'ab and Qalansiyah in the west. Furthermore, the metamorphic Hadibuh series, found in the vicinity of the main town, is made up of bedded mudstones and tuffs, with the greenschist facies locally developed due to slight re-crystallisation with reorientation of muscovite and chlorite.

Rocks of volcanic origin are exposed predominantly around the southern and eastern margins of the Hagghier mountains, where they form a distinct arc. Because they are cut by the peralkaline granite series of the main mountain range, they must belong to a much earlier phase of volcanic activity than the Miocene–Pliocene volcanism of the Aden area, and are almost certainly of Precambrian origin. Both igneous and volcanic masses are dissected by numerous minor intrusions and dykes. The granite basement was subjected to repeated marine transgressions when Socotra formed part of Pangaea, and the same applied later to Gondwana after the break-up of the super-continent, as evidenced by the substantial deposits of limestone and other sedimentary rocks surrounding the Hagghier core.

3.4 Triassic and Jurassic Strata

It is only fairly recently that Lower Mesozoic strata have been recognised on the island. In the eastern part of Socotra, the presence of fossilised conodonts, worm-like animals long extinct, indicates the occurrence of Triassic strata, for instance in the fault-bounded vicinity of Ra's Mumi. Triassic successions contain a varied group of facies, consisting mainly of sandstones, but also limestones of various degrees of purity.

In the south-eastern part of the island, Jurassic layers are preserved, unconformably overlying those of the Triassic (Samuel et al. 1997). They outcrop at Ra's Falanj, and are remarkable in that they display a well-developed herringbone cross-stratification. Late Jurassic strata consist of calcareous sandstones and siltstones, but also packstones and wackestones. At R'as Falanj, coral floatstones are also present, and reflect the proximity to ancient reef build-ups. The oldest calcareous rocks on Socotra are found in the Jurassic strata in the south-east.

3.5 Cretaceous Strata

On most parts of the Socotran Platform, Cretaceous sedimentary rocks lie unconformably above basement. On Socotra itself, the first layer above basement usually consists of sandstone and limestone with marls from the Aptian to Cenoman.

The maximum thickness of this sequence is 300 m, and occurs in the region between Ra's Falanj and Ra's Mumi on the south coast where the strata are visible on outcrops of steep escarpments or in deep gorges. The oldest and thickest Cretaceous successions, of Barremian to Aptian age, can be found on the smaller western islands, as well as in the east of Socotra. A further marine transgression during the Cretaceous is recorded by a transition from estuarine to mixed shallow-marine carbonate and siliciclastic sedimentation. Thin sedimentary sequences of Campanian to Maastrichtian age of the Late Cretaceous have been identified in two areas of western Socotra.

The Cretaceous layers form a continuation of the corresponding beds of the nearby Somalian Peninsula and the southern coast of Arabia. More specifically, they correlate in age with the Fartaq, Harshiyat and Qishn Formations of the Mahra province in southern Yemen.

3.6 Tertiary Strata

Cretaceous strata are overlain discordantly by a series of limestone layers from the Late Palaeocene to Eocene. This Palaeocene–Eocene succession is often over 600 m thick (Samuel et al. 1997). Regarding the Tertiary limestone strata, an angular unconformity is discernible only locally in the field with the underlying Upper Cretaceous deposits, so that the contact zone of the rocks of these two periods remains largely obscure.

The Palaeocene–Eocene succession on Socotra bears a strong resemblance to the Umm Ar-Radhuma strata of mainland Arabia, which are well exposed along the southern coast, and also to the Auradhu Series of Somalia (Beydoun and Bichan 1970). Othman (1996) described the rocks of this succession as light-grey, crystalline, often porcellaneous limestones that were formed by shallow sub-littoral deposition. They often comprise oolithic or pellety limestones and contain nodular cherts. Silicified or re-crystallised fossils are common. Marginal reef complexes representing predecessors of the Socotran islands can be identified within the Tertiary strata.

The Gulf of Aden rifting commenced in the Late Oligocene (Edgell 2006), and this event was followed by a new phase of deposition. A wide range of facies reflects strong tectonic controls on deposition (Samuel et al. 1997). This Early to Mid-Tertiary limestone series makes up virtually the entire plateau landscape of Socotra, reflecting the hardness and durability of the strata. Erosional processes, including karst development, have reduced their original thickness somewhat, especially along pre-defined lines of weakness, such as bedding planes and fractures.

In the west of Socotra, earlier Tertiary strata are overlain by Miocene sediments. These are the youngest and softest strata of the plateaus, and consist of chalky limestones and marls equivalent to the Shihr group of the mainland. They are the result of near-shore deposition along a pre-Oligocene coastline. Being less resistant to erosion, they are preserved mainly in structural depressions where they were subsequently covered by more resistant coarse alluvium and gravel. The Miocene strata on Socotra have so far not been precisely dated, and it is also unclear to what extent uplift in this epoch is temporally related to the central Hagghier uplift.

3.7 Quaternary Deposits

The Miocene uplift period of the entire Somali–Arabian area, including the Socotra Archipelago, ceased before the onset of the Quaternary. Since that time, the coastal plains on Socotra have become filled with fluvial sediments. A variety of facies can be encountered:

- sandy conglomeratic limestones and calcareous breccia of marine origin;
- fluviatile fan, sheet, terrace and channel deposits;
- recent scree debris and aeolian sands.

Alluvial fans extend from the limestone and granite mountains, forming wide plains with a distinct but fairly shallow slope. In the case of Wadi Thar in the west, the plain is almost flat. These piedmont fans consist of cobbles and rock detritus overlying gravelly alluvium in the vicinity of the mountains. Towards the coast, the particle size gradually decreases, and a carpet of coarse fragments such as cobbles, gravel and pebbles was formed on the surface due to wind deflation or water sorting after occasional heavy rains and subsequent flooding. In some areas, such as on the southern and northern coastal plains, coarser material was transported over low aeolian dunes in the interior. The course of the main drainage systems leading from the mountains is often marked by thin lines of vegetation and, locally, patches of sabkha.

Marine deposits that occur inland are the result of Miocene uplift, and several raised beaches occur along the southern Nuged coast. For example, horizontal terraces of sandy conglomeratic limestones and cemented conglomerates with a variety of marine organisms such as corals and bivalves are found in some inland locations. Sand dunes are a sign of more recent geological phenomena. They are well developed in the Nuged area on the south coast, where crescent dunes occur. The sand-covered slopes extending to the top of Jebel Hauwari (385 m) on the north coast (Fig. 3.3) are a striking example of the strength of the northerly winds on Socotra, which occur between May and October, and the ongoing effects of geological processes.

3.8 Brief Geology of the Ancillary Islands

The geology of the western three islands is similar to that of the western part of Socotra. The basement of the islands of Darsah and Samhah is made up of gneisses and amphibolites (Kossmat 1907), and these are capped by hard limestone series forming platforms, dating from the Cretaceous up to the Tertiary. A similar situation can also be found in parts of Abdalkuri, although basement rocks are exposed in some hilly areas of the central and eastern region (see Kossmat 1907).

Fig. 3.3 Sand dunes encroaching on hillside slopes in the north of Socotra. December 2008

3.9 Brief Tectonic and Climate History of the Region

An early phase of tectonic activity had gradually ceased during the Ordovician, and it was followed by a long period of quiescence characterised by several erosional phases on Gondwana, which on Socotra led to peneplanation of the basement. Tectonic activities resumed in the Late Jurassic/Early Cretaceous, during which the rigid landmass of the African shield became ruptured. The rifting of the Red Sea and the Gulf of Aden began in the Late Oligocene, ca. 25 Mya, and effectively led to Arabia becoming a separate tectonic plate (Edgell 2006). As detailed geological studies have shown, prior to this rifting Socotra was located directly adjacent to Dhofar in southern Oman (Samuel et al. 1997). Socotra was deeply affected by the Gulf of Aden rifting, and post-Lower Miocene faulting follows a predominantly WSW–ENE trend across the island. This tectonic activity is also ultimately responsible for Socotra, and probably Abdalkuri too, becoming stretched and assuming their current shapes. In marked contrast to the situation in East Africa and over large tracts of southern Arabia, there is no indication that this tectonic activity was accompanied by volcanism (Beydoun and Bichan 1970). In the Late Tertiary/Quaternary, tectonic activity gradually came to an end, and extensive alluvial plains were formed in the northern, western and southern coastal area of Socotra.

The precise time at which the Socotra Archipelago became detached from the African mainland is a matter of some debate. Krupp et al. (2002) suggested that this event took place at about the same time as the separation of India and Madagascar from Africa, therefore about 165 Mya. Kopp (1999) and Mies (2001) postulated that Socotra finally became detached towards the end of the Cretaceous, ca. 70 Mya, but other authors point to a much more recent separation. As most of the region was repeatedly submerged for long periods during the Tertiary up until the Miocene, this point could at first sight perhaps be of lesser relevance for the biogeography of the flora. In this context, however, a further point of contention is whether, as indicated above, the Hagghier mountains of Socotra experienced substantial Cretaceous uplift. In this case, it could follow that the Hagghier area of Socotra has been an island above sea-level since the end of the Cretaceous (65 Mya) and would therefore have to be regarded as one of the most isolated landmasses on earth since the end of this period. This would have profound biogeographical consequences, similar in importance, for instance, to the separation of Madagascar or even India from the African continent (Mies 1995a, 1998a, b; Kopp 1999). Others, such as Fleitmann et al. (2004), state that basement uplifts appear to be relatively young (Miocene) in geological terms. In line with this assumption, Naumkin (1993) and Guebourg (1998) postulated that the rocks of the Hagghier mountains were once overlain by Tertiary strata. This led them to assume that Socotra as an island above sea-level could not be older than the Miocene. However, there are no traces of Tertiary limestone or even Cretaceous rocks in the central Hagghier mountains, including in the valleys there. The absence of these rocks cannot be convincingly explained by erosional processes, as limestone caps are present elsewhere on the island (Mies 1995a, 1998a, b; Kopp 1999). Moreover, Jebel Hauwari on the north coast and the centrally orientated Tertiary series could be interpreted as a marginal reef complex comprising a steep fore-reef, a reef core and a lagoon as back-reef. This would suggest the presence of an island immediately to the south, which could have been formed only by the earlier uplifted Hagghier peaks. While sliding northwards into the Gulf of Aden rift, Jebel Hauwari assumed its present inclination.

Miller and Morris (2004) speculated on the possible existence of a recent land bridge to the Horn of Africa during the Pleistocene, when sea-levels were much lower, but Fleitmann et al. (2004) suggested that Socotra has remained isolated from the mainland for at least 15 My.

The backbone of the following account of the climate history is based on a recent contribution by Morley (2007), with other papers mentioned where relevant. Despite the obvious difficulties in reconstructing the climatic conditions and associated vegetation types for specific areas over the past 140 million years, soon after the time that angiosperms began evolving, progress has been made in recent years. However, although general trends may be discernible over such a long period in the Earth's history, our knowledge remains fragmentary. This applies in particular to the north-eastern African–Arabian region, where fossil records of both plants and animals are almost absent, as is the case with large tracts of the African continent.

During the Early Jurassic, Madagascar was sandwiched between eastern Africa and India, and both Madagascar and India were situated very close to Socotra, with the northern tip of India located just several hundred kilometres away (Storey 1995; Birse et al. 1997). Madagascar, as part of the Madagascar–Seychelles–Indian block, gradually became detached from mainland Africa some time during the Middle

Jurassic, ca. 165 Mya (Rabinowitz et al. 1983), soon after the rifting stage of Gondwana had commenced during the Early Jurassic, ca. 180 Mya (Storey 1995). Sea-floor spreading in the Mascarene Basin during the Late Cretaceous led to the separation of Madagascar and Greater India (Storey et al. 1995), with Madagascar assuming its approximate present geographical position in the Late Cretaceous (Rabinowitz et al. 1983).

During the Early Cretaceous, many parts of Africa experienced a strongly monsoonal climate, with semi-arid conditions prevailing locally. Throughout this period, the continent lay somewhat further south than it does today, with eastern Africa more so relative to the western part of the continent. In the east, the equator ran through an area that is located roughly 16° further north than currently. This means that the Horn of Africa, with Socotra attached, was situated several degrees to the south of the equator in the southern tropics. With the gradual fragmentation of Gondwana, Africa began drifting northwards in a slightly anticlockwise rotational movement, with the Horn of Africa region attaining its current geographical position in the northern tropics during the Miocene, about 16 Mya.

The hot and humid greenhouse climate that gradually set in during the Cretaceous led to lowland megathermal rainforest dominating large swathes of central Africa up until the Oligocene (Axelrod and Raven 1978), including areas further north than the Horn of Africa. The Horn of Africa itself differed climatically from the adjacent regions, because at least since the Late Cretaceous, and extending into the Middle Eocene ca. 45 Mya, the presence of evaporites suggests that arid conditions persisted there, at least intermittently, even as the region drifted northwards over the equator. During the Mid-Cretaceous, much of Socotra, along with the adjacent areas, lay submerged in a shallow sea, although it emerged intermittently during the Late Cretaceous. Rising sea-levels brought about by the warmer global climate during the Early Tertiary led to large tracts of East Africa and the Horn of Africa becoming re-submerged, witnessed by the massive limestone deposits from the Palaeocene to Early Eocene that today cover most of Socotra (Beydoun and Bichan 1970). According to Mies (2001), some of the central highlands of Socotra protruded as islands from this shallow sea due to substantial Cretaceous uplift. The arid flora of the region persisted on such islands and in other regions of northern Africa that were not so much affected by the overall wetter macroclimatic conditions.

India and Madagascar split off from Africa in the Jurassic. During the Late Cretaceous, Africa, with Arabia still firmly attached, became fully disconnected from South America. By this stage, most of the floras of the world had developed a relatively modern aspect, with 139 present-day families having fossil records dating from the Early Tertiary and Late Cretaceous (Muller 1981). The explosive radiation of the angiosperms during the Cretaceous was accompanied by a corresponding demise of gymnosperms in the vegetation (Burgoyne et al. 2005).

By the Late Eocene, eastern Africa had already completed a substantial part of its north-westwards movement. From the Middle Eocene, global temperatures began to decline, a trend that accelerated at the beginning of the Oligocene, about 33 Mya, and the decline has gradually continued until the present. The cooling of the climate was associated with an accumulation of polar ice and lower precipitation.

The resulting decline in sea-levels, accompanied by an uplifting of the Ethiopian–Arabian region that commenced during the Miocene (Edgell 2006), exposed those areas of East Africa, including Socotra, which had previously been submerged during most of the Tertiary. The species-rich megathermal rainforests of Africa, partly monsoonal in nature, slowly began to contract, but they still covered a continuous broad band from the West to East African coast, even infringing on the extreme south-west of Arabia. In the Horn of Africa and eastern Arabia, however, it seems that the climate was too arid to support closed forest.

The general aridification of the climate that has taken place since the Oligocene is continuing to have major repercussions for the distribution of species in the northern African–Arabian region and beyond. In eastern Africa, C_4-grass savanna began to expand by the Late Miocene. As the climate became harsher during the Pliocene, the savanna-type vegetation of northern Africa gradually became more sparse, leading to the formation of the Sahara desert. Throughout the Quaternary, a series of wet–dry cycles, possibly associated with the glacial–interglacial periods further north, but more probably attributable to the movements of the ITCZ and polar front (Le Houérou 1997), greatly influenced the Saharan region. During the wetter phases, Mediterranean floral elements, including trees and shrubs, expanded their range southwards, only to retreat at the onset of the intervening dry periods. Throughout these climatic shifts, it seems that mountains offered suitable enclaves allowing a number of species to persist that otherwise would have become extinct (Gams 1964; Lösch et al. 1990), similar to the case with the Afromontane flora (Hedberg 1970).

During the glacial phases of the Pleistocene, it seems that as the rainforests contracted, "arid corridors" existed between South Africa and the north-east of the continent, including the Horn of Africa (Verdcourt 1969; White 1983; Burgoyne et al. 2005). Bellstadt et al. (2008) underlined that repeated plant migrations were taking place in both directions. Bobe (2006) has postulated that arid corridor links between north-east Africa and southern Africa (including Namibia) probably first originated in the Miocene. Such corridors could help explain the present-day disjunct distributions of a number of taxa, for instance in the genera *Wellstedia*, *Moringa* and several others (see Sect. 4.2.2). Galley et al. (2007) found that with certain related species in the Cape region and the highlands of tropical Africa (the Afrotemperate flora, which also includes the highlands of Ethiopia), a vicariant hypothesis was not tenable, and that the direction of migration was strictly unidirectional from south to north, taking place over the past 17 My, consistent with the Mio-Pliocene formation of the mountains in this area.

Chapter 4
Flora and Biogeography

Abstract In the first part of this chapter, the history of botanical exploration is explored. The most important floristic work to appear in recent times is the *Ethnoflora of the Soqotra Archipelago*, outlining 828 species of vascular plants that have been recorded from the area. A striking feature is the high proportion (37%) of endemics. Palaeoendemics appear to be rather poorly represented, and recent molecular phylogenetic studies are beginning to emphasise the role of neoendemism in shaping the flora. The vascular flora has its closest affinities to the adjacent regions of the Horn of Africa and southern Arabia, but distinct links exist to other geographical areas much further afield. Recent evidence also suggests that long-distance dispersal of ancestral species has contributed substantially to the development of the present-day flora, as demonstrated by various examples. It can be assumed that with the arrival of the first settlers on the islands, many native plant communities have been subject to pronounced changes in species composition, and examples of the two major groups that have benefited from human impact, "invaders" and "colonisers", are given. Twenty-six species of fern are known from the archipelago, with only one endemic. Fairly detailed overviews now exist of the bryophyte and lichen flora. The biogeography of these two groups is of considerable interest, as they are quite ancient in geological terms and occur in comparable habitats throughout the world.

4.1 History of Botanical Exploration

In antiquity, Socotra was called "Dioscorida", and referred to as the place from which incense, myrrh and dragon's blood originated. It was only later that the Canary Islands became known as a source of dragon's blood (Casper 2000), and during the Age of Discovery, when a maritime route to India and South-East Asia was being sought, the red resin exuded from the fruit of certain rattan palm species

G. Brown and B.A. Mies, *Vegetation Ecology of Socotra*, Plant and Vegetation 7,
DOI 10.1007/978-94-007-4141-6_4,

in the Far East was also used. Although the natural products of Socotra were then well known, it seems that business with European maritime traders during this period was minimal. In fact trade in general was probably considerably lower than it had been in the first few centuries AD. The main market for the small amounts of aloe and dragon's blood that were being produced on Socotra was Arabia, when once a year dhows would take advantage of the south-west monsoon to sail there. As the economic importance of Socotra gradually diminished, its geo-strategic relevance became enhanced. For a few years at the beginning of the sixteenth century, Socotra was an important naval base for the Portuguese on their way to India. A British military expedition was undertaken to Socotra in 1824, and in the following year, Lieutenant J.R. Wellsted presented his report to the Royal Geographical Society of London. He gave a vivid account of the life of the inhabitants and their natural environment. Wellsted travelled throughout the island on foot and by camel, referring to the vegetation and the plants wherever he could by their local vernacular names. Unfortunately, he collected only very few plant specimens on his travels. So far, only one species of *Romulea* collected by him has been located in the Royal Botanic Gardens at Kew. However, his detailed description of the vegetation and aspects of land use some 180 years ago is invaluable for comparing the present-day situation and assessing the degradation that has taken place over this period. Later in the nineteenth century, Balfour (1888) followed Wellsted's routes on the island, using his map for selecting places of botanical interest.

Up until 1880, occasional visits to the islands by ships on their way to India had been the sole source of botanical information. A few specimens were gathered by anonymous collectors during the occupation of Socotra by British colonial troops from 1834 to 1839. Some of them were named by a collector referred to as "Mr. Nimmo from Bombay". Louis Hyacinthe Boivin was a botanist on the French vessel *Duconadec*, under the command of Captain Guillain, visiting Socotra for a few days in 1847, and the specimens collected during this trip were deposited in the herbarium in Paris. The Political Resident at Aden, Captain Hunter, collected a few plants in 1876 and sent them to Edinburgh (Hunter 1876). In the same year, Commodore Wykeham Perry anchored off the coast of Socotra, and used the opportunity to collect a few living specimens of a local aloe (*Aloe perryi*) from the island, which he took back with him to Britain.

On 11 February 1880, Isaac Bayley Balfour (Fig. 4.1), the gardener Alexander Scott, both from Edinburgh, and Lieutenant Cockburn of the Aden Regiment (B.C.S.) landed at Qalansiyah in the west of Socotra, where their ship had safe anchorage (Balfour 1881; Scott 1881). Within 48 days, they created the foundation of one of the outstanding floras of the nineteenth century. On 25 February, the expedition moved from the western tip of the island to Hadibuh, arriving there a few days later on the 29th. One week was devoted to field trips into the Hagghier mountains from their base camp on the outskirts of Hadibuh. They continued towards the eastern part of Socotra 6–18 March. Their last field trip took place 22–27 March, crossing the Hagghier range to Nuged, before returning to Aden on 30 March. The expedition yielded thousands of herbarium specimen, living plants and other

Fig. 4.1 Isaac Bayley Balfour

scientific collections (cryptogams, zoological and mineralogical samples). Balfour (1888) wrote:

> In exploring the island, I deemed it better, considering the short time of our sojourn, rather to attempt to cover as much ground as possible, with the view of obtaining a representative collection, than to examine in detail a limited tract of the country. ... Especially amongst the hills of the Haggier range are there valleys which would well repay a careful and intended investigation. The expedition just completed ought to be considered only preliminary; I am assured a rich harvest awaits any collector who may visit the island.

One set of specimens was given to Kew, another to the British Museum. The rest remained with Balfour during his tenures at Glasgow, Oxford and later Edinburgh. He soon published his findings of the new species (Balfour 1882a, b, 1883, 1884). Balfour had originally been a student at Edinburgh University, obtaining a DSc degree in 1875 (Anonymous 1924). Later, he took up the position of Herbarium Keeper and then that of Professor of Botany in Glasgow. The expedition to Socotra had supplied him with a theme for his botanical "dissertation" in 1883. He moved to Oxford University in 1884 to take responsibility for the Botanical Garden and the herbarium collections. In 1888, he was recalled to Edinburgh where he taught until 1922. With the support of various other botanists, including Schweinfurth, Balfour published his *Botany of Socotra* (Balfour 1888). This work listed over 600 species, of which more than 200 species and 20 genera were new to science.

In 1881, a German party consisting of two scientists, the zoologist Emil Riebeck and botanist Georg Schweinfurth (Fig. 4.2), sailed along the coast of Arabia, eventually docking in Socotra on 9 April 1881. While waiting in Aden to continue their

Fig. 4.2 Georg Schweinfurth

journey by dhow, they received a provisional list of plant species from Balfour. They camped in the northern foothills of the Hagghier range and spent most of their time in the mountains until 18 May. Schweinfurth was also a gifted writer, giving vivid accounts of the inhabitants and the natural landscape of Socotra (e.g. Schweinfurth 1884a, b, 1891a, b) (Fig. 4.3). He collected numerous botanical samples from Socotra, yielding approximately 800 herbarium sheets of flowering plants, apart from many lichens. As the main focus of his attention was the vascular plants of Arabia and Ethiopia (Schweinfurth 1894, 1896, 1899), he forwarded many samples collected on Socotra to Balfour, therefore contributing substantially to his *Botany of Socotra*. The non-vascular cryptogams were sent to other specialists, particularly noteworthy being a large collection of lichens that were forwarded to J. Müller-Argau in Geneva.

Henry Ogg Forbes from Liverpool Museum and W.R. Ogilvie Grant from the British Museum spent a total of 84 days, from 2 December 1898 until 24 February 1899, on the islands (Forbes 1899). They visited Abdalkuri for four days, but spent most of the time on Socotra itself. Initially, they set up their base camp on the Hadibuh plain, later moving on to Dahamis, presumably the location that Schweinfurth referred to as "Wadi Kishen" (Fig. 4.4), and is now known more commonly as Wadi Dihzafaq or Wadi Denegen. As they were suffering from fever, probably malaria, they shifted to the higher elevations of the Homhil area, and from there to a ridge in the Hagghier mountains above Wadi Dihzafaq called Dimele (Adho Dimele). The main purpose of the expedition was to explore the fauna of the island.

Fig. 4.3 Illustration of the endemic cucurbit *Dendrosicyos socotrana* and *Adenium obesum* (Taken from Schweinfurth 1891a)

Fig. 4.4 The camp of an expedition led by the zoologist Emil Riebeck, which included Georg Schweinfurth, in Wadi Kishen (Wadi Dihzafaq) (Taken from Schweinfurth 1891b)

Botanical samples were also collected and later passed on to Balfour, who in the meantime had been appointed Professor of Botany in Edinburgh. A book entitled *The Natural History of Socotra and Abd-el-Kuri* was edited by Forbes (1903), who requested that Balfour be again responsible for the accounts of the higher plants.

An Austrian expedition dispatched by the Academy of Science in Vienna to undertake a multidisciplinary survey of social and natural sciences arrived in southern Arabia in 1898 (Peucker 1899; Müller 1907). They landed on Socotra on 8 January 1899 and stayed there for 87 days, until 4 March. Several of the team members, namely Oskar Simony, Stefan Paulay, Franz Kossmat and David H. Müller, later travelled to Abdalkuri. Furthermore, Samhah was visited for the first time by a scientific expedition. Von Wettstein (1906) published a series of photographs taken by Kossmat that gave the first impression of the vegetation of Socotra to the outside world. Over a century later, these photographs have been used to analyse long-term population dynamics of the dragon's blood tree (*Dracaena cinnabari*) in the Hagghier mountains (Habrova et al. 2009). The phanerogams collected during the Austrian expedition were revised by J. Vierhapper (1903, 1904, 1905, 1906, 1907, 1909), who described a large number of new species. Regarding the lichens, J. Steiner (1907) described new and remarkable genera belonging to the Roccellaceae. Many of the specimens collected during the expedition are located in the herbaria of Vienna where they have remained largely untouched since, and are therefore still waiting to be revised.

For the first 50 years of the twentieth century, virtually no scientific work was carried out on Socotra, probably due to the political situation. French semitists worked on the linguistics of the Socotri language in the 1930s (Leslau 1938), but no botanical information is available from their visits. Engler (1910) and Pichi-Sermolli (1955) used the information at their disposal to describe the flora and vegetation of Socotra. George Popov rekindled scientific work on the islands after he had visited Socotra for a few weeks in 1953 in search of the breeding grounds and migration routes of locusts (Uvarov and Popov 1957). He published the first comprehensive description of the vegetation of the island (Popov 1957), as well as a list of his collections (Gillett, in Popov 1957).

Unfortunately, few botanical details were published from an expedition carried out by Oxford University in 1956. This expedition is worth mentioning, because it took place from 1 August to 3 October, and therefore included both periods of drought and storm for the first time (Botting 1957, 1958a, b; Gwynne 1968).

After a short visit in March 1964, the ethnologist Brian Doe gave material of *Adenium obesum* to John J. Lavranos to be used for his systematic work on this Arabian and African genus (Lavranos 1966). The Middle East Command Expedition from 6 December 1964 to February 1965 commenced work on a detailed map of Socotra. However, this expedition did not publish any botanical results.

Major Peter Boxhall assembled a multidisciplinary team to visit Socotra from March to late May 1967. Two of the team members, Alan Radcliffe-Smith from the Royal Botanic Gardens, Kew (Fig. 4.5), and John Lavranos from Pretoria (Fig. 4.6), carried out detailed botanical surveys on Socotra and later spent two days on Abdalkuri (Lavranos 1971a, 1974, 1994). They set up a base camp at Suq (Shiq) on

Fig. 4.5 Alan Radcliffe-Smith on Abdalkuri in 1967 (Photograph by John Lavranos)

Fig. 4.6 John Lavranos on Socotra in 1999

Fig. 4.7 *Euphorbia hadramautica* (Photographed in Dhofar, Oman)

the Hadibuh plain, from where the members explored the island. The first three days were spent on foot exploring the north coast to Qalansiyah, discovering the first specimens of *Euphorbia hadramautica* (Fig. 4.7) for Socotra. Above Qalansiyah, local endemics such as *Aloe squarrosa* and *Euphorbia arbuscula* ssp. *montana* (Fig. 4.8) were found. One of the species of *Ceropegia* that was noted probably refers to the edible *C. affinis*, which according to Miller and Morris (2004) is common in dense thickets, but easily overlooked. After returning to their base camp, they climbed the Muqadrihon ridge on 29 March, facing heavy rains and floods in the wadi beyond Jebel Rewged (Rughid). Here, the expedition recorded about 100 *Dirachma socotrana* trees (Fig. 4.9) in two separate locations (Lavranos, pers. communication). The next place of interest was the Hamaderoh plateau on the north-eastern coast of Socotra. From this field trip, *Echidnopsis insularis*, *Sarcostemma socotranum*, *Boswellia nana* and *Croton pachyclados* (now regarded by some as a form of *C. socotranus*) were described as new to science. Afterwards, the botanical party spent a couple of weeks in the Hagghier mountains. They camped on the ridge above Wadi Dihzafaq, as had the expedition of Forbes 64 years before. The botanical party discovered *Duvaliandra dioscorides* (Fig. 4.10) in rock crevices, and rediscovered the pomegranate *Punica protopunica* (Figs. 4.11 and 4.12) in the lower hills. After returning to their base camp, the party travelled by ship to Abdalkuri to obtain more herbarium material from the island, and also to collect living specimens of *Euphorbia abdelkuri*. These individuals are still propagated in botanical gardens all over the world as grafts on the base of other succulent *Euphorbia* species. Radcliffe-Smith stayed two weeks longer on Socotra than Lavranos, and described and revised a large number of species belonging to various groups from the island (see Mies 1994). Lavranos focused his research on the succulents from the archipelago;

Fig. 4.8 *Euphorbia arbuscula* ssp. *montana* on Jebel Ma'alah. February 1999

Fig. 4.9 The rare endemic *Dirachma socotrana* in flower. Muqadrihon pass, south of Hadibuh (Photograph by John Lavranos)

Fig. 4.10 *Duvaliandra dioscorides*, Adho Dimele (Photograph by John Lavranos, 1967)

Fig. 4.11 *Punica protopunica* on the limestone plateau near Dixam. December 2008

Fig. 4.12 *Punica protopunica* in flower. February 1999

notably *Aloe* (Lavranos 1969), Apocynaceae–Asclepiadoideae (Lavranos 1970, 1971a, a, 1979, 1993) and *Euphorbia abdelkuri* (Lavranos 1971b). Most of the living plants that are still grown in botanical gardens to this day have their origin from this expedition. The fig *Ficus vasta* and *Begonia socotrana*, the latter a species with orbicular leaves, are frequently cultivated, while other plants collected from the expedition are not so widely distributed. For instance, the pomegranate *Punica protopunica* collected on Socotra is currently grown only in a botanical garden on Hawaii. Other cuttings or seeds have proved difficult to keep in cultivation, or have become lost (Lavranos, pers. communication).

The independence of South Yemen and the Cold War made the islands inaccessible to scientists from the West. There are generally very few records of botanists working there during this period. In January 1971, A.S. Bilaidi from South Yemen collected plants on Socotra and sent them to the herbarium in Florence (Pichi-Sermolli 1976). A joint Aden University and Agricultural Centre Al-Qod Expedition was undertaken to the archipelago in 1985 (Wranik et al. 1986). The ethnologist Vitaly Naumkin published an excellent description of the life of the Socotran people, mentioning the wealth of plant species and some of their traditional usages (Naumkin 1993). However, he mentioned plants by their local names, and it is therefore difficult to know exactly which ones he was referring to. Naumkin also gave an overview of the geology of Socotra. Some time later, despite facing considerable bureaucratic difficulties, Quentin Cronk succeeded in reaching Socotra to search for threatened plant species (Cronk 1986).

The political changes in Yemen and the reunification of the country finally opened up the islands again to scientists in 1989. A regular flight connection now operates from Sana'a to Socotra. Since 1989, a number of scientists, most notably Friedrich

Fig. 4.13 Friedrich Beyhl on Socotra in 1999

Beyhl (Fig. 4.13), Abdulnasser Al-Gifri, Peter Hein, Norbert Kilian, Harald Kürschner, Anthony Miller, Bruno Mies and Mats Thulin, have undertaken substantial botanical research on Socotra. At present, the Royal Botanic Garden Edinburgh houses the largest collection of Socotra plants in the world, mainly due to its substantial involvement in the "Flora of Arabia" project (Miller and Cope 1996). Other important collections of plants from Socotra can be found in the Uppsala Botanical Garden (M. Thulin) and Berlin (N. Kilian, P. Hein and co-workers).

4.2 Vascular Plants

4.2.1 General Attributes of the Flora of Socotra

The most important work to appear in recent times on the flora of Socotra is the *Ethnoflora of the Soqotra Archipelago*, authored by Anthony Miller and Miranda Morris (2004). According to these authors, 828 species of vascular plants have been recorded from the Socotra Archipelago, although some have so far been identified only at the generic level. In the meantime, a number of additional species have been added, either completely new to the archipelago as a whole, or new to individual islands, but recorded from adjacent ones. Most of the latter involve species that have been found on the main island of Socotra, but were already known from one or several of the others. In accordance with Appendix 1, 842 species are now recognised as occurring or having occurred in the archipelago (a few are possibly now extinct). Ferns are represented by 30 species, gymnosperms by 1 (*Ephedra foliata*), monocots by 180 and dicots by 631. In terms of species richness, the most important families are Poaceae (100 species), Fabaceae (72), Asteraceae (57), Acanthaceae (34), Euphorbiaceae (34), Apocynaceae with Asclepiadoideae (30), Boraginaceae (29), Cyperaceae (27), Malvaceae (25), Convolvulaceae (24), Scrophulariaceae (23) and Rubiaceae (22). A striking feature of the vascular flora is the high proportion (37%) of endemics, an aspect that will be dealt with in Sect. 4.2.3. The vascular flora has its closest affinities to the adjacent regions of the Horn of Africa and southern Arabia, but distinct links exist to other geographical areas much further afield, including Madagascar, South Africa and West Africa/Macaronesia, as will be discussed below.

Islands typically have fewer species per unit area than the mainland (Whittaker and Fernández-Palacios 2007), but like the Canaries, Socotra is unusual in that it has a species diversity that is as high as the adjacent continental areas (Cronk 1997). This author lists various factors that may explain the high level of species diversity, including: (1) the close proximity to Arabia and Somalia; (2) physiographic richness; (3) geological diversity with a corresponding variety of substrates; (4) climatic complexity; (5) ecological complexity (reflecting the pronounced spatial heterogeneity of abiotic factors); (6) geological history and the close affiliations with the adjacent landmasses of Arabia and north-east Africa; and (7) Holocene stability and a rather benign land management system that has prevented the over-exploitation of natural resources, until very recently at least.

The equilibrium model of island biogeography (MacArthur and Wilson 1967) postulates that the number of species on an island is the outcome of a dynamic equilibrium between immigration and extinction, dependent on two main factors, namely island area and degree of isolation. It is clear from the number of factors influencing the processes of immigration and extinction, summarised by Whittaker and Fernández-Palacios (2007) and listed by Cronk above, that the relationship is more complex. This is also taken into account to a certain extent by the findings of Hobohm (2000), who showed that in the Canaries, which consist of seven main islands, the species-richness of endemics correlates better with island height than with area, because taller islands support more habitat types than do lower ones. A conspicuous feature of the Canaries, at least on the higher islands, is the altitudinal zonation with a remarkable diversity of strongly contrasting habitats, ranging from semi-desert to permanently moist laurel forest within a very small horizontal area. In this respect, Socotra is very similar to the Canaries, and that would help explain the relatively high number of species there.

An important question concerns the origins of the flora of Socotra, and this question will be examined in more detail in the discussion of the endemic flora (Sect. 4.2.3).

Fig. 4.14 Inflorescence of *Dorstenia gigas*. March 1996

4.2.2 Characteristic Distribution Patterns

The primary goal of biogeography is to shed light on the current distribution of organisms, and furthermore to uncover the mechanisms that have led to the observed spatial distribution patterns. The close links of the flora of Socotra with the neighbouring continental landmasses (Socotra is located ca. 232 km from the Horn of Africa and 351 km from southern Yemen – see Sect. 2.1) are currently being underlined by a spate of molecular phylogenetic studies (e.g. Cortés-Burns et al. 2004; Thiv and Meve 2007; Thiv et al. 2006, 2011). These areas are strongly influenced by a subtropical to tropical climate with spring and summer rains (Kürschner 1998). However, close affinities also exist to taxa in more distant parts of the world. In the following, examples are given of the characteristic distribution patterns relevant to the flora of Socotra, with particular emphasis on the endemics. This information will then be used to examine the origins of the flora in Sect. 4.2.4.

Socotra–Somalia

Dirachma socotrana (Dirachmaceae) is closely related to *D. somalensis* in Somalia (Link 1991). Together these are the only representatives of the isolated family Dirachmaceae (Thulin et al. 1998), which is now considered to be close to the Rhamnaceae (Baas et al. 2001; Ronse De Craene and Miller 2004).

Dorstenia gigas (Moraceae – Fig. 4.14) is very closely related to *D. gypsophila*, a species with an epigaeic caudex from northern Somalia (Lavranos 1972b).

Fig. 4.15 *Euphorbia arbuscula*, a locally common endemic tree on Socotra. Ra's Hebak, December 2008

Echidnopsis milleri (Apocynaceae-Asclepiadoideae): According to Lavranos (1993), the endemic *E. milleri* resembles *E. chrysantha* from Somalia. *E. milleri* is known from only one small area on Socotra. See below and Sect. 4.2.4.3 for further details.

Euphorbia arbuscula (Euphorbiaceae), a locally common tree on Socotra (Fig. 4.15), is closely related to the shrub *E. bariensis* from northern Somalia (Carter 1992). Recent molecular studies by Steinmann and Porter (2002) suggest a very close relationship with other members of the Tirucalli clade (*E. xylophylloides*, *E. tirucalli* and *E. gregaria*), although these authors do not appear to have included *E. bariensis* in their studies.

Xylocalyx (Scrophulariaceae): In her revision of the genus *Xylocalyx*, Carter (1962) provided descriptions of four species, two endemic to Socotra (*X. aculeolatus* and *X. asper* – Fig. 4.16), and two others known from northern Somalia (*X. hispidus*, *X. recurvus*). Thulin (1987) later added a fifth species, *X. carteriae*, also from Somalia.

Thulin (1993, 1995) mentions further examples of the Socotra–Somalian distribution type in his "Flora of Somalia" project, including ***Erythrina melanacantha* ssp. *somala*** (Fabaceae).

Socotra–Ethiopia–Somalia

Buxanthus pedicellatus (Buxaceae) is fairly widespread on Socotra (Figs. 4.17 and 4.18), and also occurs in the mountains of south-eastern Eritrea as well as on the northern mountain slopes of Somalia (Valenti 1965). On Socotra, it is found in areas that receive sufficient moisture from the winter monsoon or in the form of dewfall.

Fig. 4.16 *Xylocalyx* cf. *asper* on coastal dunes. Nuged plain, December 2008

Fig. 4.17 *Buxanthus pedicellatus*, fruiting. December 2008

Fig. 4.18 *Buxanthus pedicellatus*, showing arrangement of male and female flowers. December 2008

Echidnopsis (Apocynaceae–Asclepiadoideae) is a medium-sized genus with a typical Eritreo-Arabian distribution, apart from *E. milleri* (see above). Up until fairly recently, only one member of the genus (*E. socotrana*) had been recorded from Socotra, but Bruyns (2004) has described four further species, bringing the total to five. All are closely related amongst each other and are most probably of East African descent (Thiv and Meve 2007). See Sect. 4.2.4.3 for further details.

Socotra–Dhofar (Oman, South Arabia)

Cocculus balfourii (Menispermaceae) was formerly regarded as a Socotran endemic (Fig. 4.19) until it was found in Dhofar (Oman), as described by Forman (1980). The species is also known from south-eastern Yemen (Miller and Morris 2004).

Socotra–Dhofar–Somalia/Djibouti

Poskea socotrana (Globulariaceae) occurs on rocky terrain on Socotra. It has recently been found on mainland Yemen (Miller and Morris 2004), and previously in NE Somalia (Morucchio 1969). *Poskea* is an endemic genus to the Horn of Africa and Socotra, where, apart from *P. socotrana*, two other localised species occur: *P. africana* and *P. newbouldii* (Morucchio 1969).

Zygocarpum coeruleum (Fabaceae), the only member of this small genus to occur on Socotra, has five close relatives in north-eastern Africa and southern Arabia (Thulin and Lavin 2001), including *Z. somalense* from north-eastern Somalia as well as *Z. dhofarense* from Dhofar in southern Oman and *Z. yemense* from western Yemen. This endemic genus to the Horn of Africa and southern Arabia was formerly included in the more widespread genus *Ormocarpum*, of which an overview was given by Gillett (1966). Further details are given in Sect. 4.2.4.3.

Fig. 4.19 *Cocculus balfourii*, a widespread species at higher altitudes. December 2008

Saharo-Sindian and Mediterranean-Macaronesian

The Saharo-Sindian regional zone covers a vast area stretching from the Canaries in the far west throughout the Sahara, encompassing most of the Arabian Peninsula and extending into Pakistan, Afghanistan and western India. It is usually subdivided into three main phytogeographical units: (1) the Saharan regional subzone in Africa; (2) the Arabian regional subzone; and (3) the Nubo-Sindian regional subzone. In the west, it is in contact with the Mediterranean floral region.

Campylanthus (Scrophulariaceae) is a small genus of 18 species with a conspicuous disjunct distribution, and representatives are known from Macaronesia, north-eastern Africa, Arabia and Pakistan. *C. spinosus* occurs on Socotra (Fig. 4.20), and is also known from Somalia. At the time of the last revision (Hjertson 2003; Miller and Morris 2004), 15 species were known, but in the meantime, a further three have been added, two from the main area of occurrence in north-eastern Africa/southern Arabia, and one from northern Arabia. The last record is quite remarkable, as it is from the mountains of northern Oman (Hjertson et al. 2008), whereas all other species on the Arabian Peninsula occur ca. 800 km further south. A characteristic feature of the genus is that many of the species have a very restricted distributional range with small populations (Kilian et al. 2002).

Ochradenus (Resedaceae): The distribution of *Ochradenus* stretches from the Tibesti mountains in the central Sahara desert throughout the Saharo-Sindian belt into Arabia and Pakistan (Miller 1984). The endemic *O. socotranus*, the only species known from Socotra, is closely related to the more widespread *O. arabicus* and *O. baccatus*.

Schweinfurthia (Scrophulariaceae) is a small genus consisting of six species (Miller et al. 1982). It has its centre of distribution in southern and eastern Arabia.

Fig. 4.20 *Campylanthus spinosus* is fairly widespread on coastal plains and the limestone plateau. March 1997

Only *S. pedicellata* is known from Socotra, a species that also occurs on the immediately adjacent continental landmasses of southern Arabia and the Horn of Africa, as well as on the Comoro Islands and in Tanzania.

Further examples of this distribution type include ***Farsetia*** (Brassicaceae), ***Gymnocarpos*** (Caryophyllaceae, Petrusson and Thulin 1996), ***Lavandula*** (Lamiaceae), ***Reseda*** (Resedaceae) and ***Withania*** (Solanaceae).

Socotra–North West Africa

Hemicrambe (Brassicaceae) is a small genus consisting of just two species: (1) *H. fruticosa* on Socotra; and (2) *H. fruticulosa*, known only from Morocco, nearly 7,000 km further to the west.

Socotra–Tropical Africa–Arabia

Boswellia (Burseraceae): A total of ca. 20 species of this xerophytic genus occur in north-eastern Africa and Arabia, extending into India. Eight species have been recorded from Socotra, all of which are endemic. Some of the Socotran species have retained female reproductive organs with five carpels as a primitive feature, whereas the number has been reduced to three in the mainland species (Engler 1913; Gillett 1991; Thulin and Al-Gifri 1998).

Gnidia is the largest genus in the Thymelaeaceae, containing about 140–160 species (Rogers 2009). It is almost completely restricted to Arabia, southern Africa and Madagascar, and represented on Socotra by the endemic evergreen shrub *G. socotrana* (Fig. 4.21).

Fig. 4.21 *Gnidia socotrana*, a fairly widespread endemic evergreen shrub at medium elevations. March 1997

Hypericum balfourii and ***H. socotranum*** (Clusiaceae) were placed by Robson (1985, 1993, 1996) in the section Campylosporus, which has a general African distribution.

Maerua angolensis (Capparaceae) is a widespread species in Africa and parts of Arabia. Kers (1993) noted that the populations from Socotra and the Horn of Africa cannot be differentiated taxonomically at the species level, but could be treated as a distinct subspecies (ssp. *socotrana*). A further differentiation within ssp. *socotrana* was suggested, with Socotran populations belonging to the var. *socotrana*, and those from the adjacent regions of the Horn of Africa to var. *africana*.

Further examples include ***Allophyllus*** (Sapindaceae, but not in mainland Arabia), ***Aloe*** (Asphodelaceae), ***Anisotes*** (Acanthaceae, East Africa, Arabia), ***Carphalea*** (Rubiaceae), ***Cephalocroton*** (Euphorbiaceae), ***Euclea*** (Ebenaceae), ***Eureiandra*** (Cucurbitaceae), ***Ledebouria*** (Hyacinthaceae) and ***Melinis*** (Poaceae).

Socotra–East African Coast

Kraussia socotrana (Rubiaceae) is a Socotran member of a genus with one species, *K. floribunda*, in south-eastern Africa and a further two, *K. kirkii* and *K. speciosa*, in the coastal regions of Kenya and Tanzania (Bridson 1995). According to this author, *K. socotrana* most closely resembles the south-eastern African taxon *K. floribunda*.

Socotra–East Africa–Madagascar

The small genus ***Metaporana*** (Convolvulaceae) comprises four species in north-eastern Africa and Madagascar. Details on the phylogenetic classification of the Convolvulaceae were given by Stefanovíc et al. (2003). On Socotra, the endemic small shrub *Metaporana obtusa* is locally dominant in western parts of the island.

Socotra–Arabia–Madagascar–Tropical Africa–Asia (India)

Actiniopteris semiflabellata (Actiniopteridaceae), a typical fern of rock crevices, is known from Socotra, as well as Arabia, south-eastern Egypt, Sudan, Ethiopia, Somalia, Kenya, Uganda, Eastern Congo, Madagascar, Réunion and Mauritius.

Aerva (Amaranthaceae) is a medium-sized palaeotropical genus represented by four species on Socotra, two endemics (*A. microphylla*, *A. revoluta*) and two more widespread taxa (*A. javanica*, *A. lanata*). Other members of the genus, including several endemics, are known from Arabia, southern Africa, West Africa, north-eastern Africa, Madagascar, the Mascarenes and south Asia. See Sect. 4.2.4.3 for further discussion.

Crotalaria socotrana (Fabaceae), an endemic tree on Socotra, was initially included in the genus *Priotropis*, together with two other species, one from the eastern Himalayas and the other from eastern Tanzania. The genus was later abandoned, with Thulin (1998a) transferring the remaining species, *P. socotrana*, also to *Crotalaria*.

Cryptolepis (Apocynaceae–Asclepiadoideae): In the subfamily Periplocoideae, which has recently been transferred to the Apocynaceae, *Curroria macrophylla* and the genera *Ectadiopsis*, *Mitolepis* and *Socotranthus* are now included in the genus *Cryptolepis* (Venter and Verhoeven 1997). This genus has a very broad geographical distribution, being found throughout much of Africa (including Madagascar) as well as in parts of Arabia and Asia (Joubert et al. 2008).

The genus ***Dicoma*** (Asteraceae) comprises in its wider circumscription about 60 species of herbs, shrubs and small trees, although Ortiz (2001) and Ortiz and Netnou (2005) have recently suggested removing about half of the species from the genus and transferring them to the reinstated genus ***Macledium***. Most of the 60 species are found in tropical and southern Africa, including Madagascar, with several species also occurring in Arabia and Asia. Two species have been recorded from Socotra, the weed *D. tomentosa* (also known from Yemen and Saudi Arabia), and the endemic *D. cana*, a common species on the island, and now referred to by Ortiz and Netnou (2005) as *Macledium canum*.

Dracaena (Dracaenaceae): The genus *Dracaena* is represented on Socotra by the xeromorphic *D. cinnabari* (Fig. 4.22). Its closest relatives are found in the mountains of southern Arabia (*D. serrulata* – Fig. 4.23), Sudan (*D. ombet* – Fig. 4.24) and northern Somalia (*D. schizantha*), as well as in Madagascar, southern Africa and South Asia. Furthermore, *D. draco* (Fig. 4.25) occurs in Macaronesia and the Anti-Atlas mountains of southern Morocco (Mies 1995b; Benabid and Cuzin 1997). See Sect. 4.2.4.3 for a more detailed account of *D. cinnabari* and related species.

Fig. 4.22 *Dracaena cinnabari*. Dixam plateau, December 2008

Fig. 4.23 *Dracaena serrulata*. Jebel Samhan, Dhofar, Oman, March 2011

Fig. 4.24 *Dracaena ombet* in the mist oasis Erkwit, Sudan (Photograph by A. Elshafie)

Fig. 4.25 *Dracaena draco* (Photographed on La Gomera, Canary Islands)

Fig. 4.26 *Jatropha unicostata*, an abundant endemic shrub on Socotra. December 2008

The genus ***Dyerophytum*** (= *Vogelia*) (Plumbaginaceae) is a small genus of four species that has a distributional area covering India, Arabia and Socotra, with a further species occurring in South Africa (Baker 1948). Two of these species, both endemics (*D. pendulum* and *D. socotranum*), occur on Socotra.

The genus ***Exacum*** (Gentianaceae) was formerly regarded as being endemic to Socotra until it was reported from the Dhofar mountains in Oman (Miller and Morris 1988), Africa (Porembski 1996), Madagascar and India (Klackenberg 1985). Three species are now considered to occur on Socotra, all endemics (Miller and Morris 2004), with these authors treating *E. gracilipes* as belonging to *E. affine*. More details regarding the disjunct distribution pattern of this genus and its possible origins are given below (Sect. 4.2.4.3).

Erythroxylon socotranum (Erythroxylaceae), a dwarf shrub endemic to Socotra, is the first member of the tropical family Erythroxylaceae to be found in south-western Asia (Thulin 1996). The genus, which is also absent from Somalia, has a regional centre of endemism in Madagascar and the Comoro Islands, and contains as well a variety of species in tropical East Africa, in the Mascarene Islands, in the Seychelles and on Aldabra (Thulin 1996).

Jatropha unicostata (Euphorbiaceae), a spineless shrub and the only endemic member of this palaeotropical genus on Socotra (Fig. 4.26), seems to be most closely related to *J. mahafalensis* from the south-western part of Madagascar (Lavranos, pers. communication).

The genus ***Launaea*** (Asteraceae) contains numerous species that occur in the drier regions of northern and eastern Africa, extending through much of Arabia into Asia, with one species found in Australia (Kilian 1997). Seven species have been recorded from Socotra, of which four are endemic. Section 4.2.4.3 provides further details.

Fig. 4.27 *Tragia balfouriana* is an endemic Euphorbiaceae with stinging hairs

Scaevola socotraensis (Goodeniaceae): Most of the 11 genera of the Goodeniaceae, a family comprising ca. 400 species, are almost entirely restricted to Australia, but one genus, *Scaevola*, has dispersed and radiated throughout much of the Pacific region and Indian Ocean coasts (Howarth et al. 2003). *S. socotraensis*, a small endemic shrub, has so far been recorded only from its type locality at Ra's Qatanahan on Socotra (St John 1962), where it is found growing near freshwater springs and seepages. See Sect. 4.2.4.3 for more details on the disjunct distribution of this genus.

Tamridaea (Rubiaceae) is a monotypic genus that, on the basis of molecular studies, has been shown to be most closely related to the genera *Virectaria* and *Pseudosabicea* from tropical Africa, as well as to the pantropical genus *Sabicea* (Bremer and Thulin 1998; Bremer 2009). *Tamridaea capsulifera* is shown in Fig. 6.83.

Tephrosia socotrana (Fabaceae) appears to have its closest systematic affinities to *T. noctiflora*, a species that is widespread throughout tropical Africa and that is also known from the Seychelles, Madagascar and India (Thulin 1997). It was only fairly recently found on Socotra and described by Thulin (1997), and it is apparently still known only from its type locality above Ra's Hebak on the northern coast (Miller and Morris 2004).

Tragia (Euphorbiaceae): The section Tagira of the genus *Tragia*, characterised by twining herbs with stinging hairs, consists primarily of tropical African and Malagasy species. Outside of this main range, *T. balfouriana* occurs on Socotra (Fig. 4.27), and *T. moammarensis* in southern Arabia. Furthermore, the distribution of *T. cinnabarina* extends from tropical West Africa to Sri Lanka and the Malabar coast (Pax and Hoffmann 1931).

Further examples of this distribution type include ***Arthraxon lancifolius*** (Poaceae), ***Blepharis*** (Acanthaceae), ***Caralluma* s.l.** (Apocynaceae–Asclepiadoideae), ***Cissus*** (Vitaceae), ***Dichrostachys*** (Fabaceae), ***Ecbolium*** (Acanthaceae), succulent members

Fig. 4.28 The rare endemic succulent *Kleinia scottii*. Dixam plateau, December 2008

of the genus ***Euphorbia***, ***Gaillonia*** (Rubiaceae), ***Hybanthus enneaspermus* agg.** (Violaceae), ***Lepturus*** (Poaceae), ***Neuracanthus*** (Acanthaceae), ***Psiadia*** (Asteraceae), ***Pulicaria*** (Asteraceae), ***Setaria intermedia*** (Poaceae) and ***Tavernieria*** (Fabaceae).

Socotra–North East Africa (Arabia)–South Africa

The genus ***Gaillonia*** (including *Jaubertia*, Rubiaceae), comprising a total of nine species, has a distributional range that includes Socotra, southern Arabia, Somalia and Namibia (Thulin 1998b). Four species, all endemics, are known from Socotra, namely *G. puberula*, *G. putorioides*, *G. thymoides* and *G. tinctoria*.

Graderia (Scrophulariaceae) is a small genus consisting of four species. The small shrub *Graderia fruticosa* (see Fig. 6.123) is endemic to Socotra, with two further species found in South Africa and another in Tanzania. *Graderia* belongs to a clade (together with, amongst others, *Striga*) that has its centre of diversity in eastern and southern Africa (Wolfe et al. 2005).

Kleinia scottii (Asteraceae) is the only succulent member of the Asteraceae to be found on Socotra (Figs. 4.28 and 4.29). Halliday (1988) considered it to be very closely related to, or even conspecific with, the Somalian species *Kleinia polytoma*. Other allied species mentioned by Halliday (1988) are *K. odora* and *K. squarrosa*. *K. odora* is known from Dhofar (Miller and Morris 1988), whereas *K. squarrosa* has a tropical African distribution. *Kleinia neriifolia* is a common succulent in semi-arid to arid habitats in the Canary Islands. More recently, Rowley (in Eggli 2002) treats *K. scottii* as a subspecies (ssp. *scottii*) of *Senecio longifolius*, a widespread succulent throughout tropical Africa, Madagascar and South Africa.

Fig. 4.29 Inflorescence of *Kleinia scottii*. Dixam plateau, December 2008

Wellstedia (Boraginaceae): Because of its taxonomic isolation, the genus is sometimes treated as a family in its own right (Wellstediaceae). It is a typical example of a genus with a north-east–south-west African disjunct distribution (Hunt 1971; Thulin and Johansson 1996). Five species occur in the north-eastern region, namely *W. socotrana* on Socotra, *W. laciniata*, *W. somalensis* and the recently described *W. robusta* (Thulin 1998a) in northern Somalia as well as *W. filtuensis* from south-eastern Ethiopia (Hunt and Lebrun 1975; Thulin and Johansson 1996). *W. dinteri* is known from southern Namibia and adjacent parts of the Cape (Thulin and Johansson 1996).

On the basis of data derived from Monod (1971) and de Winter (1971), Werger (1978) provided a useful summary of remarkable disjunct distribution patterns involving taxa of the arid regions of southern and northern Africa, with the latter region also encompassing in some cases the adjacent parts of southern Europe and the Arabian Peninsula, extending into south-west Asia. Taxa relevant to Socotra, and not mentioned above, include ***Aizoon***, ***Citrullus***, ***Enneapogon desvauxii***, ***Fagonia***, ***Moringa***, ***Stipagrostis***, ***Tricholaena*** and ***Zygophyllum simplex***.

Further examples include ***Babiana*** (Iridaceae, South Africa), although *B. socotrana* has recently been excluded from *Babiana*, and referred to the new genus ***Cyanixia*** (Goldblatt et al. 2003), ***Camptoloma*** (Scrophulariaceae, South Africa), ***Euryops*** (Asteraceae, South Africa, Arabia – Nordenstam 1968a, 1968b, 1969) and ***Lasiocorys*** (Lamiaceae, South Africa).

Fig. 4.30 *Secamone socotrana*. February 1999

Socotra–Africa–New Caledonia

The genus ***Acridocarpus*** (Malphigiaceae) occurs throughout Africa, with *A. socotranus* endemic to Socotra. A single species (*A. austrocaledonicus*) is also known from New Caledonia. Although Engler and Drude (1915) doubt whether the species is native to the island, recent molecular phylogenetic studies by Davis et al. (2002) suggest a long-distance dispersal event from Madagascar to New Caledonia around 15–8 Mya.

Socotra–Madagascar–Australasia

Secamone (Apocynaceae-Asclepiadoideae) is a palaeotropical genus (Klackenberg 1992), and, with 81 species, by far the largest of the nine genera in the subfamily Secamonoideae (Bruyns 2004). The majority of species (62) occur on Madagascar and the adjacent islands. Fifteen species are known from mainland Africa, and two others are distributed from India to northern Australia. Apart from *S. socotrana* (Fig. 4.30), which has long been known from Socotra, an extremely rare and local endemic, *S. cuneifolia*, was recently described by Bruyns (2004).

Socotra–Asia–Pacific Islands

Punica protopunica (Lythraceae) is endemic to Socotra and thought to be the wild ancestor of the cultivated pomegranate *P. granatum*. It is unclear whether *P. protopunica* was once native over a much wider area, or whether trees were exported from Socotra in historical times and subsequently cultivated. *P. granatum* is now grown in many tropical countries around the world.

Fig. 4.31 *Fagonia luntii*. Ra's Hebak, January 1998

Socotra–Central America–North America–South Africa

The genus ***Chapmannia*** (previously *Arthrocarpum*) (Fabaceae) comprises seven species in total with a remarkable disjunct distribution, four endemic to Socotra, one in Somalia and two in the New World (Thulin 1999b). See Sect. 4.2.4.3 for more details.

Fagonia (Zygophyllaceae) is a genus comprising about 34 species. It shows a remarkable distribution, occurring throughout the Saharo-Sindian region and adjacent areas, i.e. across a continuous broad band from the Canaries in the west into south-west Asia, in south-western Africa, western South America, western North America and north-eastern Mexico (Beier et al. 2004). Two species occur on Socotra, *F. luntii* (Fig. 4.31) and *F. paulayana* (Fig. 4.32), neither of which is endemic.

The genus ***Thamnosma*** (Rutaceae) occurs on Socotra (represented there by *T. hirschii* and *T. socotrana*), in Somalia, Arabia and southern Africa as well as in Central America (Balfour 1888; Thulin 1999a; Thiv et al. 2011). Whereas *T. socotrana* (Fig. 4.33) is endemic to Socotra itself, *T. hirschii* is found on the adjacent landmasses of southern Arabia and Somalia.

It is possible that the New World genus ***Tiquilia***, which occurs in south-eastern North America and north-eastern South America, is a sister group of *Wellstedia* (Thulin and Johansson 1996). *Tiquilia* is usually placed in the subgenus Ehretioideae of the Boraginaceae, whereas *Wellstedia* belongs to its own subfamily, or is treated as a separate family altogether.

Fig. 4.32 *Fagonia paulayana*. Nuged plain, March 1997

Fig. 4.33 *Thamnosma socotrana*. Fieri peaks, February 1999

4.2.3 Endemism

As already indicated, endemism is a striking feature of the flora of the Socotra Archipelago and one of the principal reasons for the tremendous interest in the islands. Of the 842 recorded species of higher plants listed in Appendix 1, 309 (i.e. 37%) are considered to be restricted to the archipelago. The number of endemics continues to rise, although it is likely that some will be found in under-explored areas of the adjacent mainland. An important reason for this high degree of endemism is invariably the relatively long degree of isolation, as is the case for many oceanic island floras (Silvertown 2004). As discussed in Sect. 3.9, Fleitmann et al. (2004) estimate that Socotra has been isolated from mainland influences for at least 15 My, although it is important to bear in mind that barriers are by no means effective for all species.

The archipelago encompasses an area of roughly 4,050 km^2, and with the number of endemic plant species at around 309, this gives a figure of 0.076 per km^2, virtually identical to that (0.08) calculated by Caujapé-Castells et al. (2010) for the Canaries, and higher than that for Hawaii (0.06) and the Galápagos (0.022).

A superficial appraisal of the Socotran flora suggests that the degree of endemism is generally higher in those families whose seeds are dispersed by means other than anemochory. Furthermore, because Socotra does not lie on the main migration route for birds (Cheung and DeVantier 2006), it can be safely assumed that bird dispersal (i.e. long-distance dispersal) does not play a major role in the flora. It therefore follows that most seeds are likely to be dispersed mainly in close proximity to the mother plant. Many groups containing large numbers of endemics (e.g. Euphorbiaceae) have no obvious features to facilitate pronounced long-distance dispersal. However, there are several exceptions of anemochorous groups in the Asteraceae and Apocynaceae (including Asclepiadoideae) whose seeds are furnished with a distinct pappus, but they appear to be exclusively entomogamous, i.e. pollinated by insects. Bird-pollination, although observed in the Socotran flora, probably plays very much a subordinate role (see Sect. 5.7). Due to the lack of any detailed studies, it is unclear as to what extent pollination is dependent on certain insect species. The predominance of non-anemochorous and entomogamous plant families, and the existence of highly specific plant–insect interactions, would theoretically serve to promote geographic and genetic isolation and maintain high degrees of endemism. Further aspects of pollination ecology are dealt with in Sect. 5.7, and seed dispersal in Sect. 5.8.

4.2.3.1 Endemic Taxa

Up until fairly recently, the Dirachmaceae were regarded as a monotypic family endemic to Socotra, represented by the locally common *Dirachma socotrana*. However, since the discovery of *Dirachma somalensis* in northern Somalia (Link 1991), no endemic family is now known from the archipelago. It is nonetheless noteworthy that the two members of the Dirachmaceae have a very restricted distribution.

The 15 endemic genera of the Socotra Archipelago currently recognised are listed in Table 4.1. Most of the endemic genera are monospecific, the exceptions being *Ballochia* and *Trichocalyx* (Acanthaceae), as well as *Rughidia* (Apiaceae). Whereas some of the species have a very restricted range – for example, *Nesocrambe socotrana* appears to be confined to the dry far west of Socotra

Table 4.1 The endemic genera of the Socotra Archipelago (After Miller and Morris 2004, except *Cyanixia*)

Family	Endemic genera (with number of species)
Acanthaceae	*Angkalanthus* (1), *Ballochia* (3), *Trichocalyx* (2)
Apiaceae	*Nirarathamnos* (1), *Oreofraga* (1), *Rughidia* (2),
Apocynaceae–Asclepiadoideae	*Duvaliandra* (1), *Socotrella* (1)
Brassicaceae	*Lachnocapsa* (1), *Nesocrambe* (1)
Caryophyllaceae	*Haya* (1)
Cucurbitaceae	*Dendrosicyos* (1)
Iridaceae	*Cyanixia* (1)
Rubiaceae	*Placopoda* (1), *Tamridaea* (1)

(Miller et al. 2002) – others such as *Placopoda virgata* and *Tamridaea capsulifera* are fairly widespread and common.

A number of other taxa possibly merit distinct generic status. For example, *Cephalocroton socotranus* (Euphorbiaceae) differs from other members of the genus in that the female flowers possess entire sepals. In his monograph, Pax (1910) considered the taxon as belonging to a distinct genus, *Cephalocrotonopsis*. However, most taxonomists continue to adhere to the name originally given to it by Balfour (1883), treating *Cephalocrotonopsis* at the rank of subgenus (Radcliffe-Smith 1973).

Endemism is a very pronounced feature at the species level in the flora of Socotra. Table 4.2 summarises the number of endemic species found in the various plant families, based on the list of species in Appendix 1 (which in turn is based to a large extent on Miller and Morris 2004).

Taking those families containing ten or more species on the islands into consideration, the highest proportion of endemics is found in the Burseraceae (92%), Acanthaceae (74%), Lamiaceae (68%), Rubiaceae (68%), Euphorbiaceae (59%), Asteraceae (58%), Boraginaceae (55%), Brassicaceae (55%), Apocynaceae (53%) and Caryophyllaceae (53%), all with more than 50% endemic species in the family. Six of the 11 endemic members of the genus *Euphorbia* are succulents. The Apocynaceae (including Asclepiadoideae) contains succulent and non-succulent members, and about eight of the 16 endemics can be regarded as succulents, including both of the monospecific endemic genera *Duvaliandra* (described by Gilbert 1980) and *Socotrella* (described by Bruyns and Miller 2002).

4.2.3.2 Palaeoendemics and Neoendemics

According to the concept of Cronk (1997), endemics can be divided broadly into two main groups that represent opposite poles along a continuum of possible scenarios. (1) Palaeoendemics are relictual species whose taxonomic isolation is due to extinction in their ancestral area. Cronk (1997) infers from this definition that divergence from the ancestral form must therefore have taken place before colonisation

Table 4.2 Details of the endemic phanerogamic species in the flora of the Socotra Archipelago (Based on species list in Appendix 1)

A	B	C	D	E
Pteridophyta				
Aspleniaceae	4	1		*Asplenium* (1/3)
Monocotyledonae				
Amaryllidaceae	1	1		*Crinum* (1/1)
Anthericaceae	3	2		*Chlorophytum* (2/3)
Asparagaceae	2	1		*Asparagus* (1/2)
Asphodelaceae	5	4		*Aloe* (4/4)
Colchicaceae	1	1		*Iphigenia* (1/1)
Dioscoreaceae	1	1		*Dioscorea* (1/1)
Dracaenaceae	1	1		*Dracaena* (1/1)
Hyacinthaceae	8	6		*Dipcadi* (3/4), *Drimia* (1/1), *Ledebouria* (2/3)
Iridaceae	2	1		*Cyanixia*** (1/1)
Orchidaceae	4	1		*Holothrix* (1/1)
Poaceae	100	10	10	*Andropogon* (1/1), *Aristida* (1/3), *Dactyloctenium* (1/4), *Lepturus* (4/4), *Panicum* (2/4),*Tricholaena* (1/2)
Dicotyledonae				
Acanthaceae	34	25	74	*Angkalanthus*** (1/1), *Anisotes* (1/1), *Ballochia*** (3/3), *Barleria* (4/4), *Blepharis* (1/3), *Chorisochora* (2/2), *Dicliptera* (1/1), *Hypoestes* (1/1), *Justicia* (3/4), *Neuracanthus* (1/3), *Rhinacanthus* (1/1), *Ruellia* (4/6), *Trichocalyx*** (2/2)
Amaranthaceae	13	3	23	*Aerva* (2/4), *Psilotrichum* (1/2)
Anacardiaceae	4	3		*Lannea* (1/1), *Rhus* (2/3)
Apiaceae	7	4		*Nirarathamnos*** (1/1), *Oreofraga*** (1/1), *Rughidia*** (2/2)
Apocynaceae	30	16	53	*Cryptolepis* (4/5), *Duvaliandra*** (1/1), *Echidnopsis* (5/5), *Marsdenia* (1/1), *Sarcostemma* (1/2), *Secamone* (2/2), *Socotrella*** (1/1), *Vincetoxicum* (1/1)
Asteraceae	57	33	58	*Distephanus* (1/1), *Helichrysum* (12/13), *Kleinia* (1/1), *Launaea* (4/7), *Macledium* (1/1), *Pluchea* (2/2), *Prenanthes* (1/1), *Psiadia* (1/1), *Pulicaria* (7/7), *Vernonia* (2/2), *Volutaria* (1/2)
Begoniaceae	2	2		*Begonia* (2/2)
Boraginaceae	29	16	55	*Cystostemon* (1/1), *Echiochilon* (1/2), *Heliotropium* (10/16), *Trichodesma* (3/3), *Wellstedia* (1/1)
Brassicaceae	11	6	55	*Erucastrum* (1/1), *Farsetia* (2/4), *Hemicrambe* (1/2), *Lachnocapsa*** (1/1), *Nesocrambe*** (1/1)
Burseraceae	13	12	92	*Boswellia* (8/8), *Commiphora* (4/5)
Campanulaceae	2	1		*Campanula* (1/1)
Capparaceae	7	1		*Cadaba* (1/4)
Caryophyllaceae	17	9	53	*Gymnocarpos* (2/2), *Haya*** (1/1), *Polycarpaea* (6/8)
Celastraceae	1	1		*Maytenus* (1/1)
Cleomaceae	7	1		*Cleome* (1/7)

(continued)

Table 4.2 (continued)

A	B	C	D	E
Clusiaceae	5	5		*Hypericum* (5/5)
Convolvulaceae	24	8	33	*Convolvulus* (3/6), *Metaporana* (1/1), *Seddera* (4/6)
Crassulaceae	6	2		*Kalanchoe* (2/3)
Cucurbitaceae	10	2	20	*Dendrosicyos*** (1/1), *Eureiandra* (1/1)
Dirachmaceae	1	1		*Dirachma* (1/1)
Erythroxylaceae	1	1		*Erythroxylon* (1/1)
Euphorbiaceae	34	20	59	*Andrachne* (1/2), *Cephalocroton* (1/1), *Croton* (4/4), *Euphorbia* (11/17), *Jatropha* (1/1), *Meineckia* (1/1), *Tragia* (1/1)
Fabaceae	72	21	29	*Acacia* (2/4), *Chapmannia* (4/4), *Crotalaria* (2/6), *Dichrostachys* (1/1), *Indigofera* (3/15), *Lotus* (2/4), *Paracalyx* (1/1), *Senna* (1/4), *Taverniera* (1/1), *Tephrosia* (2/7), *Trigonella* (1/1), *Zygocarpum* (1/1)
Gentianaceae	5	3		*Exacum* (3/3)
Geraniaceae	5	1		*Pelargonium* (1/1)
Goodeniaceae	1	1		*Scaevola* (1/1)
Lamiaceae	19	13	68	*Lavandula* (1/1), *Leucas* (7/10), *Micromeria* (1/1), *Orthosiphon* (1/2), *Plectranthus* (1/2), *Teucrium* (2/2)
Lythraceae	4	1		*Punica* (1/1)
Malpighiaceae	1	1		*Acridocarpus* (1/1)
Malvaceae	25	9	36	*Hibiscus* (9/15)
Meliaceae	1	1		*Turraea* (1/1)
Moraceae	4	2		*Dorstenia* (2/2)
Nyctaginaceae	5	2		*Commicarpus* (2/4)
Plumbaginaceae	4	4		*Dyerophytum* (2/2), *Limonium* (2/2)
Polygalaceae	3	1		*Polygala* (1/3)
Portulacaceae	6	3		*Portulaca* (3/5)
Resedaceae	3	2		*Ochradenus* (1/1), *Reseda* (1/1)
Rubiaceae	22	15	68	*Carphalea* (1/1), *Gaillonia* (4/4), *Kohautia* (1/2), *Kraussia* (1/1), *Oldenlandia* (5/7), *Placopoda*** (1/1), *Pyrostria* (1/1), *Tamridaea*** (1/1)
Rutaceae	4	1		*Thamnosma* (1/2)
Sapotaceae	2	2		*Sideroxylon* (1/1), *Spiniluma* (1/1)
Scrophulariaceae	23	5	22	*Graderia* (1/1), *Lindenbergia* (1/1), *Nanorrhinum* (1/2), *Xylocalyx* (2/2)
Solanaceae	11	3	27	*Lycium* (1/1), *Withania* (2/3)
Sterculiaceae	4	1		*Melhania* (1/2)
Thymelaeaceae	1	1		*Gnidia* (1/1)
Tiliaceae	9	5		*Corchorus* (1/3), *Grewia* (3/5), *Triumfetta* (1/1)
Valerianaceae	1	1		*Valerianella* (1/1)
Verbenaceae	6	5		*Clerodendrum* (2/2), *Coelocarpum* (2/2), *Priva* (1/1)
Vitaceae	3	3		*Cissus* (3/3)

Column A = Plant family that contains at least one endemic species; Column B = Number of species in the family in the Socotra Archipelago; Column C = Number of endemics in the family in the Socotra Archipelago; Column D = Percentage of endemic species in family (for $n \geq 10$ species); Column E = Genera with endemics, indicating the number of endemics followed by the number of species in the Socotra Archipelago (**endemic genus)

Fig. 4.34 The arborescent endemic cucurbit *Dendrosicyos socotrana*. Ra's Hebak, December 2008

of the island or island-like habitat in which they currently occur. However, it is also conceivable that a species may have survived in a very small part of its original distribution area without diverging from its original form, and still be considered a palaeoendemic. (2) In contrast, neoendemics are species whose ancestral forms have dispersed to a new area (island or island-like habitat) where they have undergone pronounced radiation, i.e. they have evolved *in situ*. Neoendemics retain close relatives in their ancestral areas and can reveal considerable information on the process of colonisation and evolution, especially in the wake of recent advances in molecular phylogenies.

Palaeoendemics appear to be rather poorly represented in the flora of Socotra. Prominent examples are probably the dragon's blood tree, *Dracaena cinnabari* (see Sect. 4.2.4.3 for more details), and the arborescent cucurbit *Dendrosicyos socotrana* (Fig. 4.34), but in both cases, fossil records are absent. The palaeoendemics of Socotra and their remote disjunct relatives would therefore appear to support the view that they are remnants of a dry subtropical southern Tethys flora, and are thus vicariants.

Recent molecular phylogenetic studies are beginning to emphasise the role of neoendemism in shaping the flora of Socotra, as has been suggested for the flora of the Canary Islands (Francisco-Ortega et al. 1996; Silvertown 2004). In these latter islands, many presumed palaeoendemics that were until quite recently thought to be relicts of a more widespread Tertiary flora now extinct in the adjacent continental land regions have been shown by molecular studies to be neoendemics, the majority apparently originating from a single colonisation event of the archipelago followed by radiation (see Silvertown 2004 and references therein). The availability of vacant habitat niches appears to be a primary prerequisite facilitating radiation. The fact that the origins of many recent Socotran endemics can be traced back to about 20–5 Mya suggests that as the shallow seas covering much of Socotra receded during the Miocene, large tracts of habitat became available for colonisation within a relatively short time-frame. As indicated in Sect. 4.2.4.3, the evidence from molecular studies suggests that many endemics have their closest relatives on adjacent mainland areas, and probably reached Socotra by dispersal, rather than being of a vicariant origin. This in turn suggests that any land areas of Socotra that may have remained above sea-level during the pre-Miocene Tertiary, almost invariably granitic basement, did not provide the source for many present-day Socotran taxa, and this includes the two closely related endemics of the genus *Aerva* (*A. microphylla* and *A. revoluta*) mentioned in Sect. 4.2.4.3.

Silvertown (2004), citing the work of Schluter (2000), summarises the ecological theory on which adaptive radiation is based, and the twofold role of competition in this process: intraspecific competition for resources provides the "push" for diversification, whereas the absence of interspecific competition for resources in uncolonised habitats facilitates a "pull" of species that must adapt to the new environmental conditions. It is therefore possible that the driving force behind speciation in *A. revoluta* was a "pull" into a granite habitat due to the presence of vacant niches.

4.2.4 Origins of the Socotra Flora and Colonisation of the Islands

Apart from its close links with the floras of the adjacent continental areas, it is also apparent that unique processes have been operating in the Socotra Archipelago resulting in the high degree of endemism. No fossils of vascular plants have so far been discovered on Socotra, and the same applies to much of the wider region. Evidence relating to the origins of the Socotran flora can therefore be gleaned only from a grasp of the climatic and geological history of the islands in a regional context, as well as from a fundamental understanding of the ecology of the various species involved. However, as indicated in Sect. 3.9, it is clear that key aspects of the geological history of Socotra highly relevant to the origins of its flora cannot be determined with absolute certainty. For instance, whereas Miller and Morris (2004) speculate on the possible existence of a land bridge to the Horn of Africa during the Pleistocene, when sea-levels would have been much lower, Fleitmann et al. (2004) suggest that Socotra has been isolated from the mainland for at least 15 My. Furthermore, Mies (2001) suggests that the granite core of Socotra represents an ancient island, and that this is supported by the high number of endemic species growing there, including various members of the Euphorbiaceae (*Andrachne schweinfurthii*, *Croton sulcifructus*, *Euphorbia hajhirensis* – Fig. 4.35, *E. leptoclada*, *E. schweinfurthii*) and *Leucas kishenensis* (Lamiaceae).

Fig. 4.35 *Euphorbia hajhirensis*. Adho Dimele, January 1998

However, recent molecular studies have called into question the age of some endemics. For instance, *Aerva revoluta*, cited by Mies (2001) as an example of such an ancient taxon, has been shown by molecular studies to have evolved from its progenitor much later, long after much of the limestone part of the island had emerged from the sea (see Sect. 4.2.4.3), and probably arrived by dispersal (Thiv et al. 2006). Furthermore, some of the species have also been found growing on limestone (e.g. *Andrachne schweinfurthii, Croton sulcifructus*). How long and to what extent Socotra has been isolated from the mainland are questions of central importance for assessing the relative roles of vicariance and dispersal in shaping the current flora of the islands, a controversial issue that has divided scientists dealing with the floras of other parts of the world. This issue will be examined in more detail in Sect. 4.2.4.1. A brief overview of the tectonic and climate history relating to the East African–Arabian region over the past ca. 150 My, the period during which angiosperms radiated and dominated the floras of the world, was given in Sect. 3.9. An overview of the role of palaeo- and neoendemism, which is also central to understanding the origins of the flora, appeared in the preceding Section (Sect. 4.2.3.2).

4.2.4.1 Dispersal and Vicariance

As a broad simplification, the existing flora of a given region is a result of two main processes, namely immigration (dispersal) and vicariance. The relative importance of these two processes has generated heated controversy amongst biogeographers at least since the time of Charles Darwin, a controversy often referred to as the dispersal–vicariance debate. The sometimes highly dogmatic arguments have been fuelled not just by the absence of conclusive scientific evidence, such as accurately dated fossils, but also by a failure to incorporate existing data from all relevant sources.

Isolated islands such as Socotra are discrete entities, and therefore offer a unique opportunity to shed more light on fundamental biogeographical questions such as the roles of vicariance and dispersal. As noted in Chap. 3, the islands comprising the Socotra Archipelago represent ancient fragments of continental rock that were once part of the southern supercontinent Gondwana, and that became separated as a result of plate tectonic processes. In accordance with the classification of Wallace (1911), such continental fragments can be distinguished from continental islands that are separated from the mainland by shallow waters and are usually of recent origin (i.e. postglacial sea-level rise), as well as from the oceanic islands that are the product of submarine volcanic activity. In terms of their origin, therefore, the Socotra islands have more in common with Madagascar than, for instance, with the Canary Islands, which are typical oceanic islands. It is apparent that the nature of the process by which an island has come into existence will have implications for the biogeography of its biota. One obvious difference between a continental fragment island and an oceanic one is that the former will begin its existence as an island with a subset of species sampled from the mainland from which it has become detached, unless catastrophic events such as marine transgression have wiped out the biota. In contrast, oceanic islands will start off with no species whatsoever.

The paucity of adequate data continues to beset current biogeographical studies and efforts to interpret species distributions. However, over the past decade, the rapid development and improvement of molecular phylogenetic approaches, including molecular dating, have begun to revolutionise our knowledge and thinking on many issues. When accurately calibrated, phylogenetic trees derived from DNA sequence data allow the inference of powerful information, including the time and origin of monophyletic groups and when lineages arrived in different geographical areas, even, as is often the case, when the fossil record is incomplete (Pennington et al. 2004). However, it is important to bear in mind that the methods are still being developed and are prone to various shortcomings, summarised in Milne (2006). For example, the problem of accurate molecular dating remains, with different methods producing highly conflicting node ages. Furthermore, some of the assumptions that underlie the advanced statistical techniques used to reconstruct ancestral areas are currently too restrictive, for instance in ignoring the fact that there may be a strong asymmetry in the direction of dispersal rather than an equal probability (Cook and Crisp 2005). In other words, dispersal from area A to area B may be much more likely to occur than vice versa. It can be expected that in the coming years, as methodological issues are improved, significant advances will be made in explaining distributional patterns.

Once the plate tectonic theory was finally validated in the 1960s and up until very recently, disjunct distribution patterns were generally attributed to vicariance, rather than to dispersal processes. Before this period, dispersal had been the more widespread explanation, including for transoceanic disjunctions. With the advent of molecular phylogenetic techniques during the last few years, the importance of dispersal is being reinstated in a dramatic fashion. In fact the resurrection of dispersal, just a few decades ago viewed as a rare phenomenon, is one of the most remarkable turn-arounds in the recent history of biogeographical thinking, and striking examples of oceanic dispersal in both plants and animals are given by de Queiroz (2005). This author even goes as far as suggesting that vicariance as an *a priori* explanation for present-day distribution patterns should possibly be discarded. However, vicariance has the virtue of being more amenable to falsification than does dispersal, and can therefore serve as a useful null hypothesis.

Several plant groups (e.g. *Dracaena*, *Euphorbia balsamifera*) have disjunct distribution patterns between the Macaronesian islands, Morocco, East and South Africa, southern Arabia and Socotra, and these disjunctions are regarded as evidence for a continuous flora that once existed across northern Africa in the Late Miocene (e.g. Bramwell 1985; Mies 1995b). Molecular phylogenetic studies carried out by Andrus et al. (2004) on *Vierea* and *Pulicaria* section Vieraeopsis suggest that their morphological similarities are due to morphological convergence, and that the two genera do not share a common ancestor, thus weakening the vicariance hypothesis. Furthermore, these authors suggest that many of the putative relationships mentioned for other species are not supported by molecular phylogenies, although some notable exceptions do exist (e.g. *Campylanthus* – see above, *Aeonium*).

4.2.4.2 Long-Distance Dispersal

Regarding the origins of the Socotran flora, studies on the dispersal capability of both widespread taxa and endemics could be very instructive. The mechanisms of long-distance seed dispersal offer a fascinating field of research, despite, or perhaps due to, the methodological hurdles involved. No detailed studies or analyses have been carried out on the seed dispersal ecology of Socotran plants (for the sake of convenience, the term "seed" is used to denote the dispersal unit or diaspore). General information is forthcoming on some of the more widespread taxa of the archipelago, but very little is available in respect of the endemic flora. Such investigations are urgently required because of the importance of seed dispersal in population and metapopulation dynamics, in community structure and not least for assessing the role of dispersal in plant colonisation of the islands. The majority of seeds tend to be dispersed in the immediate vicinity of the mother plant (Willson 1993), but increasingly, long-distance dispersal across a wide range of scales is being recognised as both important and overlooked in plant community ecology (Nathan 2006; Levey et al. 2008). Although long-distance dispersal events may be extremely rare, they are possibly of enormous significance in promoting the spread of populations, also to previously unoccupied habitats (Nathan et al. 2008). A wide variety of seed dispersal mechanisms exist in plants, and these have been summarised by various authors. Perhaps the most detailed classification of dispersal syndromes relevant to Socotra (in that a number of the species or their close relatives are found on the islands) is that published by Gutterman (1993). This and other dispersal classification systems are typically based on the agent of dispersal, which in turn is inferred from seed morphology. At the most fundamental level, dispersal can be achieved through the abiotic factors wind and water, or through biotic agents, in particular animals, including humans, or even the plant itself. Some plants produce atelechoric dispersal units, i.e. are not dispersed and remain at the location of the mother plant (e.g. *Emex spinosa* – see Sect. 5.8).

Despite the obvious attractions of such a seed morphology-based classification system, Levin et al. (2003) cautioned against its over-general application. These authors pointed out that dispersal is rarely mediated by a single dispersal agent and is not restricted to the primary movement of the seeds from the plant to the substrate. Higgins et al. (2003) found only a poor relationship between morphologically defined dispersal syndromes and long-distance dispersal. Against this background, they suggested that non-standard means of dispersal could play an important role in long-distance dispersal events. Although it is relatively straightforward to implicate long-distance dispersal for species with seeds that are easily transported (including by wind, oceans, animals), this is evidently not the case with species whose seeds are heavy, such as those of *Dracaena cinnabari*, or are both heavy and toxic (*Euphorbia* spp.). Species like *Dendrosicyos socotrana*, *D. cinnabari* and *Euphorbia balsamifera* appear therefore to be good examples of palaeoendemics. But it is also clear from the occurrence of *Draceana draco* and *E. balsamifera* in the Canary Islands (and other Atlantic islands) that long-distance dispersal must have played a role in these species, as the islands are of volcanic origin and there is no evidence of any land bridge ever connecting them to the African mainland. Lems (1960) stressed that the entire flora of the Canary Islands could be attributed to long-distance dispersal. Guzmán and Verdas (2009) have recently published a paper describing the long-distance colonisation of the western Mediterranean by the shrub *Cistus ladanifer*, despite the absence of specialised dispersal mechanisms. The results of Nogales et al. (2001) provide evidence that gulls could have played a major role in the dispersal of seeds to and between the individual islands of the Canaries, being possibly capable of transporting seeds for up to 650 km.

In summary, therefore, research into long-distance dispersal faces considerable challenges, because events are so rare and unpredictable. Nathan et al. (2008) point out that traditional research focusing on the seed itself is unlikely to shed more light on vital long-distance dispersal mechanisms. A much broader approach is required to identify vectors most probably associated with long-distance dispersal, as well as to take into consideration other relevant aspects such as landscape structure, migratory routes, stop-over sites, etc. Where long-distance dispersal has been implicated, it can be assumed that for many widespread species, multiple colonisation attempts rather than a single chance colonisation event may be needed to ensure that a viable population finally becomes established. This should not be confused with speciation processes in some endemics, which, where dispersal rather than vicariance has been implicated, are thought to be the result of a single colonisation event (e.g. Juan et al. 2000; Silvertown et al. 2005).

In Sect. 4.2.4.3, the overwhelming evidence suggests that long-distance dispersal of ancestral species has played a major role in shaping the flora of Socotra, as has

Fig. 4.36 *Aerva microphylla* on low cliffs. Wadi Ireh, December 2008

been the case on other islands around the globe, such as New Zealand (Pole 2001), the Canaries (de Nicolás et al. 1989; Whittaker and Fernández-Palacios 2007) and Hawaii (Nathan 2006), and, as was suggested by Hedberg (1970), for the isolated high mountains of East Africa with their characteristic Afroalpine flora. From the assumption that during the Early Tertiary, much of the landmass of Socotra was submerged, the fact that just over 60% of the Socotran flora is non-endemic suggests that a sizeable portion of these ca. 520 species has probably dispersed at least once to the island some time since about the Miocene, but has not undergone speciation processes to any significant extent. This is probably due to sporadic gene flow between the island and mainland populations.

4.2.4.3 Examples of Dispersal and Vicariance in the Flora of Socotra

In the following, a number of examples supporting both vicariant and dispersal explanations for the occurrence of specific taxa on Socotra are presented in more detail. These explanations are often based on the results of recent molecular phylogenetic studies.

***Aerva* (Amaranthaceae)**

Aerva is a medium-sized palaeotropical genus with species found in various parts of Africa, Madagascar, the Mascarenes, Socotra, Arabia and South Asia. Molecular studies by Thiv et al. (2006) have revealed that the two Socotran endemics (*A. microphylla* – Fig. 4.36, *A. revoluta*) are sister taxa that arose about 5 Mya. Ecologically,

Fig. 4.37 *Barleria tetracantha*. Ra's Hebak, March 1997

they show contrasting habitat preferences, with *A. revoluta* restricted to the granites at higher elevations of the Hagghier mountains, whereas *A. microphylla* is characteristic of limestone at lower altitudes. Both endemics appear to be closely related to *A. artemisioides*, an endemic on limestone in southern Yemen, and all three have stronger phylogenetic affinities to *A. javanica* than to *A. lanata*. The apparently relatively young ages of the Socotran endemics (<10 My) in relation to the separation of Socotra (at least 15 Mya) led Thiv et al. (2006) to assume that *Aerva* colonised the islands via dispersal from an Eritreo-African ancestral area.

***Barleria* (Acanthaceae)**
Barleria is a mainly palaeotropical genus comprising ca. 300 species, only one of which (*B. oenotheroides*) is present in the New World tropics (Balkwill and Balkwill 1998). Four species have been recorded from Socotra (*B. aculeata*, *B. argentea*, *B. popovii* and *B. tetracantha* – Fig. 4.37), three, and possibly all four, of them being endemic to the island. One of the main centres of diversity of this genus is tropical East Africa. An infra-generic classification of *Barleria* allows seven distinct sections to be distinguished (Balkwill and Balkwill 1997). Three of the sections are represented by species on Socotra. Section Barleria has its main centres of distribution in South and East Africa, and two species belonging to this section, *B. aculeata* and *B. popovii*, are found on Socotra. The map in Balkwill and Balkwill (1998) is somewhat misleading, as it shows only one species for Socotra, but Balkwill and Balkwill (1997) list both species mentioned above as belonging to the section Barleria. The section Prionitis has it main centre of distribution in East Africa, as well as to a lesser extent in South and South-East Africa. One species of this section, *B. tetracantha*, occurs on Socotra. Section Somalia has its main centres

of distribution in South and East Africa. Only one species of this section, *B. argentea*, occurs on Socotra, even though the map in Balkwill and Balkwill (1998) indicates two. This is due to the confusion with the section Barleria mentioned above.

Two conspicuous biogeographically relevant features of the genus are the high levels of regional endemism and a strong geographic concentration of species belonging to the same infra-generic section. In this context, the fact that *Barleria* has a topochorous seed dispersal mechanism, i.e. the seeds are dispersed very close to the mother plant, is a particularly noteworthy feature. This would strongly suggest a vicariance explanation for the disjunct distribution pattern of the genus.

***Chapmannia* (Fabaceae)**

The genus ***Chapmannia*** (previously *Arthrocarpum*) comprises seven species in total with a remarkable disjunct distribution, four endemic to Socotra, one in Somalia and two in the New World (Thulin 1999b). Both the Somali–Masai and Florida–Mesoamerican clades are restricted to a seasonally dry tropical forest type vegetation (Lavin and Beyra Matos 2008).

In a biogeographical study involving dalbergioid legumes (Fabaceae), which include the genus *Chapmannia*, Lavin et al. (2000) initially suggested a vicariant biogeographical relationship between tropical North America and Africa. This assumption was based on very close molecular phylogenetic ties between the Florida–Mesoamerican and Horn of African–Arabian–Socotran clades. These and other dalbergioid disjunct distribution patterns indicated the importance of a Tertiary North Atlantic land bridge, present from about the Early Eocene to Mid-Miocene, in allowing range expansion of tropical and warm-temperate species. In fact Lavin et al. (2000) postulated that this land bridge was the principal historical reason why legumes are so concentrated in Africa and the Neotropics. Earlier, Liston et al. (1989) had argued that stochastic oceanic dispersal events during the Quaternary offered good explanations for the close phylogenetic relationship between the highly disjunct desert annual *Senecio flavus* of African–Arabian deserts and the North American *Senecio mohavensis*. More detailed molecular analyses by Lavin et al. (2004) led to the rejection of their vicariance hypothesis of transatlantic genera such as *Chapmannia*, due to the young age (ca. 8–16 My) of the clades involved, i.e. younger than the existence of the land bridge. In the light of these results, Lavin et al. (2004) suggested explanations for the disjunct taxa that centred on dispersal (i.e. ecological processes) from their ancestral area.

***Dracaena* (Dracaenaceae)**

The genus *Dracaena* consists of at least 60 species that are found in the Old World tropics, especially in Africa. Most of the species are geophytes or shrubs, but six arborescent taxa with very restricted occurrences are found scattered across northern Africa and southern Arabia. All six tree species occur in thermo-sclerophyllous communities. One of these, *D. tamaranae*, recently described from Gran Canaria, Canary Islands (Marrero et al. 1998), is extremely closely related to the more widespread species of the Canaries and north-western Africa, *D. draco*. *D. serrulata* is found in parts of southern Arabia, *D. ombet* and *D. schizantha* are known from localised occurrences in north-eastern Africa, and *D. cinnabari* (Fig. 4.38) is restricted to Socotra, where it grows in mist-affected areas of the island on both

Fig. 4.38 Flowers of *Dracaena cinnabari*

limestone and granite, usually above 300 m. This disjunct distribution has been interpreted as a relict of a continuous Late Miocene flora throughout northern Africa, i.e. it is therefore a case of vicariance.

***Echidnopsis* (Apocynaceae-Asclepiadoideae)**

Echidnopsis is a medium-sized genus that belongs to the more-or-less leafless stem succulents referred to as "stapeliads" (Bruyns 2000). Members of the genus are found in north-eastern Africa and Arabia (including Socotra). In a generic revision, Bruyns (1988) recognised 19 species with just one (*E. socotrana*) occurring on Socotra. More recently, four further endemics to Socotra have been described (Bruyns 2004). The results of a molecular phylogenetic study of the genus (Thiv and Meve 2007) suggest that the genus comprises 28 species. *Echidnopsis* has a typical Eritreo-Arabian distribution with species known from Tanzania, extending into the Horn of Africa and southern Oman. As with other stapeliads, the most probable ancestral area is north-eastern Africa (Bruyns 2000). Rather than being of vicariant origin, Thiv and Meve (2007) provide molecular evidence to suggest that the genus reached Socotra from East Africa by a single long-distance dispersal event.

***Exacum* (Gentianaceae)**

The genus *Exacum* (Gentianaceae) comprises about 64 species and shows a typical palaeotropical disjunct distribution (Klackenberg 1985, 2002; Thulin 2001). Madagascar is home to the majority of species, 38 in total. Fourteen species occur in the southern part of India and on Sri Lanka, a further seven are known from other parts of South-East Asia (including a small area in the far north of Australia), two from mainland Africa and three from Socotra and the southern-most part of the Arabian Peninsula (Yuan et al. 2005). These authors list *Exacum affine* (Fig. 4.39)

Fig. 4.39 *Exacum affine*. Homhil, December 2008

and *E. caeruleum* as occurring on Socotra and in southern Arabia. The third species they refer to is presumably *E. gracilipes*, which they examined in previous molecular studies (Yuan et al. 2003). Balfour (1883) originally described four species from Socotra alone, *E. affine*, *E. caeruleum*, *E. gracilipes* and *E. socotranum*. It appears as if the variable *E. affine* and *E. gracilipes* cannot be reliably separated, and probably constitute the same taxon, so Miller and Morris (2004) list *E. affine*, *E. caeruleum* and *E. socotranum* as occurring on Socotra, all endemic to the archipelago. *E. affine* and *E. gracilipes* were initially thought to occur in the Dhofar region of southern Oman, but it seems as if these populations represent two distinct endemic species (Thulin 2001).

Irrespective of the taxonomic details of the Socotran–Arabian taxa, Klackenberg (1985, 2002) suggested that the present-day distribution of the genus was a typical example of vicariance attributable to the break-up of Gondwana. If this were the case, then the assumption must be that *Exacum* is a very old genus, at least 130–150 Mya to be more precise. However, molecular studies by Yuan et al. (2003) indicate that the Gentianaceae diverged as a separate family only about 50 Mya, with the tribe Exaceae, to which the genus *Exacum* belongs, being about 40 My old. Such a relatively young age of the genus would appear to rule out firmly a vicariance hypothesis to explain the present-day distribution of *Exacum*.

Recent molecular phylogenetic studies by Yuan et al. (2003, 2005) show a high degree of congruency with the *Exacum* clade structure of Klackenberg (1985) that was proposed on the basis of morphological and anatomical features. However, the molecular studies point to a much closer relationship of Socotran–Arabian taxa with the Indomalesian clade than with any other, whereas Klackenberg's (1985) morphological phylogenetic analyses suggested a stronger affiliation with the African–Madagascan clade.

Given that the vicariance explanation of the current distribution pattern of *Exacum* cannot be invoked from recent molecular phylogenetic studies, a dispersal scenario seems to be the only viable alternative. The results of Yuan et al. (2005) suggest that *Exacum* originated in Madagascar during the Eocene at the earliest. From this centre of origin, long-distance out-of-Madagascar dispersal events first led to the colonisation of Sri Lanka and southern India. Whether this occurred directly or via stepping stones, such as the "Lemurian stepping stones" proposed by Schatz (1996), is unclear. The apparent absence of *Exacum* from the Seychelles would appear to weaken support for the Schatz hypothesis. Wind-mediated long-distance dispersal can probably be ruled out due to the distances involved, even though the seeds of *Exacum* can be very light. A more likely explanation is zoochory, especially as migratory birds have been strongly implicated in the long-distance dispersal of some species, such as woody Hawaiian violets (*Viola*), which, rather than being related to South American groups, were found to be strongly affiliated to Alaskan and eastern Siberian populations (Ballard and Sytsma 2000). Establishment of *Exacum* in southern India and Sri Lanka was followed by a substantial range expansion, including into Arabia and eastern Africa, accompanied by radiation. The current disjunct distribution is explained by a major range contraction leading to the extinction over large areas formerly occupied by members of the genus. This was probably triggered by Late Tertiary aridification. Isolated populations were able to survive in favourable locations such as Socotra, and within these populations, secondary diversification was initiated, leading to the formation of the closely related taxa present in today's flora.

***Fagonia* (Zygophyllaceae)**

As noted above, *Fagonia* comprises about 34 species with occurrences throughout an extended Saharo-Sindian region (with *F. luntii* and *F. paulayana* occurring on Socotra), in south-western Africa, western South America, western North America and north-eastern Mexico. On the basis of molecular studies, Beier et al. (2004) concluded that the Tertiary North Atlantic land bridge is the most probable explanation for the transatlantic disjunction, ruling out long-distance dispersal across the Atlantic from Africa to South America, as was suggested by Raven and Axelrod (1974).

***Launaea* (Asteraceae)**

The Old World genus *Launaea* has its centre of diversity in Africa and South-West Asia, including the Arabian Peninsula (Kilian 1997). The majority of species (ca. 90%) are found across northern Africa, extending through the Arabian Peninsula towards India, and from the Arabian Peninsula southwards along the East African coast. Six species of the genus are known from Socotra, of which one still remains to be described, and they are *Launaea crassifolia* (Fig. 4.40), *L. crepoides*, *L. massauensis*, *L. rhynchocarpa*, *L. socotrana*, and *L.* "sp. A" (see Miller and Morris 2004). Of these, four (*L. crepoides*, *L. rhynchocarpa*, *L. socotrana*, and *L.* "sp. A") are endemic to the Socotra Archipelago. The non-endemic Socotran species are mainly found in the north-eastern African–southern Arabian region, in the case of *L. massauensis* extending into southern Iran.

Fig. 4.40 *Launaea crassifolia*. Nuged plain, December 2008

Kilian (1997) assumes a South-West Asian origin of the genus. Some of the prominent disjunctions observed in the genus, mainly at the section level and particularly between East Africa and Asia, would initially appear to suggest a vicariant explanation for their distribution. This is also supported by the low diversity of species on islands such as Madagascar, even though the adjacent mainland African area is relatively rich in species. However, it is clear that some species of the genus have considerable powers of dispersal, as indicated by the distribution pattern of *L. sarmentosa*, which is found in East Africa, Madagascar, the Mascarenes, India, eastern Asia and Australia. It also occurs in these regions on many islands and islets, and is mainly associated with coastal vegetation types, strongly suggesting that the sea has played a major role in its dispersal.

***Scaevola socotraensis* (Goodeniaceae)**

Most of the 11 genera of the Goodeniaceae, a family comprising ca. 400 species, are restricted to Australia, but one genus, *Scaevola*, has dispersed and radiated throughout much of the Pacific region (Howarth et al. 2003). Forty of the ca. 130 species of the genus occur outside of Australia. *S. socotraensis*, a small endemic shrub, has so far been recorded only from its type locality at Ra's Qatanahan on Socotra (St John 1962), where it grows near freshwater springs and seepages. Whereas the Australian members of the genus have dry fruits, all of those known from outside Australia have fleshy ones that have presumably enabled the species to become more widely dispersed by sea. As a consequence, some of the species are typical of strand habitats, although others are more characteristic of inland locations.

The two most widespread dispersers are *Scaevola taccada* and *S. plumieri*, and both of these pantropical strand species are characterised by the ability of their fruits to remain viable after floating in seawater for several months (Howarth et al. 2003). St. John (1962) regarded the coastal *S. coriacea* from the Hawaii Islands as the most similar to *S. socotraensis*, but more recent phylogenetic analyses show that the latter is most closely related to *S. plumieri* (Howarth et al. 2003). This would fit in well with the hypothesis that many of the endemic island species of *Scaevola* originated from prominent colonisers (Howarth et al. 2003). Although *S. plumieri* does not occur on Socotra, it is known from the coasts of Africa and India, as well as from much further afield (but interestingly, not from the Hawaiian islands).

***Zygocarpum* (Fabaceae)**

Zygocarpum is an endemic genus to the Horn of Africa, Socotra and parts of southern Arabia. On Socotra it is represented by one species, *Z. coeruleum* (see Fig. 6.102). *Zygocarpum* was once included in the genus *Ormocarpum*, but molecular studies supported the separation into a distinct genus (Thulin and Lavin 2001). However, it does belong to the *Ormocarpum* group, a monophyletic clade comprising three Old World genera and two from the New World (Thulin and Lavin 2001). The transatlantic disjunct distribution was attributed by Lavin et al. (2000) to Tertiary vicariance, i.e. a formerly continuous distribution that became fragmented with the severing of the North Atlantic land bridge, although subsequent analyses (Lavin et al. 2004) strongly support a dispersal explanation. Thulin and Lavin (2001) presented molecular evidence suggesting that the distribution of *Zygocarpum* on either side of the Gulf of Aden is also due to vicariance with the rifting separation of Arabia from Africa during the Late Tertiary. The very restricted occurrences of populations within the present-day distribution of the genus point to a low degree of dispersability, therefore implicating vicariance. However, given the conclusions of Lavin et al. (2004) dismissing vicariance as an explanation of continental disjunct distributions in certain legumes, it appears as if a dispersal scenario cannot be ruled out.

4.2.4.4 Dispersal and Vicariance in the Fauna of Socotra

With respect to Socotra, the dispersal–vicariance debate is by no means restricted to plants. One of the regionally most salient examples of dispersal in animals is the occurrence of chameleons in South Asia and on some of the Indian Ocean islands. Socotra harbours one endemic species, *Chameleo monachus* (Fig. 4.41) (Wranik 1999). Rather than being attributable to a Gondwanan vicariant history, strong evidence has emerged that their distribution in South Asia and on Indian Ocean islands is due to long-distance oceanic dispersal from Madagascar, their putative centre of origin (Raxworthy et al. 2002). This conclusion is even more remarkable given the fact that chameleons are essentially terrestrial organisms, and that several out-of-Madagascar dispersal events must have taken place. How the chameleons managed to disperse across the ocean remains open to speculation. However, reptiles are known to cling to driftwood occasionally, allowing them to be transported for

Fig. 4.41 *Chameleo monachus*. Wadi Ayhaft, December 2008

considerable distances across open bodies of water (for examples, see de Quieroz 2005). It has been speculated that during its northwards movement from Madagascar, India came much closer to Arabia than previously assumed (Briggs 2003). Taking this into consideration and from phylogenetic data, Macey et al. (2008) favour this relatively short-distance dispersal hypothesis over a formerly continuous Gondwanan distribution followed by wide-scale extinction to explain the occurrence of *C. monachus* on Socotra.

Dispersal has also been implicated for colubrid snakes, of which Socotra is home to two species, *Hemerophis* (formerly *Coluber*) *socotrae* and *Ditypophis vivax*. Both genera are endemic and monotypic. The results of Nagy et al. (2003), although not conclusive, appear to indicate a multiple colonisation of Socotra by colubrids, possibly from Madagascar. Because the time at which Socotra became detached from mainland Africa is still disputed (with suggestions ranging from ca. 14 Mya to 70 Mya – see Sect. 3.9), and molecular dating suggests a fairly recent Cenozoic origin of the Socotran colubrids, a vicariant history cannot strictly be ruled out, even though this is unlikely given that most of Socotra was submerged until quite recently in its geological history.

4.2.4.5 Invaders and Colonisers

It can be assumed that with the arrival of the first settlers from Arabia, India and Africa from about 5,000–11,000 years ago, many native plant communities of Socotra have been subject to pronounced changes in species composition. It is possible to distinguish two distinct groups of species that have benefited from human impact, and following the terminology of Grime (2001), these are referred to as "invaders" and "colonisers". Invaders are introduced species that have expanded their former range, often quite dramatically, whereas colonisers are species that were already native to the region but have increased their presence in the vegetation cover as land-use patterns have changed. Although the distinction between these two groups is clear in theoretical terms and some obvious examples exist, it may not be quite as straightforward to assign any particular species to one of the two categories, mainly because the original flora before the arrival of humans is unknown.

Invaders (adventives) include not only pantropical or cosmopolitan weeds found in date palm groves and other cultivated areas, but also weeds that persist and have become naturalised in native plant communities. Examples of invaders are *Acanthospermum hispidum*, *Achyranthes aspera*, *Ageratum conyzoides*, *Amaranthus dubius*, *A. graecizans*, *Argemone mexicana*, *Bidens biternata*, *B. pilosa*, *Capsella bursa-pastoris*, *Chenopodium murale*, *Misopates orontium*, *Mollugo nudicaulis*, *Prosopis juliflora*, *Ricinus communis*, *Veronica polita* and *Zaleya pentandra*. Some of these species, for instance *Bidens pilosa* or *Capsella bursa-pastoris*, remain quite rare and do not appear to constitute any serious threat to the native flora, but others, such as *Ageratum conyzoides*, *Argemone mexicana* and *Prosopis juliflora*, are not so benign. The last of these in particular has wreaked havoc in some parts of the Arabian Peninsula where it now dominates at the expense of native plant communities (such as in eastern parts of the UAE, in damp wadis in parts of Oman and Yemen). Furthermore, invaders are not restricted to annuals, but are represented by various plant functional groups. Most, if not all, have in common the ability to produce large quantities of propagules within a very short time.

A good example of a coloniser is the widespread herb *Tephrosia apollinea*, which, due to massive overgrazing, has been able to increase its presence in the vegetation cover dramatically, especially at lower altitudes. The same probably applies to the endemic *Jatropha unicostata*, and also to *Adenium obesum* (see Sects. 7.3.1 and 7.3.2). All of these species contain toxic compounds, and are therefore not grazed to any extent by domestic livestock. It is unclear whether *Calotropis procera* is originally native to the archipelago. Popov (1957) noted that on Socotra it was observed most commonly in sand dunes on the Nuged plain. This shrubby tree is generally regarded as a weed and an indicator of overgrazing in many parts of its range. For instance, it dominates locally on sand sheets in Dubai Emirate (Brown, pers. observation), where high camel densities are present and sufficient moisture is available, or on the rolling hills above the escarpment woodland in Dhofar, southern Oman (see Sect. 7.3.8). It is also found as a ruderal in urban sites in some parts of mainland Arabia. On Socotra, we did not observe it in the habitat mentioned by Popov (1957), but found it to be abundant in neglected areas of the capital Hadiboh, where it was growing as a ruderal. Furthermore, young plants were noticed at regular intervals beside the newly constructed road at lower elevations, indicating that anthropogenic disturbance is contributing to the spread of this species, as is the case with railway lines, riverbanks and other such corridors in other parts of the world. In some instances, individual older *Calotropis* trees were found within native vegetation in wadis (Fig. 4.42), but did not appear to be spreading there.

Fig. 4.42 *Calotropis procera* in a wadi near Hadibuh. December 2008

Davis et al. (2000) have recently proposed an ecological theory that underlies the intermittent nature of invasibility in most plant communities, identifying resource availability as the main controlling factor. This theory can be summarised as follows: a plant community becomes more susceptible to invasion when there is an increase in the amount of unused resources. From this theory, it follows that if competition for resources is reduced, for instance due to overgrazing, other non-resident species can become established. Furthermore, it can also be inferred from this theory that arid habitats are generally less at risk from invasion than more mesic ones. On Socotra, therefore, habitats that are theoretically most at risk from introduced aliens are in particular damp wadi beds and the climatically more favourable locations at higher altitudes.

4.3 Ferns

Twenty-six species of fern are known from the Socotra Archipelago (Miller and Morris 2004). Only one of these, *Asplenium schweinfurthii*, is endemic. Most of the others have a wide distribution throughout the Old World Tropics, with the Adder's Tongue *Ophioglossum reticulatum* being pantropical, and *Pteridium aquilinum* (bracken), frequent on some of the high mountain peaks on Socotra, cosmopolitan. Despite this broad distribution, some of the species are known only from single or a handful of locations on Socotra, although they may well have been overlooked.

Fig. 4.43 *Actiniopteris semiflabellata*, a widespread fern

Fig. 4.44 *Actiniopteris semiflabellata* when dry and physiologically inactive

The majority of species inhabit shaded locations amongst rocks and boulders or in rock crevices. The endemic *Asplenium schweinfurthii* is fairly common in montane thickets on Socotra itself, and was even recorded from Abdalkuri in the nineteenth century. *Actiniopteris semiflabellata* is a widespread fern in the mountains, on both limestone and granite (Figs. 4.43 and 4.44). It is a typical example of a poikilohydric plant: one that can tolerate extreme desiccation of its aerial organs, but rapidly becoming fresh green and resuming normal photosynthetic activity when moisture becomes available. As such, it is one of the most drought-tolerant of ferns in the archipelago, along with *Onychium divaricatum.* Conversely, *Marsilea coromandeliana* is a semi-aquatic fern of moist muddy locations, and *Ceratopteris cornuta* is the only truly aquatic fern species known from Socotra, inhabiting water-holes on the Hadibuh plain.

4.4 Non-Vascular Cryptogams

Over the past decade, there has been an upsurge in studies dealing with some groups of the non-vascular flora of Socotra, so that a good overview now exists in particular of the bryophytes and lichens. Information is also forthcoming on the marine algae (e.g. Schils and Coppejans 2003, and Schils in Cheung and DeVantier 2006). Perhaps the most detailed overview of the marine organisms and their habitats was undertaken by Klaus and Turner (2004).

The biogeography of the bryophytes and lichens is of considerable interest, as these groups are quite ancient in geological terms and occur in comparable habitats throughout the world. Kappen and Straka (1988) have provided experimental evidence for the long-distance transport of both phanerogam pollen and cryptogam spores, thus implicating the role of spores in particular in long-distance dispersal of the relevant organisms. The results of Heinrichs et al. (2006) suggest that the global distribution of the liverwort genus *Plagochila* can in some cases also be attributed to rare long-distance dispersal events.

4.4.1 Bryophytes

Up until quite recently, knowledge of the bryophyte flora of the Socotra Archipelago was very limited, with few published papers. However, the situation has now changed markedly with the intensive work carried out over the past decade by Harald Kürschner and co-workers (e.g. Al-Gifri et al. 1995; Frey and Kürschner 1988; Kürschner 2000a, b, 2003a, b, 2004a, b, 2006b; Kürschner and Ghazanfar 1998; Kürschner and Ochyra 2004; Kürschner and Sollman 2004), culminating in the publication of a bryophyte flora for the main island of Socotra (Kürschner 2006a). Apart from providing a key with which to identify all currently known species (listed in Appendix 2), this paper also presents a phytogeographical analysis of the bryophyte flora of the island, and gives an overview of the history of bryophyte exploration.

The first account of the bryophytes of the island, written by Mitten (1888), appeared in Balfour's 1888 work on the vascular flora. A further overview was given by Balfour (1903). Georg Schweinfurth collected samples in 1881, which were later identified by Müller (1901), and eventually compiled by Long (1986). Townsend (1969) added further taxa collected during the 1967 expedition that included Radcliffe-Smith (see Sect. 4.1). By this time, the total number of species had risen to 27: 19 mosses and eight liverworts. Crosby et al. (1983) revised the bryophyte flora of the Indian Ocean area. Mies (1994) and Al-Gifri et al. (1995) listed 11 liverworts (Hepaticae) and 25 mosses (Musci) in total, among them new taxa for the islands and the endemic *Sematophyllum socotrense*. It seems as if there has been taxonomic confusion surrounding *Fossombronia crispa*, which at the time was regarded by these authors as new to the bryophyte flora of South-West Asia. According to Kürschner (2006a), the specimens recorded for the island most probably refer to *F. caespitiformis*.

The following overview of the bryophyte flora is based to a large extent on the work of Kürschner (2006a). According to this author, 80 species are now known from Socotra: 48 mosses, 31 liverworts and one hornwort (Appendix 2). Six species are considered as endemics, although the taxonomy of one, *Barbula schweinfurthiana*, remains unclear. The other five endemics are *Schlotheimia balfourii*, *Sematophyllum socotrense*, *Tortella smithii*, *Weissia artocosana* and *W. socotrana*. These six endemics, which underline the bryophytogeographical significance of Socotra, represent about 7.5% of the total number of species, a markedly lower fraction than the vascular flora.

The highest proportion of species (ca. 48%) are typical tropical elements, and Kürschner (2006a) further divides them into four subgroups: (1) pantropical species, i.e. occurring throughout the tropics; (2) species with an African–SE Asian–Australian distribution pattern; (3) Afromontane species, a group comprising species restricted to evergreen forests on Socotra at higher altitudes, and with the same or closely related species also occurring in the mountainous regions of eastern Africa; and (4) species with an African–Socotran distribution, representatives of which, in contrast to the previous group, are usually found in woodlands at lower altitudes on Socotra, and can be regarded as typical xerotropical African elements.

About 13% of the species on Socotra are cosmopolitan, including the hornwort *Anthoceros punctatus*, which has been recorded from a variety of moist habitats, especially at lower altitudes.

Northern taxa account for about 9% of the bryophyte flora, and are derived from species with a former Pangaean distribution, but away from the xerothermic and arid regions. These taxa are today often characteristic of boreal and temperate areas with higher rainfall. They were formerly able to migrate southwards along both eastern and western routes into southern Arabia and Socotra during periods of higher precipitation.

On the drier limestone plateaus and at lower elevations, bryophytes of xerotherm-Pangaean or circum-Tethyan origin predominate. A number of thallose Marchantiidae species, including *Exormotheca pustulosa* and *Oxymitra incrassata*, belong to the first category, which accounts for about 6.5% of the bryophyte species of the island.

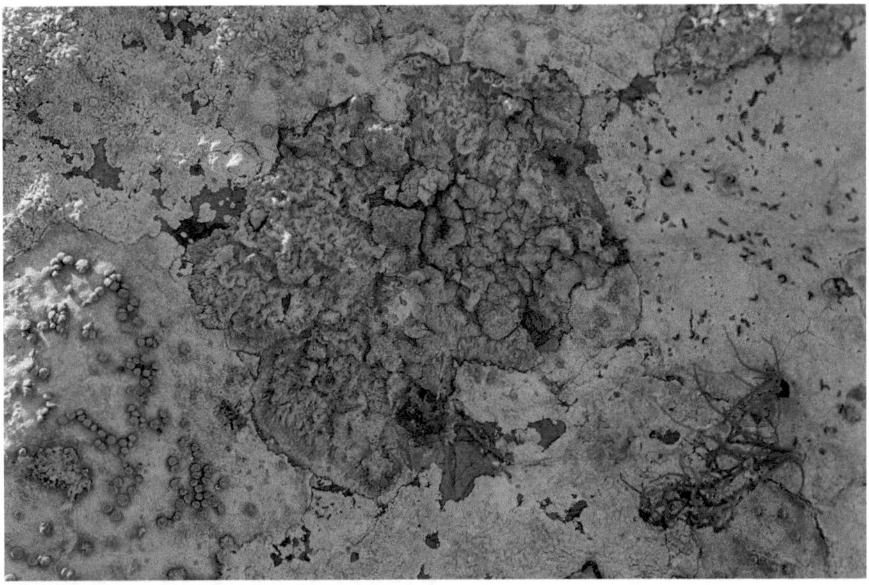

Fig. 4.45 *Roccellographa cretacea*. Limestone rock, Hamaderoh, March 1997

The proportion of circum-Tethyan species is larger (ca. 16.6%), and this group contains species that were formerly distributed along the northern and southern coasts of the Tethys Sea. Both groups are much better represented in the bryoflora of mainland Yemen, where they are found in the inland desert regions (Mies 2001).

4.4.2 Lichens

Müller-Argau (1882a, b, 1887, 1888) and Darbishire (1903) recorded 154 species of lichens from the Socotra Archipelago. Müller-Argau also examined the lichen samples he had received from I.B. Balfour and G. Schweinfurth. The genus *Minksia* (Opegraphaceae) was first described from Socotra by Müller-Argau (1882a), but further species were later discovered in Chile, Central America and the Andaman Islands in the Indian Ocean. Darbishire (1903) also studied material collected during the expedition of H.O. Forbes in 1898/99, assigning specimens to species on the basis of descriptions provided by Müller-Argau. Lichens obtained during an Austrian expedition that was undertaken at the same time as that of Forbes were examined by J. Steiner (1907). Samples collected from Samhah allowed him to describe the genera *Roccellographa* (Fig. 4.45) and *Simonyella* (Fig. 4.46) as new to science. Each of these genera are monotypic. Another monotypic genus, *Feigeana* (Fig. 4.47), was described by Mies et al. (1995) with its sole representative *F. socotrana*. This species has so far been recorded only from the type locality on limestone rocks in the

Fig. 4.46 *Simonyella laevigata*. Limestone rock, Mumi, January 1999

Fig. 4.47 *Feigeana socotrana*. Limestone rock, Mumi, January 1994

Fig. 4.48 *Roccella hertelii*, *Pertusaria socotrana* (a yellowish crustose species, and covering most of the rock in this image) and presumably a few small patches of *Roccellographa cretacea*. Dixam plateau, December 2008

Mumi area of eastern Socotra. As *Roccellographa* and *Simonyella* have been found in coastal mountain ranges around the north-western Indian Ocean in recent years (Tehler 1990; Kürschner and Ghazanfar 1998), where they are fairly common representatives of a characteristic saxicolous lichen association, it is conceivable that *Feigeana* will also be discovered there in the future.

Many more lichen than bryophyte species have been recorded from Socotra (Appendix 3), and the current number stands at 254 (Schultz and Mies 2003; Mies and Schultz 2004), including some species recently described by Schultz (2005) and Breuss and Schultz (2007). The most common species belong to the mycophytic order Arthoniales (including Opegraphales), which have *Trentepohlia* as their phycobiont. This is concordant with the situation in most other arid subtropical and maritime regions of the world (Mies and Printzen 1997). Even so, it can be assumed that many more species are likely to be discovered, especially at higher altitudes where species of the wet tropics occur, as indicated by the recent discovery of the epiphyllic Ectolechiaceae *Tapellaria epiphylla* by Mies and Schultz (2004). About 67 species and five infra-specific taxa are thought to be endemic.

The high number of genera of the Roccellaceae on Socotra (Fig. 4.48), some of them endemic to the islands and the adjacent coasts, is of considerable biogeographical interest. Table 4.3 shows the genera of Roccellaceae currently recognized, along with their respective known ranges (Mies and Printzen 1997). Seven of the 19 genera occur on Socotra. Two genera, *Roccellographa* and *Simonyella*,

Table 4.3 Lichen genera of the Roccellaceae and their distribution

Genus, author and date	Number of species	Distribution
Roccella DC. 1805	36	Tropical, subtropical (incl. Socotra)
Dirina Fr. 1825	10	Tropical, subtropical (incl. Socotra)
Combea De Not. 1846	1	S. Africa
Schizopelte Th. Fr. 1875	1	S. California
Minksia Müll.-Arg. 1882	6	± Pantropical (incl. Socotra)
Dendrographa Darb. 1895	2	Mexico, California
Ingaderia Darb. 1897	3	Chile, Peru, Mediterranean
Pentagenella Darb. 1897	1	Chile
Reinkella Darb. 1897	1	Peru
Roccellaria Darb. 1897	1	Chile, Peru
Roccellina Darb. 1898	25	Tropical, subtropical (incl. Socotra)
Roccellographa J. Steiner 1907	1	Socotra, Somalia, Masirah/Oman
Simonyella J. Steiner 1907	1	Socotra, Masirah/Oman
Roccellodea Darb. 1932	1	Galápagos
Dolichocarpus R. Sant. 1949	1	Chile
Gorgadesia Tav. 1964	1	Cape Verde
Hubbsia W.A. Weber 1965	1	S. California, Guadaloupe
Sigridea Tehler 1993	4	± Tropical, subtropical
Feigeana Mies, Lumbsch and Tehler 1995	1	Socotra
Follmanniella Peine and B. Werner 1995	1	Chile

that are currently known only from Socotra, Masirah Island (off the eastern coast of Oman) and Cape Guardafui (Somalia) appear to have a very restricted range where they are part of a coastal saxicolous community on limestone cliffs.

A similar high diversity within the Roccellaceae can be found only in the coastal areas of Central and South America. The climate of the coastal mountain ranges in these three regions is characterised by aridity, but also occasional dewfall. At first sight, it appears as if the specific climatic conditions are responsible for this disjunct distribution. However, other regions of the world have a broadly similar climate, for instance the escarpments flanking the Red Sea (which are relatively close to Socotra), the south-western part of Madagascar, the south-western coasts of Africa and the coastal Campo Caatinga in north-eastern Brazil, but they have a low diversity of Roccellaceae, with *Roccella* itself often being the only genus present. It is therefore apparent that other factors need to be taken into consideration. For example, tectonic events could have played a major role. This point is supported by the observation that many of the genera are monotypic and quite evidently palaeoendemics, being restricted to a very limited area. Conversely, the high number of currently accepted species of *Roccella* in Macaronesia and the Galápagos can most probably be attributed to recent speciation events within the genus, similar to the situation in certain vascular plant groups.

Tehler (1983) explained the present-day distribution of some *Roccellina* taxa with the fragmentation of Gondwana. Sipman (1983) used a similar argument to interpret the distribution of species belonging to the Megalosporaceae. Kärnefelt (1979) dated the segregation of Cetrarioid taxa back to the time when Pangaea separated into Laurasia and Gondwana, and the subsequent fragmentation of the latter. Schultz et al. (1999) believe *Paulia*, a genus belonging to the ancient lichen family Lichinaceae, also to be of a Pangaeic age. Phylogenetic analyses of fresh and recently collected herbarium material were used to examine the origins and relationships within the order Lichinales, an order that almost exclusively possesses cyanobacteria as the primary photobiont and to which the Lichinaceae belong (Schultz et al. 2001). The Lichinaceae not only from Socotra, but also worldwide and especially from Arabia, have been the subject of a series of recently published papers by Schultz and co-workers (e.g. Schultz 2003, 2005; Schultz and Büdel 2002; Schultz and Mies 2003; Schultz et al. 2000, 2001). Galloway (1988) assumed that the close association of several macrolichen genera with *Nothofagus* phorophytes from the austral cool temperate zone, as well as the occurrence of certain genera in shrublands, sub-alpine bogs and grasslands, can be attributed to their Gondwanan distribution. As the tectonostratigraphy throughout the Southern Hemisphere is quite complicated, he was unable to give a simple interpretation of lichen distribution. For instance, the southern island of New Zealand is composed of at least nine terranes. In contrast, Wedin (1995) refutes a tectonostratigraphic argument for the Southern Hemisphere genera of the Sphaerophoraceae in Australia, New Zealand and southern Chile, preferring the explanation of long-distance dispersal facilitated by the prevailing westerly winds. In the case of Socotra, the archipelago has occupied a climatologically and geologically relatively isolated position since at least the Cretaceous, and its simple geological history would appear to support a vicariant explanation of lichen distribution.

Many biogeographical aspects of the Socotran lichen flora correspond well with the palaeogeographical findings from bryophytes. Arvidsson (1982) suggested that the genus *Coccocarpia* once occupied a continuous moist tropical belt throughout Gondwana, possibly even further south than it is found today. The present isolated occurrences of the genus are interpreted as relicts from a warmer period. Vicariance can also be invoked as an explanation for the present disjunct distribution patterns of many Socotran lichen taxa, which are known to have been previously more widespread. For instance, the fact that the Roccellaceae have centres of endemism in Central and South America and Socotra cannot be explained by long-distance dispersal, but can more convincingly by tectonic events.

Chapter 5
Ecology and Adaptive Strategies

Abstract Most of the Socotra Archipelago is occupied by desert or semi-desert vegetation types, and deserts impose severe restrictions on plant growth. Few detailed studies have been carried out on adaptive features of Socotran plants to their environment, and the information that is forthcoming mainly relates to plants of the desert plains in both the north and south of the island. In this chapter, some key adaptations to arid desert environments are discussed, followed by a brief overview of the adaptations to saline and high montane ecosystems. The Raunkiaer life-form classification system is of limited use for predominantly desert ecosystems, because it is rather restricted in its structure–function relationships relevant to desert life. The five main types are briefly discussed, followed by a more detailed overview of the major growth-forms. Other classification systems place greater emphasis on the role of physiological adaptations, and "plant functional types", which reduce the complexity of species diversity in biological systems to a limited number of key types, can facilitate a better understanding of important ecosystem processes. Several of the more common functional classification systems are outlined, and their application to the flora of Socotra is indicated. Short accounts of topics such as pollination ecology, including the potential significance of bird-pollination, seed dispersal and germination are given.

5.1 Introduction

Grime (2001) defines stress as the external constraints that limit the rate of dry matter production of all or part of the vegetation. As outlined in Chap. 6, with the notable exceptions of shaded wadis and high montane areas, most of the Socotra Archipelago is occupied by desert or semi-desert vegetation types. Desert environments impose severe restrictions on plant growth, at least at certain times of the year, and in this context, the environmental parameter most affecting plant life is low, variable and unpredictable water availability, often coupled with excessive

G. Brown and B.A. Mies, *Vegetation Ecology of Socotra*, Plant and Vegetation 7,
DOI 10.1007/978-94-007-4141-6_5, © Springer Science+Business Media Dordrecht 2012

levels of solar irradiation. Other stresses to which plants are frequently exposed include low relative humidity and extremely high potential evapotranspiration, as well as high air and soil surface temperatures (Gibson 1996). Furthermore, elevated substrate salinity can be a stress factor in some desert areas (Orshan 1986), and this is also the case on Socotra locally. Xerophytes, plants inhabiting ecosystems characterised by drought, comprise a diverse group of species with a wide variety of sometimes strongly contrasting morphological or physiological adaptations.

At high altitudes, especially above the cloud zone, plants are subject to a variety of environmental stress factors, including high rates of insolation, temporary drought, strongly contrasting diurnal temperatures and high winds, the last of these having the potential to cause severe mechanical damage.

Despite, or perhaps partly because of, these stresses, plant communities on Socotra often contain a remarkable diversity of species, many endemic, and, with respect to the predominantly arid ecosystems, biomass production can be surprisingly high.

It should be emphasised that very few detailed studies have been carried out on adaptive features of Socotran plants to their environment, and information that is forthcoming mainly relates to plants of the desert plains in both the north and south of the island. Some of this information can be inferred from studies on the same or closely related species from comparable habitats in adjacent mainland regions. In the following, some key adaptations to arid desert environments will be discussed, followed by a brief overview of the adaptations to saline and high montane ecosystems.

5.2 Adaptation of Plants to Desert Ecosystems

5.2.1 Photosynthetic Pathways

The three main pathways for photosynthesis (C_3, C_4 and CAM) are all represented in the flora of Socotra. Most of the drought-deciduous species that are inactive during the hot, dry period are likely to be typical C_3 plants, as are the majority of annuals. This is probably also the case with the evergreen species at higher altitudes, although even superficial studies are so far lacking. A substantial proportion of the plants that are active during the hot summer months have either the CAM or the C_4 pathway, with the former being often characteristic of succulent species of dry habitats. Plants inhabiting coastal marshland, including the succulents, are usually of the C_3 or C_4 type, and physiological aspects of these halophytes are dealt with in Sect. 5.3. C_4 metabolism is generally regarded as an important device that allows plants to maximise rates of photosynthesis at ambient temperatures that are supra-optimal for C_3 plants (Rundel and Gibson 1996). This is underlined by the fact that most C_4 species occur in the warmer regions of the world (Larcher 2003). Apart from having a higher temperature optimum for photosynthesis, C_4 species have a superior water-use efficiency (WUE) to C_3 species, often nearly twice as high (Barbour et al. 1998), i.e. they lose less water for equivalent amounts of carbon acquired. Although plant

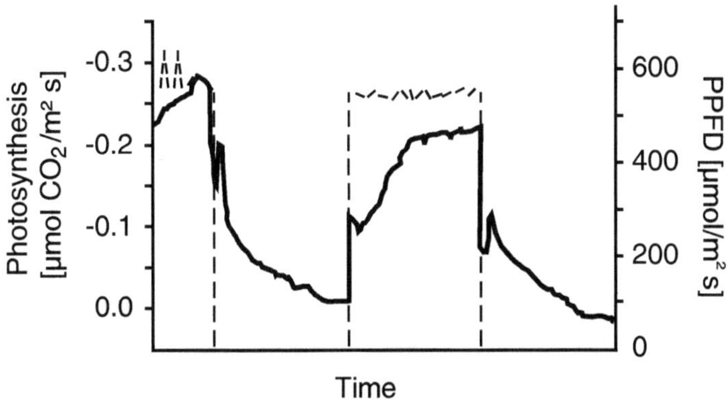

Fig. 5.1 Net rate of CO_2 exchange (*solid line*) in *Dracaena cinnabari* exposed to several weeks of drought to demonstrate the occurrence of CAM

species show a considerable acclimation potential of photosynthesis to temperature, which varies greatly with plant species and ecotype, the C_3 mode of photosynthesis is overall more efficient than the C_4 type at lower temperatures, conditions that are prevalent during part of the year throughout much of Socotra.

Results of Mies et al. (1996) show that the spartinoid succulents of the genus *Euphorbia* in the Canary Islands, namely *E. aphylla* and the introduced *E. tirucalli*, are characterised by CAM. However, if there is an adequate supply of water, the small leaves of these species that develop at the beginning of the rainy period display the C_3 mode of photosynthesis. This pattern is also likely to be found in corresponding plants in Socotra, for instance in *E. arbuscula* and the widespread *E. schimperi*. Initial experimental findings suggest that *Dracaena cinnabari*, a plant that produces very large, tough leaves, is a facultative CAM plant. With adequate provision of water, the leaves exhibit typical C_3 photosynthesis, but they gradually change to CAM when exposed to increasing levels of drought (Fig. 5.1). Figure 5.2 shows that species such as *Aloe perryi*, *Caralluma socotrana* (Fig. 5.3) and *Cissus subaphylla* (Fig. 5.4) are typical CAM plants, as indicated by the more acidic sap during the very early morning. The same applies to the leaves of *Jatropha unicostata*, although this species is not strictly a succulent. Leaves of *Dendrosicyos socotrana* and *Adenium obesum*, both species with a succulent stem but non-succulent leaves, are characterised by the C_3 mode of carbon fixation.

5.2.2 Plant Water Relations

5.2.2.1 Leaf and Stem Adaptations to Counteract Low Water Potentials

Whereas some species have small, narrow leaves to avoid the potentially damaging effects of high radiation loads (see Sect. 5.2.3.1), others characteristic of hot, dry environments on Socotra have relatively large leaves. In general, these species

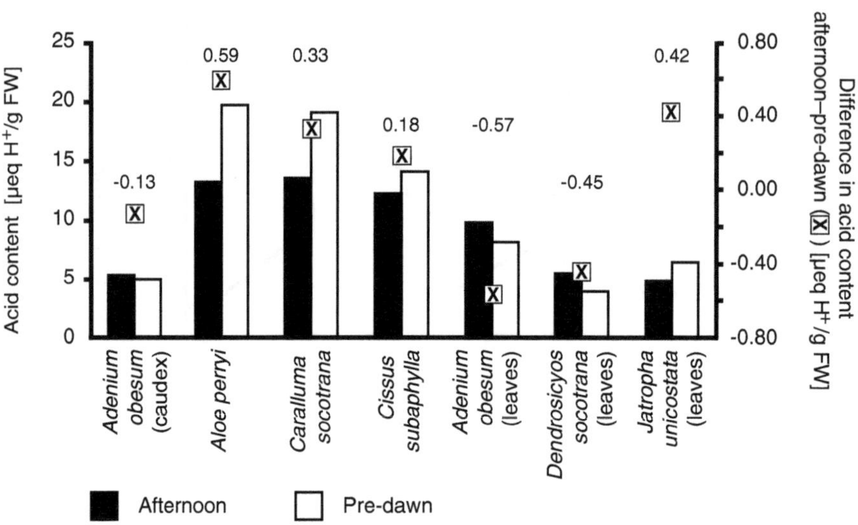

Fig. 5.2 Titratable acid content in the green organs of various succulent species measured in the afternoon (*black columns*) and pre-dawn (*white columns*)

Fig. 5.3 *Caralluma socotrana*. Dixam plateau, December 2008

tend to discard their leaves during the summer months and remain dormant, i.e. they display a typical drought-avoidance strategy. This is, for instance, the case with the widespread *Croton socotranus*, but it also develops smaller leaves in response to the onset of drought, or in locations with reduced water availability,

Fig. 5.4 *Cissus subaphylla*. Hadibuh plain, December 2008

before finally discarding them almost completely. *Jatropha unicostata* also disposes of its leaves when confronted with acute water shortage, although the apical ones may remain active. As there is often pronounced small-scale variation in water availability on the plains and slopes on Socotra, it is still possible to see *Jatropha* with a full complement of leaves well into the summer, for example in shallow drainage water channels, but dormant plants without leaves on the adjacent plain proper. Large-leaved phreatophytes, deep-rooted terrestrial species that obtain a regular supply of water from the water table or other sources, are dealt with in Sect. 5.5.2.4.

Most succulents on Socotra display the same characteristic physiological adaptations to drought that have been described in these plants from other parts of the world (e.g. Rundel 1982; von Willert et al. 1992). During the day, the plants close their stomata to minimalise transpirational water-loss, as exemplified by the CAM plant *Euphorbia arbuscula* (Fig. 5.5). Notable exceptions are the C_3 plant *Adenium obesum*, and presumably the morphologically similar *Dorstenia gigas*, which both have a woody, strongly succulent trunk, but lack a green cortex, so photosynthesis is restricted to the non-succulent leaves.

5.2.2.2 Root Systems

During periods of extended drought, which often correspond to very high surface temperatures, soil-water potentials in typical desert ecosystems may remain below

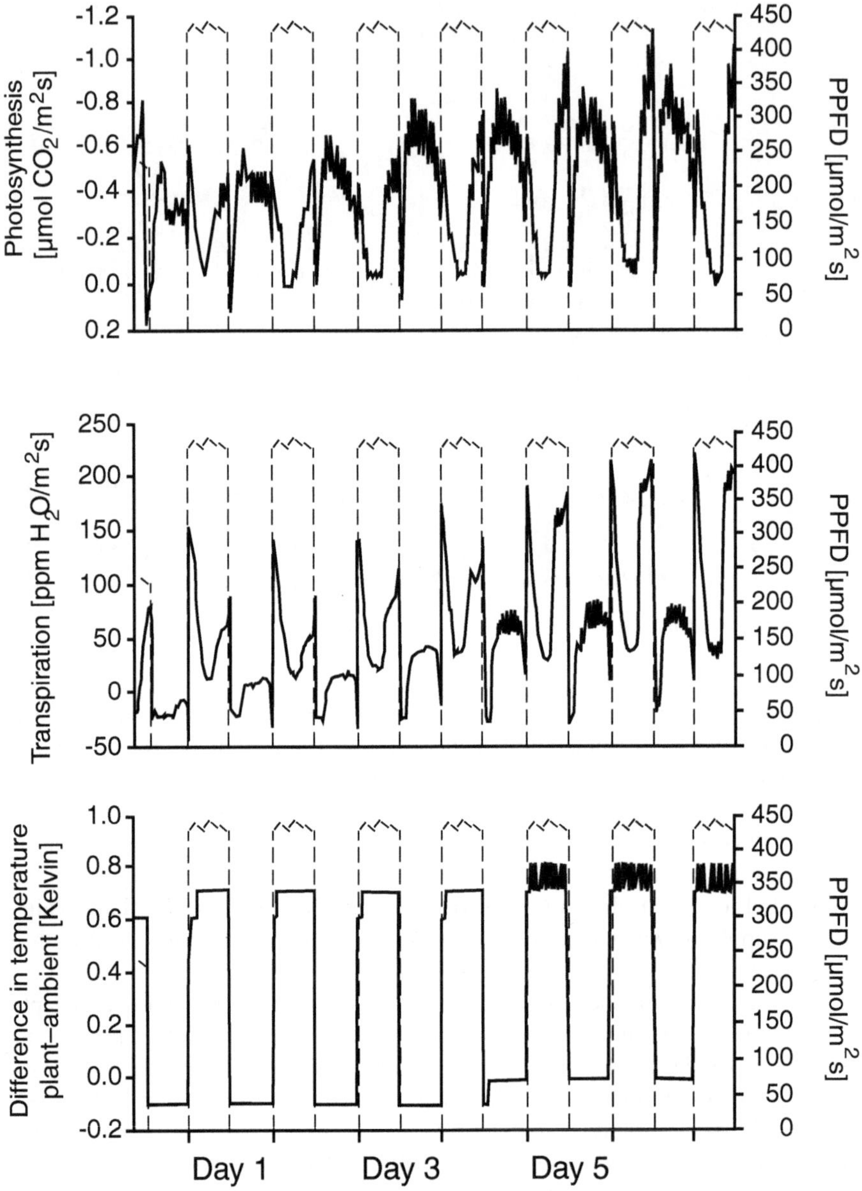

Fig. 5.5 Net rate of CO_2 exchange, transpiration and difference in plant surface temperature in *Euphorbia arbuscula* under artificial lighting conditions (12-h day–night regime) after exposure to drought

−10 MPa (Gibson 1996). At such times, the ability of plants to maintain a positive water balance can be severely tested. Little specific information is forthcoming on the rooting architecture of plant species on Socotra, and most of what is known relates to the more widespread species. Many of the desert dwarf shrubs presumably possess a well-developed system of lateral roots with one or several deeper taproots, as described for corresponding species in other arid regions of the world (e.g. Rundel and Gibson 1996). Some evidence has been presented to indicate that the regular spatial patterning sometimes reported in desert shrubs, and also occasionally observed on Socotra, is attributable to underground competition for water (e.g. Beals 1968; King and Woodell 1973), but Barbour (1973) and Ebert and McMaster (1981) regarded this evidence as inconclusive. The role of competition in water-limited ecosystems in vegetation patterning has since been reiterated by several authors (e.g. King and Woodell 1984; Prentice and Werger 1985; Allen et al. 2008).

Some of the more common small succulents often give the impression of being fairly scattered in the landscape. Closer inspection reveals that this applies to the aerial organs only, as the roots of different individuals may be in intimate contact with each other and competing for resources. Most succulents tend to possess shallow root systems. This may in part be a direct consequence of the specific habitat conditions, which are often characterised by shallow, stony and well-drained substrates with low water-holding capacity. However, shallow root systems enable the plants to exploit even small amounts of rainfall and early morning dewfall that may not penetrate the ground to any great degree. After rainfall, many succulents rapidly develop an extensive fine-root system near the surface, sometimes up to several metres in length, only to discard it during periods of drought, leaving the thicker roots remaining.

Rhizosheaths are structures composed of persistent root hairs, mucilage secreted from the roots of the plants and sand particles that form a cylinder around the actual root. It has been suggested that a chief role of rhizosheaths is to promote water absorption and to protect the roots from desiccation (Volkens 1887; Price 1911). It has also been determined that rhizosheaths provide a favourable "nanohabitat" for soil bacteria, including some that are able to fix atmospheric nitrogen (Wullstein et al. 1979). The majority of species possessing rhizosheaths are perennial graminoids of nutrient-poor desert dunes and sand sheets (Danin 1996). However, these structures have also been recorded in other monocot families (e.g. Liliaceae), several dicots and a limited number of phreatophytes (see Danin 1996). Several desert species from the Arabian mainland possess rhizosheaths, including the graminoids *Centropodia forsskaolii*, *Cyperus conglomeratus*, *Lasiurus scindicus*, *Panicum turgidum*, *Pennisetum divisum*, *Stipagrostis drarii* and *S. plumosa* (Danin 1996 – Fig. 5.6). Some of these and other closely related species occur on Socotra, and they could play an important role in certain communities on the plains in maintaining natural soil fertility.

Fig. 5.6 Rhizosheaths of *Stipagrostis plumosa* from Kuwait. The actual root is visible in the centre of the image where the rhizosheath has been removed

5.2.3 Energy Balance

5.2.3.1 Energy Balance and Leaf Morphology

A characteristic feature of some species of the desert plains is microphylly, i.e. the development of small, narrow leaves (Gibson 1996). Leaves less than 10 mm in width have been shown to have at least two distinct benefits for the plant. (1) They remain close to ambient air temperatures without the requirement for additional features to assist cooling. As temperatures rarely exceed 42°C during the hottest months on Socotra, this means that plants are not often exposed to lethal cellular temperatures (above ca. 44–46°C). (2) Apart from preventing high temperatures, a second adaptive advantage of narrow leaves is that transpirational cooling is much reduced, enabling the plants to conserve internal water supplies (Gates et al. 1968). Examples of species with narrow leaves on Socotra include the woody-based herb *Convolvulus hildebrandtii*, the widespread small shrub *Cryptolepis intricata*, the dwarf shrub *Euryops arabicus* (locally abundant at higher altitudes, where narrow leaves may also reduce the threat of mechanical damage arising from strong winds – Fig. 5.7), the locally abundant evergreen shrub *Gnidia socotrana* and many others. Examples of species that rely heavily on transpirational cooling to prevent overheating of their large leaves are given in Sect. 5.5.2.4.

Fig. 5.7 *Euryops arabicus* has very narrow leaves

5.2.3.2 Energy Balance and Water Cycling

The succulent shrub *Adenium obesum* (Fig. 5.8) reaches a height of up to 6 m in favourable locations and is able to store large amounts of water. Most of this water is retained deep inside its massive trunk. A network of vascular elements is also located close to the cortex. Measurements by Mies (1998c, 1999b) suggest that *Adenium* possesses an effective water-cycling mechanism within its trunk to protect the sensitive outer layers from overheating. In the example shown in Fig. 5.9, sun shining on the swollen trunk during the morning caused it to warm up well above ambient temperatures, a process enhanced by its red-brown colour. In contrast, temperatures 20 cm inside the trunk remained virtually the same as those of the ambient air. However, on two consecutives days of measurements, a remarkable change took place around midday: stem surface temperatures suddenly began declining to just above ambient, and this was accompanied by an equally dramatic increase in internal plant temperatures of up to ca. 8°C above those of the trunk surface, and about 10°C above ambient. During the afternoon, with declining external temperatures, internal temperatures dropped just as dramatically as they had previously risen to mirror those of the trunk surface and ambient air by the early evening. This phenomenon, although widely cited, has so far been only superficially examined and requires more detailed studies. Mies (2001) was not able to document this water-cycling mechanism for cooling in the succulent trunk of *Dendrosicyos*.

Fig. 5.8 *Adenium obesum*. Ra's Hebak, December 2008

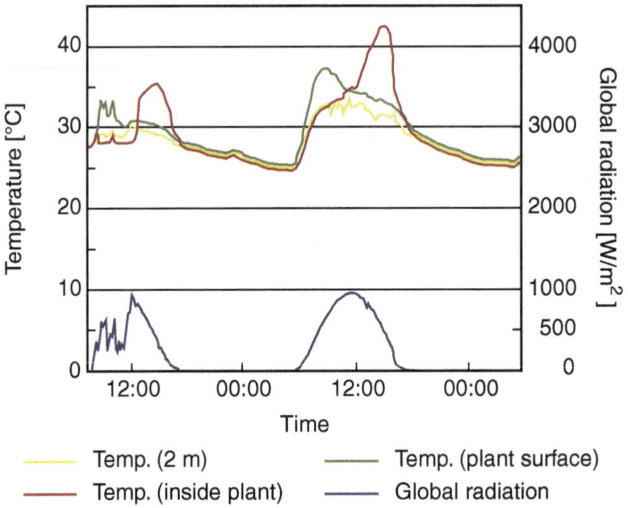

Fig. 5.9 Global radiation, air temperature, plant surface temperature and temperature inside *Adenium obesum* in the field at Ra's Hebak, 22–24 March 1997

5.2.3.3 Leaf Anatomy, Reflectance, Orientation

Although plants require light to drive photosynthesis, absorption of excessive amounts of solar irradiation can be detrimental, and this applies in particular to desert plants during the hot, dry period. The potential negative effects of photoinhibition have not been investigated in Socotran plants. However, it is well established that an increase in leaf temperature generally leads to a reduction in the rate of photosynthesis (Nobel 2009). Apart from microphylly, plants have developed a variety of morphological and anatomical features to prevent overheating of the leaves caused by high insolation, and such features abound in plants on Socotra. Reflective, often white or shiny surfaces are developed or modified to suit the appropriate environmental circumstances for the time of year. However, some degree of caution may be warranted, because as emphasised by von Willert et al. (1992), it appears that most of the reflected radiation is in the visible spectrum, not so much in the infrared wavelength region. In the case of trichomes, which are often developed as a highly reflective leaf pubescence, it has been suggested that their primary role is specifically to reflect infrared radiation, as has been shown by Ehleringer and various co-workers for the American desert shrub *Encelia farinosa* (e.g. Ehleringer and Björkman 1976; Ehleringer et al. 1976), rather than to create a boundary layer around the leaf to reduce transpiration directly. In *Encelia*, a reduction in leaf temperature of 5°C increased photosynthetic rates by up to 60%, while concomitantly conserving substantial amounts of water due to a 23% decline in transpiration rates (Ehleringer and Mooney 1978). This chief role of the trichomes appears to be corroborated by more recent studies on various members of the New World tropical and subtropical genus *Tillandsia*. The mainly epiphytic species, which usually grow in exposed situations and can be regarded as physiological xerophytes, have a suite of adaptations to minimise potential drought stress, including CAM. However, the results of Benz and Martin (2006) have shown that trichome-enhanced boundary layers have virtually no effect on *Tillandsia* gas exchange, suggesting that the trichomes serve primarily to reflect radiation.

On Socotra, plants with a distinct pubescence on sun-exposed organs include a number of dwarf shrub species of the coastal plains and low-altitude slopes. Two widespread examples are *Indigofera pseudointricata*, with the whole plant being silvery throughout and possessing much-reduced leaves, and *Campylanthus spinosus*, which has very small, linear, silvery leaves. *Chapmannia sericea*, a shrub of exposed limestone slopes, has small leaflets that are usually covered with dense, silky hairs. The endemic *Lachnocapsa spathulata*, a cushion plant occurring on coastal plains and exposed limestone slopes, has densely tomentose, fairly large leaves. It is a quite conspicuous fact that leaves of cushion plant species occurring mainly in the mid-altitudinal range are often very small with silvery hairs on their upper surface, and examples include the *Macledium canum* (Fig. 5.10), *Pulicaria dioscorides* and *Wellstedia socotrana*, as well as the more widespread *Echiochilon pulvinata*.

Mies and Aschan (1995) described the radiation regime and temperatures within the canopy of the hemispherically shaped shrubs of *Euphorbia balsamifera* in the

Fig. 5.10 *Macledium canum*. Hamaderoh, March 1997

Canary Islands. The high reflectance values of this shrub, which occurs in the Socotra Archipelago on Abdalkuri (see Sect. 6.3.4.2), are similar to those of alpine cushion plants. In addition, the leaves, which can be shed during conditions of acute water shortage, are small and narrow, and as a consequence, leaf temperatures rarely exceed 2–3°C above ambient.

Leaf orientation can be used by plants to reduce or increase light interception, depending on the circumstances. It is well known that leaves consistently exposed to the sun are inclined at a much steeper angle to the horizontal than shade leaves of the same species (e.g. Valladares and Pearcy 1998). On Socotra, several small shrub species have leaves that are usually held perpendicular to the ground, thus reducing light interception, such as the endemic *Trichocalyx obovatus*, as well as the more widespread *Buxanthus pedicellatus* (Fig. 5.11) and *Dodonaea viscosa*.

5.3 Adaptation of Plants to Highly Saline Ecosystems

Most dominant species inhabiting coastal salt-marsh or drier saline habitats on Socotra are halophytic perennials. Succulent dwarf shrubs, either leaf succulents with woody stems or stem succulents, constitute the predominant growth-form, and members of the Chenopodiaceae, as well as to a lesser extent the Zygophyllaceae, are well represented. Trees, however, are largely absent with the exception of the mangrove *Avicennia marina* and the tamarisk *Tamarix nilotica*. The latter grows in

Fig. 5.11 *Buxanthus pedicellatus*, with leaves held perpendicular to the ground

moist locations that have a high salt content, or on coastal dunes where the trees could be in contact with the saline water in the subsoil.

Both C_3 and C_4 photosynthetic pathways are found in plants associated with highly saline environments, and the occurrence of the particular pathway provides insight into the degree of water availability, a key environmental parameter. Species with the C_3 pathway are usually hydrohalophytic, such as *Arthrocnemum macrostachyum* and *Avicennia marina*, whereas the C_4 pathway occurs mainly in xerohalophytes. In contrast to the latter, hydrohalophytes do not need to conserve water to any significant degree, as it is readily available. The C_4 pathway confers a distinct advantage on plants that regularly face water deficiency, in that it facilitates a greater water-use efficiency, as the stomata need be open only slightly to carry out effective CO_2 fixation (von Willert et al. 1992). Most chenopods on Socotra occur on substrates that are primarily dry. As a consequence, they tend to be C_4 species (*Atriplex farinosa*, *A. griffithii*, *Seidlitzia rosmarinus*, *Sevada schimperi*, *Suaeda monoica*, *S. vermiculata*), as determined by Akhani et al. (1997), or probably C_4 (*Anabasis ehrenbergii*, *Cornulaca ehrenbergii*, *Lagenantha cycloptera*, *Salsola spinescens*). These latter species were not examined specifically by Akhani et al. (1997), but members of the same genus are exclusively or mainly C_4. Apart from *Arthrocnemum macrostachyum*, which inhabits salt-marsh, the only other definite C_3 chenopod known to occur on Socotra is the widespread weed *Chenopodium murale*.

C_3 plants growing in wet saline environments face a particular problem in connection with their high rates of transpiration, namely the accumulation of potentially toxic concentrations of salts in their aerial organs. This is because the plants cannot

Fig. 5.12 The dwarf shrub *Zygophyllum qatarense* has succulent cylindrical leaflets and petioles. North coast, Mumi, December 2008

reduce the uptake of dissolved salts to any significant extent. Two main mechanisms have evolved in hydrohalophytes to deal with the problem. The first is succulence, which is moderately developed in *Arthrocnemum macrostachyum*, and which serves to dilute salt concentrations in the cell and to moderate their toxic effects. The leaves of *Arthrocnemum* are inconspicuous, and photosynthesis is undertaken mainly by the young succulent stems. These become woody with age as the succulent assimilatory cortex is shed from older parts of the branch as a result of suberisation of the inner layers. Whole stem segments can also be discarded occasionally, and both this and decortication result in the removal of salt from the plant. *Zygophyllum qatarense* is a dwarf shrub with succulent cylindrical leaflets and petioles (Fig. 5.12), growing on a range of weakly to moderately saline substrates, from moist to dry. When growing on intermittently moist or wet ground, such as on sabkha-like substrates, the species is able to shed its leaves entirely as the substrate dries out and increases in salinity. The plants can remain dormant for many years until the next rainfall occurs, which leaches salts from the root zone, enabling the shrubs to resume growth for a limited period and develop new leaves (Brown et al. 2008).

The second important mechanism is the possession of salt-excreting glands, as found in the trees *Avicennia marina* and *Tamarix nilotica*, as well as in many halophytic grasses. Examples of the latter on Socotra include *Aeluropus lagopoides* (Fig. 5.13) and *Sporobolus iocladus*. Salt-excreting glands are also present in the two species of *Limonium* known from the island, both endemics, *L. sokotranum* (Fig. 5.14), with its cylindrical succulent leaves, and *L. paulayanum* (Fig. 5.15), which has spatulate leaves.

Fig. 5.13 *Aeluropus lagopoides*, a widespread grass with salt-excreting glands. Nuged plain, December 2008

Fig. 5.14 The endemic dwarf shrub *Limonium sokotranum*. North coast, west of Ghubbah, December 2008

Fig. 5.15 The endemic *Limonium paulayanum* has spatulate leaves. North coast, Mumi, December 2008

Avicennia is also adapted to waterlogging and soil anoxia by producing characteristic pneumatophores (aerial roots) that display negative geotropism and protrude up through the muddy substrate. In some instances, probably where the hydrology of the locality has changed and the substrate is not regularly inundated any more, these pneumatophores may not be developed. This is the case with some of the isolated trees observed on the north coast (see Fig. 6.20).

Short-lived salt-succulents include the widespread *Zygophyllum simplex* (annual or short-lived perennial), typically found on dry coastal plains and sand sheets (Fig. 5.16), and the very common prostrate annual herb *Aizoon canariense*. Like this latter species, the hemiparasitic *Xylocalyx* cf. *asper*, a locally common species on coastal dunes in the south, has fleshy succulent leaves that are sometimes covered with small salt crystals (see Fig. 4.16).

Germination can be problematic for plants of hypersaline habitats, as the seedlings of many species are particularly sensitive to elevated salinity. As a consequence, the seeds of most species germinate after heavy rainfall, when soil-salt concentrations are temporarily reduced. Germination can also take place at higher temperatures. For instance, mass germination of *Zygophyllum qatarense* was observed after heavy rain in May 2003 on the edge of the Rub Al Khali in the UAE, when diurnal temperatures had already reached 45°C (Brown 2006). Khan (2003) found that optimum germination in four halophytic grass species from the Pakistani coast, all of which occur on Socotra (*Aeluropus lagopoides*, *Halopyrum mucronatum*,

Fig. 5.16 *Zygophyllum simplex* is a widespread, short-lived salt-succulent. Nuged plain, December 2008

Sporobolus iocladus and *Urochondra setulosa*), was achieved in distilled water, with increasing salt-treatments leading to a reduction in germination success. In respect of temperature, seeds of all these grasses showed best germination with a 30:20°C (light:dark) temperature regime.

5.4 Adaptation of Plants to High-Montane Ecosystems

The growth-form most characteristic of the exposed high altitudes above the main forest zone are small shrubs up to about 1 m in height (nano-phanerophytes), almost invariably with elliptic to linear, very narrow leaves (less than 12 mm). Typical examples are *Hypericum scopulorum* (locally dominant), as well as the very local *H. fieriense* and *H. balfourii* (the latter up to 3 m high). Furthermore, the locally abundant, but otherwise very rare, endemic *Helichrysum nimmoanum* has a similar growth-form.

Cushion plants are present at all altitudes on Socotra, but they are a particularly characteristic feature of the high montane areas where they are more conspicuous physiognomically. This growth-form can be regarded as an adaptation to thermal extremes, mechanical stress and desiccation, the latter two often brought about by the extreme windy conditions, apart from the paucity of available moisture. The dense, cushion-like growth-form enables the plants to reduce the impact of wind and create

Fig. 5.17 The cushion-forming woody endemic *Helichrysum aciculare*. March 1996

Fig. 5.18 *Helichrysum rosulatum*. March 1996

their own microclimate, thus preventing mechanical damage and mitigating the effects of pronounced differences in temperature and humidity that are a daily occurrence at higher altitudes. Dense cushions also enable the accumulation of plant remains and detritus that would otherwise be blown away. In the interior of the cushion, with its more moderate microclimate, patches of humus develop, which presumably also provide a microhabitat for various small animals and microorganisms.

Apart from their growth-form, cushion plants often display various other characteristic adaptations. For instance, the needle-like leaves of *Helichrysum aciculare* (Fig. 5.17), a dense, cushion-forming woody herb up to about 20 cm in height, are dark green above and have white hairs below, thus serving to reduce transpirational water-loss. Similar features are also found in *Helichrysum rosulatum* (Fig. 5.18), another common cushion-forming woody herb or dwarf shrub at higher altitudes, often growing together with *H. aciculare*. *Thamnosa socotrana* is a locally common evergreen dwarf shrub that shows a transition to the cushion growth-form in that the narrow leathery leaves are congested at the apex of branches. Furthermore, the margins of these leaves are strongly inrolled to reduce surface area and therefore loss of water.

5.5 Life and Growth-Forms

As outlined in the previous section, plants growing under particular environmental conditions have evolved adaptations, often comprising a suite of traits, that enable them to cope with the predominant environmental stresses to which they are exposed. Various different systems have been proposed to classify plants according to morphological or functional types. In respect of climate, Raunkiaer (1904, 1907, 1934) developed the widely used ecological system of life-form classification based on the morphological adaptations of plants to survive the unfavourable season (winter cold or summer drought). These adaptations are primarily linked with the protection afforded to the sensitive renewal buds (apical meristems). This system was first conceived for northern temperate regions, but it has since been expanded somewhat to allow its global application (e.g. Müller-Dombois and Ellenberg 1974), although it could be argued that some of these additional groupings are more concerned with growth-form, rather than life-form *sensu* Raunkiaer.

By itself, the Raunkiaer life-form classification system is of limited use for predominantly desert ecosystems, because it is rather restricted in its structure–function relationships relevant to desert life (Gibson 1996). In the following, the five main types as proposed by Raunkiaer are briefly discussed, with the way they relate to the flora of Socotra, followed by a more detailed overview of the major growth-forms.

5.5.1 Raunkiaer Life-Form Classification

Five main groups were initially recognised by Raunkiaer (1934): (1) phanerophytes; (2) chamaephytes; (3) hemicryptophytes; (4) geophytes (cryptophytes); and (5) therophytes.

Phanerophytes are perennial woody plants with their renewal buds well above (>50 cm) the surface of the ground. Mesophanerophytes have their buds higher than

5 m above the ground. A fairly common representative of this subcategory on Socotra is the tropical chestnut *Sterculia africana*, found mainly as an emergent in succulent shrubland. Those phanerophytes with their buds between 2 and 5 m above the surface are more specifically referred to as microphanerophytes, with a third group, with buds below 2 m but higher than 0.5 m, as nanophanerophytes. The phanerophytes encompass all trees and shrubs on Socotra, including many of the leafless tall shrubs with green, often cylindrical or broom-like stems. A specialised growth-form within the phanerophytes is that of the bottle trees, in particular the microphanerophytes *Dendrosicyos socotrana*, *Dorstenia gigas* and *Adenium obesum*. Given the prominence of bottle trees and other succulents in the flora of Socotra, they will be dealt with in a separate section below.

Chamaephytes are perennial plants that rarely exceed 50 cm in height, but have their renewal buds well above the surface of the ground. Many examples exist in the flora of Socotra, and this group is particularly well represented by dwarf shrubs and many woody-based herbs. With the latter, some species often produce most of their renewal buds near the base of the plant, such as the short-lived perennial and facultative annual *Cleome scaposa*, tending therefore towards the hemicryptophyte life-form; others, such as *Aerva lanata*, are more obviously chamaephytes. The chamaephytes also include various stem succulents, such as *Euphorbia spiralis*.

Hemicryptophytes are perennials with their renewal buds located on, or very close to, the surface of the ground, and these buds are protected by dead shoot remains in some species. They are usually herbaceous species, but some may become lignified at the base to a certain extent as the plant matures. Typical examples on Socotra are *Corchorus depressus* and *C. erodioides* (Fig. 5.19), as well as various tufted perennial grass species such as *Cymbopogon jwarancusa*, *Lepturus pulchellus* and *Panicum rigidum*. The small succulent herb *Euphorbia hadramautica* can also probably be included in this category.

Geophytes (cryptophytes) are perennials that survive the harsh season as bulbs, corms, rhizomes or some other subterranean organ, as outlined in Sect. 5.5.2.13.

Therophytes are short-lived plants, often referred to as annuals, that complete their life-cycle from seed to seed production within one growth season. Section 5.5.2.13 discusses this category of plants in more detail.

5.5.2 *Growth-Form Classification of Socotran Plants*

Whereas Raunkiaer life-forms are related to the position of the renewal buds during the unfavourable growth season, the growth-form categories outlined below are primarily based on criteria relating to life-span, degree of succulence and leaf traits. The categories follow in part the generalised classification of warm desert plants outlined by Gibson (1996) and Rundel and Gibson (1996), but have been slightly modified and expanded to include non-desert ecosystems that are also found in the Socotra Archipelago. Like many other biological categories, this growth-form

Fig. 5.19 *Corchorus erodioides*, a typical hemicrytophyte. Dixam plateau, December 2008

classification is occasionally less than adequate, and it is possible that some species can be assigned to more than one category.

5.5.2.1 Drought-Deciduous Trees

Most of the tree species on Socotra can be classified as being drought-deciduous, shedding their leaves when water becomes a strongly limiting factor. Some species may dispose of their mainly mesomorphic leaves (i.e. ones that show no specialised adaptations to extreme conditions) rapidly with the onset of drought; the leaves of others persist much longer into the dry period, and in seasons of favourable moisture availability, the trees may remain green throughout the year. Some of the more conspicuous representatives of this group are *Boswellia* spp. (frankincense) and *Commiphora* spp. (myrrh). *Allophylus rubifolius* is a widespread drought-deciduous tree at higher altitudes. In contrast to the evergreen *Ficus cordata*, *F. vasta* is a deciduous species, probably because of the limited water availability in its typical habitats, which include cliffs, rocky slopes and huge boulders.

5.5.2.2 Drought-Deciduous Shrubs and Dwarf Shrubs

Two of the most common shrub species on Socotra, *Croton socotranus* and *Jatropha unicostata*, are drought-deciduous species, but when water does become available, they respond very rapidly by producing new foliage. This is also the case when rainfall is received during the hot summer months, so that both species are probably facultative evergreens. The leaves of *Jatropha* are remarkably large (10–15 × 4–6 cm), and like those of many evergreen species, are somewhat leathery, whereas the leaves of *Croton* are much smaller and highly variable in both size and shape, in part probably due to the effects of the local environmental conditions. However, this variability in leaf size in *Croton* may not always be directly linked to water stress, as has been observed in other desert species (e.g. Cunningham and Strain 1969), but in some cases appears to be a characteristic feature of specific locations.

Most of the dwarf shrubs on the plains can probably be classified as drought-deciduous species. Many have very narrow leaves or leaflets that are rapidly shed with the onset of drought conditions.

5.5.2.3 Evergreen Trees and Shrubs

A number of evergreen tree and shrub species occur on Socotra, of which the succulent representatives will be dealt with separately below. One of the perceived benefits of the evergreen growth-form is that it allows primary productivity to recommence without delay once water becomes available. Many of these species have specialised features to prevent excessive loss of water, such as strongly scleromorphic leaves. Investing more energy in the construction of tougher, longer-lived leaves also protects them to a greater degree from mechanical damage and makes them less attractive to small herbivores.

Evergreen species occur mainly at higher altitudes, usually above 400–500 m, although they may thrive much lower in shaded locations, particularly in wadis. One of the most conspicuous evergreen representatives is the dragon's blood tree, *Dracaena cinnabari*, whose distribution, dealt with in more detail in Sect. 7.5.1, coincides with areas receiving frequent moisture in the form of drizzle or mist. Other common evergreen species are small shrubs with leathery leaves, such as *Buxanthus pedicellatus*, *Dodonaea viscosa*, *Euclea divinorum* and *Gnidia socotrana*. In contrast, *Salvadora persica* is an example of an evergreen shrub that usually occurs well below 500 m.

Ficus cordata is an evergreen species with large leathery leaves growing mainly along the banks of wadis, although occasionally it is found in exposed localities on cliffs, scree and other rocky substrates. *Tamarix nilotica* on the other hand is a small evergreen tree of saline substrates at low altitudes, whose small, scale-like leaves have salt-secreting glands (see Sect. 5.3). *Avicennia marina* is an evergreen species with somewhat leathery leaves that also possess salt-secreting glands.

5.5.2.4 Phreatophytes

Phreatophytes are species that require more-or-less constant access to water, usually by virtue of deep rooting systems that enable the plants to tap into subterranean water supplies, either the groundwater itself, or water in the valdose zone.

Fig. 5.20 *Citrullus colocynthis* (Photographed near Al-Ain, UAE)

Furthermore, it is likely that many of these plants are able to exploit water temporally available in the upper surface layers by means of their extensively branched rooting systems. In this context, hydraulic lift, in which water absorbed by the roots in the subsoil is released by the roots into the dry upper surface layers once transpiration has ceased (usually at night), to be then absorbed the following day when transpiration recommences, has been documented from various desert tree and shrub species around the world (Ward 2009). Examples of phreatophytes on Socotra include a number of species, some of which may be drought-deciduous (e.g. *Acacia edgeworthii*, *Ziziphus spina-christi*), others evergreen (e.g. *Tamarix nilotica*), often inhabiting wadi beds or slight depressions where water accumulates underground. The C_3 desert vine, *Citrullus colocynthis*, a prostrate, creeping plant that can form dense mats up to several metres in diameter (Fig. 5.20), is an example of a short-lived perennial plant on the plains that can remain active, and even flowering, during the hot summer months. This is quite remarkable, as the leaves show none of the typical xeromorphic structures or adaptations (such as small leaves) to environments characterised by drought and high temperatures (Sen 1973). In fact the leaves are quite large, measuring up to 6×5 cm, and are located in close proximity to the ground. In the summer, the substrate surface can reach temperatures probably well in excess of 50°C. Transpirational cooling (profuse "sweating") is the means by which *Citrullus* comes to terms with these conditions (Lange 1959; Brown 2001). Lange (1959) also reported that a temperature depression of as much as 15.3°C could be achieved compared with ambient air temperatures. This loss of water requires that the plants have permanent access to soil-moisture, and to ensure this, the plants develop a long taproot with a fibrous lateral root system located just above the water table (Althawadi and Grace 1986). *Citrullus* can therefore be regarded as a phreatophyte. From Kuwait it is well known that its cooling mechanism is so effective that it is exploited by desert larks, which stand or sit on the plants with their wings slightly open during periods of intense heat, as described in detail by Cowan and Brown (2001) and Brown (2009). Similar observations were briefly mentioned by Shobrak (1998) regarding the prostrate herb *Corchorus depressus* from the Saudi Arabian desert, a species that also occurs on Socotra, as well as the very similar *Corchorus erodioides*. The small tree *Salvadora persica*, an evergreen that has a scattered occurrence on the coastal plains of Socotra, is an example of a microphanerophyte that transpires large amounts of water to cool its large leaves (Sen et al. 1972), so that leaf temperatures largely mirror the ambient. The leaves are amphistomatic, but our observations suggest that stomatal density is up to 50% higher on the upper leaf surface. The remarkably large leaves of the evergreen shrub *Calotropis procera* are also amphistomatic, presumably for the same reason, but appear to have equal stomatal densities on both sides of the leaf.

5.5.2.5 Aphyllous Spartinoid Shrubs

There are only a few examples of aphyllous spartinoid shrubs on Socotra. In this group, the stem assumes the main responsibility for carrying out gas exchange. In some species, a distinct palisade chlorenchyma is developed below the epidermis. One of the more common representatives is the Apocynaceae *Periploca visciformis*, a species that produces on its stems minute leaves, which are rapidly shed. This species is fairly widespread on rocky substrates in succulent shrubland, especially in the northern part of the island. Its palisade chlorenchyma is situated several layers of undifferentiated parenchyma below the epidermis. Another member of the same family, *Sarcostemma socotranum*, is superficially similar in growth-form to *Periploca*, but is somewhat succulent. There is no distinct palisade chlorenchyma developed, but chloroplasts are contained in rather undifferentiated cortex cells.

Ochradenus socotranus is a small shrub with spartinoid growth-form, and closely related to the more widespread *O. arabicus* that occurs on mainland Arabia. It produces very narrow, small, linear leaves that are often rapidly discarded, although some usually remain on the plant.

The very rare *Calligonum comosum*, known only from Abdalkuri (but common in NE Africa, mainland Arabia and extending into Pakistan), is another example of an aphyllous shrub. It is a conspicuous fact that there are no spartinoid representatives of the Fabaceae on Socotra, not even at higher altitudes.

5.5.2.6 Stem and Leaf Succulents

Von Willert et al. (1992) define a succulent as a plant possessing at least one succulent tissue. A succulent tissue is a living tissue that, apart from other possible functions, provides a temporary storage of utilisable water. This allows the plant to

Fig. 5.21 *Adenium obesum* in flower. December 2008

be independent of the external water supply for lengthy periods when soil-water conditions have deteriorated to such an extent that the root is no longer able to provide the aerial organs with the necessary water. Furthermore, the plants, which invariably have the CAM photosynthetic pathway, can reduce transpirational water-loss to nearly zero. Succulent species are therefore able to persist in habitats with long periods of drought between fairly regular pulses of water availability, as noted by Werger (1986) for succulents in the Karoo and southern Kalahari.

Succulence is developed in a number of different plant families in the flora of Socotra, but especially in the Apocynaceae (including Asclepiadaceae), and to a lesser extent Euphorbiaceae. Furthermore, succulence is also found as an adaptation to high substrate salinity ("salt succulence"), where, despite the often wet habitats, physiological water deficiency occurs. These plants, belonging mainly to the Chenopodiaceae and Zygophyllaceae, are discussed in Sect. 5.3.

Some succulents are completely leafless (or virtually so), except at the seedling stage; others produce leaves during the favourable season, i.e. are drought-deciduous. For instance, the microphanerophytes *Dendrosicyos socotrana*, *Dorstenia gigas* (see Fig. 6.69) and *Adenium obesum* all produce leaves during the wet period (in the case of the latter two, in apical clusters). In contrast, the microphanerophyte *Euphorbia arbuscula*, a small tree up to 5 m in height with a dracoid habitus (Beyhl 1996a), has numerous pencil-like, apical stems that are responsible for photosynthesis. Only the very young shoots possess minute leaves, but these are soon shed (Beyhl 2000). *Adenium* (Fig. 5.21) is a good example of a hysteranthous species, i.e. is one that can produce flowers at different times from its leaves. This is presumably facilitated by its pronounced succulence. Beyhl and Mies (1999) gave a brief overview of the biology of this species on Socotra.

Dendrosicyos socotrana, the only arborescent member of the Cucurbitaceae (Olson 2003), is a locally common species on Socotra, and many individuals can be seen on the rocky slopes at Ra's Hebak on the northern coast. When leafless, it is conspicuous for its thick white trunk, which serves to reflect sunlight and thus protects the internal tissues from overheating to some extent. The trunk is very soft and easily cut. Olson (2003) carried out a detailed investigation of the stem and leaf anatomy of the species (based on second-generation individuals from the University of California at Davis that were obtained from mother plants originally collected by J. Lavranos in 1967). The most conspicuous feature of the stem in connection with the bottle-tree habit is the large quantity of ray and conjunctive parenchyma. This is linked to water and starch storage as a means for surviving prolonged drought. The genera that appear to be most closely related to *Dendrosicyos* are *Zygosicyos* and *Xerosicyos* on Madagascar, as well as *Corallocarpus* and *Kedrostis* in Africa and Arabia, all of which develop a giant tuber on the soil surface. Olson (2003) found certain anatomical traits of *Dendrosicyos* to be characteristic of lianas, lending support to the more general hypothesis that pachycauls have developed from lianas in a number of dicot families independently. The young branches of *Dendrosicyos* have a pronounced chlorenchyma that enables them to carry out photosynthesis, even when the tree is leafless. According to Pfanz (2008), corticular photosynthesis is driven by stem-internal CO_2 derived mainly from mitochondrial activity, and can be very effective in many species. By generating O_2, corticular photosynthesis may additionally play an important role in preventing stem-internal anaerobiosis.

Socotra is home to various cactoid succulent spurges and stapeliads, the latter of which are particularly well represented. *Euphorbia spiralis* (including *E. septemsulcata*) (Fig. 5.22) is a chamaephyte with a cactoid growth-form (a typical example of convergent evolution, morphologically strongly reminiscent of neotropical cacti), possessing thick succulent stems with rows of paired spines arranged along the longitudinal ridges. It is a common species in succulent shrubland, often occurring on rocky substrates. Some succulent *Euphorbia* individuals, probably belonging to the *E. spiralis* group, have a distinct cushion-like growth-form, with the stems very densely packed. *Euphorbia schimperi* (including *E. oblanceolata*) and *E. abdelkuri* (the latter endemic to a small area on Abdalkuri; see Lavranos 1971b) are chamaephytes that, like the phanerophyte *E. arbuscula*, have numerous cylindrical stems. These are pencil-like in *E. schimperi* (as well as in various other members of the section Tirucalli), but much thicker in *E. abdelkuri*. Having no thorns and yellow, rather than white, sap, the latter species is an exception within the Section Diacanthium.

The small (usually less than 10 cm) erect herb *Euphorbia hadramautica* is an example of a succulent hemicryptophyte. It appears to have a rather short life-span of 3–5 years, although possibly longer under favourable conditions. The apical tuft of leaves dies off during extended periods of drought, making the plant even more difficult to find among small rocks. Because its succulent caudex extends underground, it is also able to regenerate from there after particularly dry years, or when

Fig. 5.22 *Euphorbia spiralis*. Ra's Hebak, December 2008

the above-ground part has been damaged (e.g. through trampling). The same probably applies to members of the genus *Echidnopsis*, with their finger-like succulent stems, partly creeping along the surface and rooting along their length, as well as to *Edithcolea grandis* (Fig. 5.23) and *Socotrella dolichocnema*. These species are often very inconspicuous when not in flower.

Caralluma socotrana is one of the more widespread stapeliads on Socotra, occurring in open situations on the coastal plain up to the medium-elevation plateaus. This leafless species, which grows in dense clumps, has large attractive orange to dark red-brown flowers. The only succulent member of the Asteraceae on Socotra is *Kleinia scottii* (see Figs. 4.28 and 4.29), a small, shrubby plant with pencil-like stems forming dense clumps, often between boulders in very open woodland and scrub on limestone hillsides at higher elevations in the Hagghier area.

5.5.2.7 Caespitose Graminoids

In terms of species, caespitose graminoids, either loosely or densely tufted, are fairly well represented in the flora of Socotra, but they do not usually dominate the vegetation. Most tussock-forming grasses (Poaceae) are found on the coastal plains and mid-elevation limestone plateaus, but they also feature in the anthropogenic grasslands of forest clearings at higher altitudes.

Panicum turgidum, a rare species on Socotra, but common on the adjacent mainland areas of both Africa and Arabia, is found on coastal plains close to sea-level. The

Fig. 5.23 *Edithcolea grandis*. Homhil, 1967 (Photograph by John Lavranos)

leaves are usually quite small (i.e. imitating the aphyllous spartinoid growth-form), and may be tightly inrolled and pungent in individuals of more extreme situations. As with a number of perennial desert grasses, the leaf sheaths and branched culms contribute significantly to photosynthesis, and in addition, the species is characterised by the C_4 mode of carbon fixation. Whereas *Panicum* is a rather loose tussock grass, especially on more sandy substrates, *Stipagrostis sokotrana*, another species of coastal plains at very low altitudes that is also widespread in the Arabian Peninsula, is densely tufted, and has strongly inrolled leaf-blades. *Urochondra setulosa* is a halophytic caespitose grass growing in large dense clumps that produces short rhizomes. It is locally common on sand dunes on the west coast of Socotra, and also on Abdalkuri. *Halopyrum mucronatum* grows in dense clumps in similar situations, except that it also develops long runners, often exceeding several metres in length.

Other caespitose grasses found on the coastal plains extending onto the low elevation slopes include *Bothriochloa insculpta*, *Cenchrus ciliaris* and *C. setigerus*, all species known from the adjacent continental landmasses, as well the endemics *Lepturus pulchellus* and *Panicum rigidum*. Another widespread species, *Cymbopogon jwarancusa*, is abundant in the near-coastal vegetation at Ra's Haulaf. *Pennisetum setaceum*, *Tetrapogon villosus* and the endemic *Lepturus tenuis* are examples of tussock grasses occurring mainly on the mid-elevation limestone slopes and plateaus. Common species of the woodland clearings and grasslands at higher elevations are *Heteropogon contortus*, *Themeda triandra* and *Tricholaena teneriffae*.

A number of sedges (Cyperaceae) also display a distinct caespitose growth-form, including *Cyperus conglomeratus*, a species of sandy coastal plains, and several members of *Fimbristylis*, such as *F. cymosa*.

Fig. 5.24 *Capparis cartilaginea*, a widespread cliff-dwelling perennial. Ra's Hebak, December 2008

5.5.2.8 Stoloniferous Perennial Graminoids

Stoloniferous perennial graminoids are common on Socotra. *Cladium mariscus* forms tall, very dense stands by wadis, whereas other Cyperaceae, such as *Cyperus dilatatus* and *Cyperus rotundus*, are more obviously stoloniferous. *Cynodon dactylon* (Bermuda grass) is a good example of a stoloniferous Poaceae, forming quite extensive stands locally, especially in disturbed places. *Paspalidium geminatum* and *Paspalum vaginatum* are two stoloniferous grasses found mainly in mud by permanent water. Many tufted graminoids also produce short and occasionally longer rhizomes or runners, presumably depending in part on the specific habitat conditions.

5.5.2.9 Cliff-Dwelling Perennials

On Socotra, *Capparis cartilaginea* (Fig. 5.24) is an evergreen trailing shrub growing on cliffs and rocky slopes. It has thick leathery leaves that reduce transpirational water-loss, allowing the plant to grow in quite exposed conditions. The endemic *Cryptolepis socotrana* (Figs. 5.25 and 5.26), whose leaves are slightly leathery, is usually found in more shaded locations than *Capparis*. *Ficus cordata* can also occasionally be observed hanging from cliffs, presumably where there is some seepage water. It is interesting to note that the two species of *Limonium*, both endemic and commonly found along the coast, also occur as cliff-dwelling plants, often far inland and away from the influence of salt (see Sect. 6.3.5.1). Other characteristic cliff-dwelling perennials include *Scaevola socotraensis*, a very rare small shrub in the south of the island, and the fairly widespread endemic woody-based herb *Leucas spiculifolia*, which usually grows out of the reach of marauding goats.

Fig. 5.25 *Cryptolepis socotrana* on cliffs. Wadi Ireh, December 2008

Fig. 5.26 *Cryptolepis socotrana* in fruit. Wadi Ireh, December 2008

5.5.2.10 Cushion and Mat-Forming Plants

Cushion plants occur at all altitudes on Socotra, but are best represented in the high montane areas. They are discussed in Sect. 5.4. At lower altitudes, representatives of this growth-form include the endemics *Lachnocapsa spathulata*, *Pulicaria dioscorides* and *Wellstedia socotrana* (see Fig. 6.106). *Teucrium balfourii* (Fig. 5.27) is a widespread mat-forming endemic on rocky, mid-elevation slopes, especially in the west of the island. The leaves of this species are linear, with strongly inrolled margins.

Fig. 5.27 *Teucrium balfourii*, a widespread mat-forming endemic. Hamaderoh, March 1997

5.5.2.11 Lianas

Lianas are plants that typically root in the ground, but depend on others for support. Further subdivisions are possible, and an obvious one relevant to the flora of Socotra is that between the succulent lianas (e.g. *Cissus subaphylla* and *C. hamaderoensis*) and the non-succulent ones, such as *Cocculus pendulus*. Whereas the endemic vine *Cissus paniculata* (Vitaceae) has large, persistent leaves, and round, only slightly succulent stems (Fig. 5.28), the other two endemic members of the genus, *C. subaphylla* and *C. hamaderohensis*, have small, deciduous leaves and large, grey-green stems, flattened in the former, and rectangular in the latter. Tendrils are absent or non-functional in *C. subaphylla*, but present in *C. hamaderoensis* (Figs. 5.29 and 5.30).

Most members of the Menispermaceae are woody climbers, including *Cocculus pendulus*, an uncommon species on Socotra, but widespread in eastern Africa, Arabia and extending into India. However, *Cocculus balfourii*, a species restricted to Socotra and the southern-most part of Arabia, is remarkable in that it is a dense shrub with sharply pointed, bilaterally flattened cladodes (see Fig. 4.19).

5.5.2.12 Epiphytes

True epiphytes are poorly represented in the flora of Socotra, the most prominent example being the rare orchid *Angraecum dives*. Another species is the minute

Fig. 5.28 *Cissus paniculata*. Muqadrihon pass, south of Hadibuh, December 2008

Fig. 5.29 *Cissus hamaderohensis*. Dixam plateau, December 2008

Fig. 5.30 *Cissus hamaderohensis*, detail of stems. Homhil, December 2008

ephemeral *Utricularia striatula*, which grows on mossy branches in woodland at the higher altitudes. Other species may be facultative epiphytes, in that they are normally rooted in the ground, but occasionally grow on trees where suitable substrate has accumulated.

5.5.2.13 Ephemerals

Ephemerals comprise two distinct groups of mainly non-woody plants that show certain similarities, in that both are inactive and largely invisible during the unfavourable growth season: (1) annuals or therophytes; and (2) geophytes.

Therophytes, short-lived plants that complete their life-cycle within one growth season, survive the unfavourable period of the year as seed in the ground. Their entire life-span may be as little as a few weeks, but it usually lasts for several months. In particularly favourable conditions, some annuals may grow as facultative short-lived perennials. Conversely, it is known that some desert perennials may be facultative annuals, and examples from mainland Arabia include the grasses *Centropodia forsskaolii*, in fact mainly an annual in the UAE (Brown, pers. observation), and *Stipagrostis plumosa* (Danin 1996).

The most common desert annual dicots on Socotra are the facultative ephemeroid *Zygophyllum simplex* and *Aizoon canariense*, both widespread species

Fig. 5.31 *Ledebouria grandifolia*. Muqadrihon pass, south of Hadibuh, January 1998

throughout much of the Arabian Peninsula and beyond, and often found on somewhat saline substrates.

Geophytes are fairly well represented on Socotra, and this category includes species with bulbs growing in succulent shrubland, such as members of the genus *Dipcadi* (including the three endemics), but also species with bulbs typically occurring in heavily shaded woodland and thickets. Examples of this latter group are *Drimia porphyrostachys* and species of the genus *Ledebouria*. In wet years, the bulbs of *Ledebouria grandifolia* (Fig. 5.31) produce one or two large fleshy leaves that can form dense carpets in drought-deciduous woodland communities, whilst in drier years, the bulbs may remain completely dormant.

The endemic *Dorstenia socotrana*, a very local species on seaward-facing limestone cliffs in NE Socotra, is a perennial stemless herb with an underground tuber, and can therefore be classified as a geophyte, in contrast to the cliff-dwelling microphanerophyte *D. gigas*.

A number of perennial grasses are primarily geophytes in that they possess subterranean rhizomes from which new shoots can develop. In addition, some species, a good example being *Panicum turgidum*, are also capable of developing new shoots from seemingly dead culms, i.e. aerial organs, and therefore display a transition to the chamaephytes.

5.5.2.14 Parasitic Plants

Socotra is home to a number of holo- and hemiparasitic plants, many of which have a widespread distribution from northern Africa/southern Europe to Asia and beyond.

Fig. 5.32 *Cuscuta planiflora* (Photographed in the Musandam, northern Oman)

There are no endemic representatives in the archipelago. They can all be accommodated in one of two families, the Convolvulaceae (*Cuscuta*) or Scrophulariaceae (*Cistanche*, *Orobanche*, *Striga*, although some taxonomists include these genera in the Orobanchaceae, a family not recognised by Heywood et al. 2007). The fairly common *Cuscuta planiflora* (Fig. 5.32) and morphologically similar *C. chinensis* are parasitic twining herbs that are devoid of chlorophyll. They are mainly found at lower altitudes, where they parasitise on other herbs. The most widespread and conspicuous parasitic member of the Scrophulariaceae is *Cistanche phelypaea* (= *C. tubulosa*). This species, with its large, pyramid-shaped inflorescence of yellow flowers, is a common parasite on *Zygophyllum qatarense* and other dwarf shrubs belonging to the Chenopodiaceae that grow on sandy, including weakly halophytic, substrates. It is also occasionally found on other hosts. The minute seeds may remain dormant in the ground for many years, before germinating on a root of the host plant. The seedling then produces a haustorium that penetrates the root, cementing the parasitic association. Older plants possess a massively swollen underground stem. *Orobanche cernua* and *O. minor* have so far been mainly recorded at medium elevations in open woodland. *Striga gesnerioides* (Fig. 5.33) is the most common of the three members of the genus that have been recorded from Socotra, and it usually occurs at lower to mid-altitudes. Its host plants include various grasses and herbs.

Fig. 5.33 *Striga gesnerioides*. Wadi Dizyaf (Wadi Shi'faar), December 2008

5.5.3 Life-Form Analysis

Mies (2000) conducted an analysis of the Raunkiaer life-forms in the flora of Socotra, and the results are shown in Fig. 5.34. Despite a few recent additions to the flora of the archipelago, the results nonetheless provide an accurate reflection of the relative proportions of the different life-forms. It is clear that therophytes are the predominant group, encompassing 37.9% of the flora. Hemicryptophytes and chamaephytes are also fairly well represented, accounting for about 17.4% and 18.6%, respectively. However, a striking fact, not evident from Fig. 5.34, is that nearly 90% of the endemic species of the Socotra Archipelago are chamaephytes or phanerophytes. Their proportion is therefore extremely high, in marked contrast to the low number of endemic therophytes. Lems (1961) also noted the conspicuously high proportion of chamaephytes and shrubs (phanerophytes) in the endemic flora of the Canary Islands. Shmida and Werger (1992) underlined the fact that annuals contribute little to the endemic flora of the Canary Islands (6.7%), in contrast to their large representation in the entire flora (37.6%).

Remarkably, the proportion of annuals in the flora of Socotra (37.9%) is virtually identical to that of the Canary Islands. However, the more arid islands of that archipelago (Lanzerote and Fuerteventura) have a much higher percentage (58% and 59%, respectively; Shmida and Werger 1992). The apparent paucity of desert annual species on Socotra is somewhat surprising, as they typically account for a high

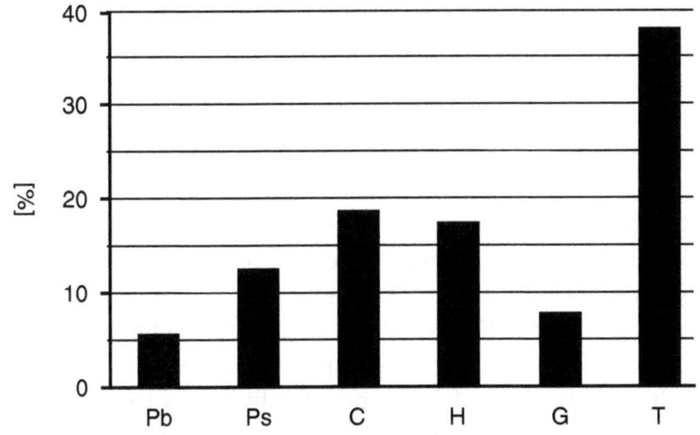

Fig. 5.34 Raunkiaer life-form spectrum of Socotran plants (Based on data from Mies 2000). *Pb* large phanerophytes (>3 m), *Ps* small phanerophytes (<3 m), *C* chamaephytes, *H* hemicryptophytes, *G* geophytes (cryptophytes), *T* therophytes

proportion of total species in desert floras, even though they comprise a relatively small percentage of community biomass (Gibson 1996). In an apparent contradiction to this statement, an analysis of Raunkiaer life-forms in the flora of Israel in relation to the environment found that therophytes were less well represented in arid ecosystems, but displayed an optimum between 200 and 500 mm annual precipitation (Danin and Orshan 1990). Socotra appears to be relatively poor in desert annual dicot species, although a fair number of annual grass species occur, and some of them can be locally abundant (e.g. *Melanocenchris jacquemontii*). This relative paucity of annuals also contrasts to other parts of the Arabian Peninsula with comparable rainfall characteristics (e.g. Kuwait, northern Qatar, northern Saudi Arabia, north-eastern UAE). In Kuwait, for example, about 69% of all species are annuals (Brown 2001). With regard to specific taxa, whereas only one annual member of the genus *Astragalus* is known from the Socotra Archipelago (*A. hauarensis*, and only on Abdalkuri), at least six species are common in the northern Gulf region, including *A. hauarensis*. One possible explanation for the low number of annuals could be the fairly high temperatures during the winter on Socotra – the other areas mentioned above receive winter rainfall and tend to be cooler at that time of the year. Whether the isolated island effect plays a role is uncertain, but Shmida and Werger (1992) noted that the proportion of annuals in the flora of the Canary Islands was very low when compared to that of mainland Spain.

5.6 Adaptive Strategies

Classification systems based on plant morphology, such as the one developed by Raunkiaer, will continue to provide significant insight into certain aspects of plant ecology. However, they do not offer a complete picture of vegetation and ecosystem processes, and how individual species come to terms with the specific constraints of their environment. Other classification systems have placed more emphasis on the role of physiological adaptations. In recent decades, a wealth of studies have been carried out on so-called "plant functional types" (PFTs). These are derived from traits based on species morphology, anatomy, physiology and/or life history, and represent groupings of species that show close similarities in their response to environmental factors (Barbour et al. 1998; Grime 2001). As non-phylogenetic groupings, functional classifications often cut across taxonomic classifications and are generally accepted to be more meaningful in terms of plant response to environmental factors than specific taxonomic units such as individual species (Smith et al. 1997). PFTs therefore reduce the complexity of species diversity in biological systems to a limited number of key types, facilitating a better understanding of main ecosystem processes. A major challenge has been to identify a manageable number of PFTs that distil the wide range of plant responses into a readily applicable universal system. In the following, several of the more common functional classification systems are briefly given, with an account of how they relate to the flora of Socotra.

5.6.1 Relating Net Carbon Gain to Environmental Stress in Arid Ecosystems

Perhaps the simplest way for plants to cope with environmental stress in desert ecosystems is by avoiding the period of the year when factors limiting growth are at their most severe. Growth, productivity and flowering are therefore usually tightly linked to the brief periods in which adequate soil moisture is available. In response to these selective pressures, warm desert plants have undergone profound evolutionary processes and developed a diversity of physiological, morphological and life-cycle adaptations that allows them to maximise net carbon gain during periods of high soil-moisture availability and to facilitate survival throughout extended drought. Survivorship and reproductive success of plants are therefore assumed to be correlated with traits that maximise net carbon gain, and Gibson (1996) presented a modified simple conceptual model, developed by Ehleringer (1985), for arid ecosystems, relating net carbon gain to environmental stress (Fig. 5.35). In this model, annuals are assumed to be little tolerant of stress (i.e. they represent typical stress-evaders), and are characterised by high levels of net carbon gain during the short favourable period. In stark contrast, long-lived perennials, with the exception of geophytes, which are similar in their behaviour to annuals, are shown to be highly tolerant of environmental stress, but display low levels of carbon gain per unit time. Short-lived perennials occupy an intermediate position, and as such have intermediate rates of carbon gain per unit time over their life-span. This simple scheme is easily applied to plant species on Socotra.

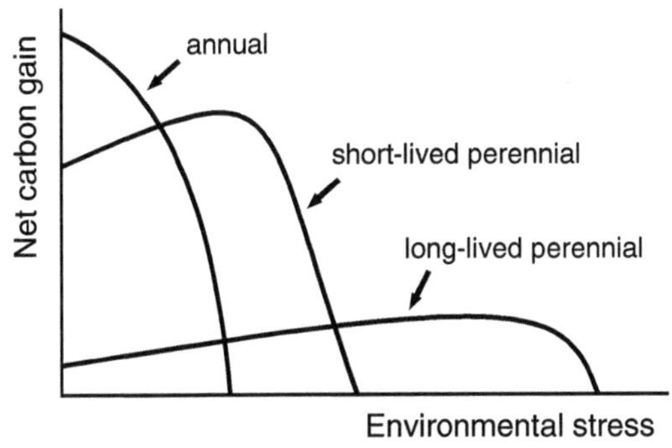

Fig. 5.35 Simple conceptual model relating net carbon gain to environmental stress and plant growth-form in arid ecosystems (Adapted from Ehleringer 1985 and taken from Gibson 1996, Figure 1.3). Reproduced with kind permission of Springer

5.6.2 Plant Strategies in Arid Ecosystems

Various authors, including Shantz (1927), Evenari et al. (1982) and Evenari (1985), have developed a system to classify desert plants according to their autecological adaptations and growth-forms for surviving periods of drought. This system, summarised in Larcher (2003), is highly relevant to the flora of Socotra.

1. Arido-active plants are ones that remain active throughout the year and delay desiccation by various mechanisms, such as deep rooting systems (e.g. phreatophytes), succulence, and reduction of transpirational loss (epicuticular waxes, reflective leaf surfaces, reduction of leaf size).
2. Arido-tolerant plants can tolerate desiccation of their organs without suffering any apparent physiological damage. This feature is dependent on the capacity of the protoplasm to tolerate severe water-loss, which for the majority of species is lethal, and also to be able to rehydrate in a coordinated manner once water does become available. On Socotra, this group of plants is represented by cryptogams, namely lichens, bryophytes and poikilohydric ferns.
3. Arido-passive plants spend the unfavourable season in a metabolically inactive state, and are generally regarded as "stress-evaders". They survive either as seed, as with desert annuals, or as subterranean storage organs (i.e. bulb, corm or rhizome), as in the case of geophytes.

5.6.3 r/K-Selection Theory and the Three-Strategy CSR Classification System of Grime

A fairly early approach to developing clearly defined strategies in the context of ecosystem structure and dynamics was proposed by MacArthur and Wilson (1967), and involved the two-model "r/K-selection" theory. Accordingly, organisms could be placed somewhere along a continuum between the two extremes of r- and K-selected species. The r-selected species are typically short-lived, uncompetitive ones that exploit temporary, but highly productive habitats. In the context of arid ecosystems, this group would include desert annuals, where "productive" refers to the temporal availability of plentiful water during the favourable growth season. The habitats are temporary in the sense that drought (i.e. disturbance) causes the death of the annuals at regular intervals. K-selected species are longer-lived plants that are highly competitive, occupying the same patch for many years.

The category of K-selected organisms is somewhat problematic, because it contains two very different sets of species: (1) fast-growing species of highly productive habitats; and (2) slow-growing species of less productive ones. Recognising this dilemma, Grime (1977) devised a three-strategy model based on the two attributes, stress and disturbance, which limit the amount of biomass production in a given habitat. If the lowest and highest levels of these two attributes are examined to determine the extreme positions, then four different permutations of stress and disturbance are theoretically possible. In practice, however, plants occupying habitats characterised by high disturbance and intense levels of stress do not exist, according to Grime, so that three distinct types remain: (1) competitors: plants of highly productive habitats, but low intensity of disturbance; (2) ruderals: plants of highly productive, but highly disturbed habitats; and (3) stress-tolerators: plants of habitats with a low intensity of productivity (due, for example, to the poor nutrient status of the soil) and disturbance. Again, these three groupings represent extreme cases, and species are more likely to occupy intermediate positions. It should be noted that when this model is applied to desert ecosystems, drought is regarded as being of high disturbance for one category of plants (arido-passive ones), but low for another (arido-active stress-tolerators). The stress-tolerators probably include many C_3 drought-deciduous shrubs characteristic of the plains and limestone plateaus, where essential nutrients are in short supply. These plants safeguard metabolic processes by utilising resource (water) pulses to invest primarily in unpalatable tissues (in particular wood) that promote endurance and long-term survival in a dormant state. Grime's CSR classification system is not without its critics, but as indicated by a number of authors (e.g. Wilson and Lee 2000; Pierce et al. 2005), as a comprehensive and coherent theory, it is useful for understanding a number of community processes.

5.7 Pollination

Little information is forthcoming on the pollination ecology of plant species on Socotra. A superficial appraisal of the flora suggests that insect-pollinated species predominate, followed by wind-pollinated plants. An interesting field of research is the potential significance of bird-pollination, especially in respect of certain endemic species. Specialist nectarivorous birds are found mainly in tropical and subtropical regions. On Socotra, it seems that one primary nectarivorous species occurs, the endemic Socotra sunbird (*Nectarinia balfouri* – Fig. 5.36). This common passerine

Fig. 5.36 Socotra sunbird (*Nectarinia balfouri*). December 2008 (Photograph by AbdulRahman Al-Sirhan)

Fig. 5.37 Flowers of the endemic *Trichocalyx obovatus*. December 2008

can be seen feeding on a number of different plant species that possess typical ornithophilous traits, such as a red, yellow or orange corolla and the production of copious nectar. The most frequent plant species visited by this bird appears to be the endemic shrub *Trichocalyx obovatus* (Fig. 5.37), but it can also be observed on pomegranate (*Punica protopunica*) with its conspicuous red flowers. R. Porter (pers. comm.) has observed Socotra sunbirds visiting the flowers of the following species: *Aloe* spp. (Fig. 5.38), *Buxanthus pedicellatus*, *Calotropis procera*, *Commiphora* sp., *Croton* sp., *Jatropha unicostata*, *Lycium sokotranum* and *Ziziphus*

Fig. 5.38 Flowers of the distinctive *Aloe perryi*. December 2008

Fig. 5.39 Flowers of *Adenium obesum*. December 2008

spina-christi. Mies and Zimmer (1993) observed Socotra sunbirds visiting the large salverform flowers of *Adenium obesum* (Fig. 5.39), and this bird could therefore be implicated in the pollination process. The pollinators of *Adenium* remained somewhat of a mystery for early visitors to the island, as for instance in 1899, when Oskar Simony speculated that the pollinators must be insects, even though he was unable to find any insects on the flowers (Vierhapper 1907). Rowley (1980) gave a first detailed account of the pollination process in *Adenium*, and implicated large bees. From adjacent continental regions, it is known that insects with a long slender

Fig. 5.40 Abyssinian white-eye (*Zosterops abyssinica*) visiting the flowers of *Trichocalyx obovatus*. December 2008 (Photograph by AbdulRahman Al-Sirhan)

proboscis, for example hawkmoths, are important pollinators, and they could also play a role on Socotra.

Apart from the Socotra sunbird, at least one generalist passerine species, the Abyssinian white-eye (*Zosterops abyssinicus*), can be observed visiting flowers as an opportunistic nectarivore, in particular those of *Trichocalyx obovatus* (A. Al-Sirhan and G. Brown, pers. observation – Fig. 5.40). In the Canary Islands, which display many ecological parallels to Socotra and where ornithophily has also evolved to a limited extent, no specialist nectarivorous birds are known, but at least six generalist passerine species have been recorded visiting flowers (Valido et al. 2004), none of which is found on Socotra (except as possible vagrants). Whether the resident endemic Socotra warbler (*Cisticola incanus*) assumes a similar role to that of the four species of warbler that visit the flowers of certain plant species in the Canaries requires clarification.

Flies are probably the primary pollinators of most stapeliads on Socotra, as is the case in other parts of the distributional range of these plants. For instance, the common endemic *Caralluma socotrana* produces short-lived (one to several days at most), large, bright-red to orange flowers (Fig. 5.41) that develop a rather unpleasant odour. In contrast, the small flowers of *Echidnopsis* species do not appear to emit any strong smell, but could still attract insects by the bright colours of their fleshy flowers.

Given the high proportion of endemics in the flora, studies on the pollination ecology of different species could prove highly rewarding, especially in the context

Fig. 5.41 Flowers of *Caralluma socotrana*

of plant–pollinator mutualistic networks. A review of studies carried out at the community level, rather than concentrating on individual species, has revealed that pollination processes are often more generalised than previously assumed (Olesen and Jordano 2002). The connectance level indicates the percentage of all animal–plant interactions within a pollination network that are actually established, and species-poor communities typically have a high connectance level, i.e. a low degree of specialisation (Whittaker and Fernández-Palacios 2007). If the situation on Socotra is compared to that of the Canary Islands, then, from the results of Olesen and Jordano (2002), it can be expected that the degree of connectance will be much higher in the relatively species-poor, arid, lowland ecosystems than in the species-rich forest communities. However, even in the species-poor communities, it is possible that endemic pollinators are involved that, as "super-generalists", serve many plant species. Conversely, a few super-generalised plant species may exist that can be pollinated by many different insect species.

Dendrosicyos socotrana appears to be a mainly dioecious species in the wild (but not necessarily so in culture under optimum growth conditions), although occasionally both male and female flowers are also developed on one and the same tree (Fig. 5.42). This mechanism serves to reduce the chance of self-pollination substantially, especially as monoecious plants appear to be self-compatible. The delicate flowers of *Dendrosicyos* open towards the evening and have usually wilted by the following midday. Crepuscular or nocturnal moths are therefore possible pollinators, but diurnal insects could also play a role, as is the case with most other members of the family. Our own initial observations show that viable seeds are produced

Fig. 5.42 Female flowers of *Dendrosicyos socotrana*. Hadibuh plain at mouth of Wadi Dihzafaq, December 2008

in less than 20% of fruits. Before any conclusions are drawn on the apparent low viability of the seeds, it is important to assess whether this percentage, even if it represents the norm, is sufficient to ensure survivability of the population under natural conditions, as is probably the case.

5.8 Seed Dispersal and Germination

Effective seed dispersal, i.e. the dispersal of seeds that results in the successful establishment of new individuals, is a vital prerequisite for the long-term survival of most plant species, and especially so for desert annuals that have to rejuvenate from seed at regular intervals, and where the local physical environmental conditions may vary substantially from one year to another. For instance, a microhabitat that enabled germination and growth in one year may become covered with sand in the next. It is widely accepted that only few desert plants species show any obvious adaptations for long-distance dispersal (Zohary 1937; Ward 2009). In fact according to data available from the Israeli flora (Ellner and Shmida 1981), it seems that desert species are primarily characterised by atelechoric dispersal types. However, most desert annuals and Orobanchaceae (now included in the Scrophulariaceae) produce numerous small seeds, a feature that not only protects the seeds from potential

Fig. 5.43 *Xylocalyx asper*. Members of the genus *Xylocalyx* display hygrochasy

predators, but also facilitates wind dispersal, i.e. telechory, to varying degrees. Some species, especially various composites, possess a pappus, a parachute-like structure that enables the fruits to glide through the air, and seeds of many Apocynaceae/Asclepiadaceae have long cottony hairs for the same purpose.

On Socotra, the Polygonaceae *Emex spinosa*, a widespread annual throughout much of the Arabian Peninsula and parts of the Mediterranean, is the classic example of a species displaying heterocarpy, i.e. the production of morphologically different fruits (which in *Emex* act as the dispersal unit) on the same plant. In fact it has a remarkable form of heterocarpy, namely amphicarpy: the subterranean fruits are large with a poorly developed pappus, in contrast to the aerial fruits, which are much smaller with a well-developed pappus (Gutterman 1993). The subterranean fruits are antitelechoric, germinating *in situ* from the base of the dead mother plant, whereas the aerial ones are atelechoric, characterised by short-distance dispersal.

Members of the small genus *Xylocalyx* (Scrophulariaceae) display hygrochasy (Thulin 2006), with water causing specialised tissues to swell, leading to the opening of the fruits and the release of a portion of the seeds (Fig. 5.43). This feature is probably also found in members of the genus *Blepharis* (Acanthaceae) on Socotra, especially as it is known from the widespread *B. ciliaris* (which does not occur on Socotra). Hygrochasy is also characteristic of the widespread *Aizoon canariense* (Aizoaceae), one of the more common annuals on Socotra. The ecological significance of hygrochasy has been discussed by various authors, including Gutterman (1993).

Apart from generalisations that can be inferred from neighbouring regions, very little specific information is forthcoming on the dispersal mechanisms and seed germination in the flora of Socotra. Such studies are vital, because, as underlined by Schupp et al. (2010), effective seed dispersal is critical to many ecological questions, including, to name but a few, plant population dynamics, community structure, maintenance of biodiversity in fragmented habitats, ecological restoration and conservation biology in general.

Studies on germination and establishment success under field conditions are urgently required for the rare endemics that are threatened with extinction, as well as for keystone perennials that are not regenerating well in their natural habitat. In many cases, germination under greenhouse conditions appears to be unproblematic (e.g. *Dracaena* – see Sect. 7.5.1, *Dendrosicyos socotrana* and *Dorstenia gigas*), a fact that offers some hope for *ex situ* conservation at least.

Desert annuals often display highly specific germination requirements because these plants are, to a much larger extent than perennials, dependent on germination and successful establishment at regular intervals to maintain the long-term survival of the population. A good deal of information on the germination characteristics of the desert annuals of Socotra can be inferred from studies that have been undertaken in the wider region, in particular by Gutterman (summarised in Gutterman 1993). An important mechanism that prevents germination occurring at the wrong time of the year is dormancy, and this is developed in both annuals and perennials. Most species of the desert plains on Socotra appear to be winter annuals, i.e. germinating during the cooler part of the year when rainfall is received. This is also the case with many desert species on the Arabian mainland, although there are a number of exceptions. Cues that trigger germination are presumably lower temperatures and sufficient moisture, parameters that probably elicit species-specific germination responses.

Predictive germination, defined as germination that is directly sensitive to environmental factors associated with conditions favourable for immediate seedling growth (Smith et al. 2000), is an important strategy that allows plant populations, of both perennials and annuals, to thrive in highly variable, harsh environments. The amount and temporal distribution of available soil moisture are primary environmental variables on which predictive germination is based (Gutterman 1993).

A second important mechanism that enables desert plants to persist in their harsh environment is fractional (= delayed) germination (Baskin and Baskin 1998). With fractional germination, only a certain proportion of germinable seeds will actually germinate, even under ideal conditions. This strategy buffers a population from the consequences of near or complete reproductive failure in unfavourable years, and has been well documented from American desert annuals (Venable and Lawlor 1980; Philippi 1993). Fractional germination is also known from desert annuals of the Middle East, such as *Spergularia diandra* (Gutterman 1996). It appears as well to be a widespread mechanism in many desert annuals in the north-east of the Arabian Peninsula (Brown 2001).

Chapter 6
Vegetation

Abstract Extending from sea-level to just over 1,500 m, the compression of climatic zones along an elevational gradient as well as small-scale differences in exposure, rainfall and substrate are responsible for a broad spectrum of different habitats, ranging from deserts to evergreen forests within a relatively small area on Socotra. A concise overview of the vegetation of the islands along an altitudinal gradient is presented in this chapter. As a considerable proportion of the native flora is endemic, most of the natural plant communities are unique to Socotra, although physiognomically, they are often very similar to vegetation units of comparable habitats in the adjacent mainland areas. A highly distinctive feature of the mid-elevation plateaus in the eastern part of Socotra is the open dragon's blood tree (*Dracaena cinnabari*) vegetation, which is of outstanding conservation value. Widespread grazing by domestic livestock is greatly modifying the natural vegetation cover, leading to serious degradation over many areas. The role of inaccessible cliff ledges and remote plateaus is highlighted as a refuge for many palatable plant species, and, with respect to the potential natural vegetation, such habitats could provide a reliable indication as to the composition and structure of vegetation that previously existed over large tracts in the more accessible areas.

6.1 General Comments on the Vegetation

In accordance with the terminology of Zohary (1973), Socotra and the adjacent areas belong phytogeographically to the Eritreo-Arabian subregion of the Sudanian region. White (1983) referred to the same area as the "Somali-Masai regional centre of endemism" (Fig. 6.1). The characteristic vegetation of this phytogeographical unit is *Acacia–Commiphora* shrubland, dominated by a wealth of tree and small shrub species in which the genera *Acacia* and *Commiphora* (myrrh) are conspicuous elements. However, it should be noted that *Acacia* is poorly represented on Socotra, and does not play a dominant role in the vegetation. *Boswellia* (frankincense) is

G. Brown and B.A. Mies, *Vegetation Ecology of Socotra*, Plant and Vegetation 7,
DOI 10.1007/978-94-007-4141-6_6, © Springer Science+Business Media Dordrecht 2012

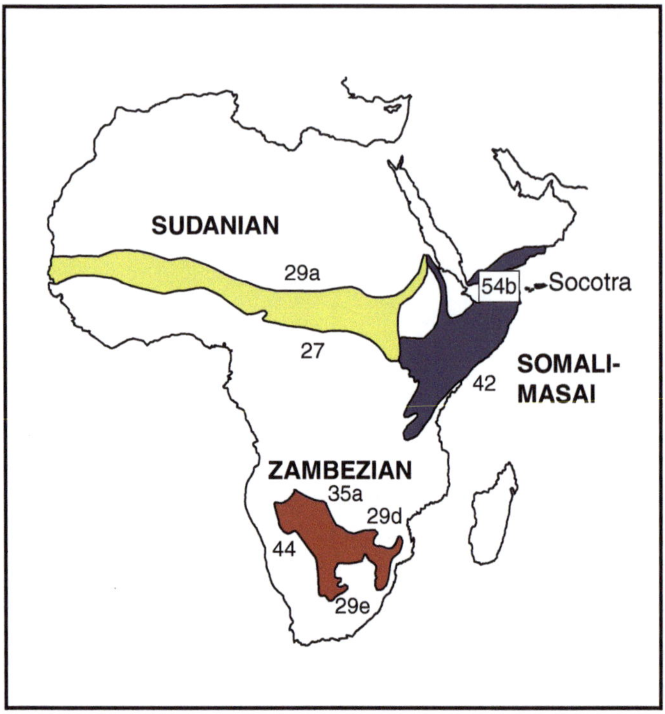

Fig. 6.1 Distribution of *Acacia–Commiphora* shrubland in Africa and Arabia and related vegetation types in Africa in accordance with White (1983). SOMALI-MASAI: 42=*Acacia-Commiphora* deciduous bushland and thicket; 54b semi-desert grassland and shrubland; SUDANIAN: 27=woodland; 29a=undifferentiated woodland; ZAMBESIAN: 29d/e=South Zambezian undifferentiated woodland and wooded grassland; 35a=transition from undifferentiated Zambezian woodland to *Acacia* deciduous bushland and wooded grassland; 44=Kalahari deciduous *Acacia* bushland and wooded grassland

another example of a genus of shrubs or small trees typically found in north-eastern Africa and southern Arabia, with most of the ca. 18 known species occurring in this region, including eight, all endemics, recorded from Socotra.

In their global vegetation scheme, Walter and Breckle (2004) classified Socotra as a transitional zone, referring to it as the "Ceno-Ecotone II/III". This category lies between the "Tropical Cenobiom II", typical of a humid-arid climate type (summer rain), and the "Cenobiom III", comprising arid subtropical deserts. Other areas within the Ceno-Ecotone II/III include the Sahel Zone, Cape Verde Islands and the arid regions of south-western Madagascar. As with this last area, a prominent feature of the flora of Socotra is the host of succulent plant species.

Extending from sea-level to just over 1,500 m, the compression of climatic zones along an elevational gradient, sometimes referred to as "telescoping", has led to the creation of a wide variety of environmental conditions within a relatively small area, particularly on Socotra itself, with which this chapter is mainly concerned. In accordance with the general observations of Bruijnzeel (2005) from other parts of

Table 6.1 Classification of deserts and semi-deserts according to Shmida (1985)

Extreme desert:	Below 70 mm/year	No dwarf shrub cover
Desert:	70–120 mm/year	Dwarf shrubs less than 10% cover
Semi-desert:	120–350 mm/year	Dwarf shrubs ca. 20–50% cover

the world, the local occurrence of woodland at relatively low elevations on Socotra can probably be attributed to the high humidity of the oceanic air, which promotes cloud formation. The same phenomenon is also responsible for the development of woodland along the Dhofar escarpment on the southern coast of Oman. To what extent the "Massenerhebung effect" (Brockmann-Jerosch 1919; Troll 1948) plays a role in the observed compression of vegetation zones at these relatively low altitudes is open to question, but, consistent with other islands from around the world, it is noticeable that cloud forest is developed at much lower altitudes on Socotra than on mainland areas away from the maritime influence.

The overall arid macroclimate of the island has been substantially modified by the close proximity to the sea, although this influence is less pronounced in the rain shadow in the south. Another important factor leading to more diverse environmental conditions is the markedly varied geology, with exposed limestone and granite layers occurring in the immediate vicinity of each other. In summary, therefore, differences in altitude, exposure, rainfall and substrate are responsible for a broad spectrum of different habitats, ranging from deserts to evergreen forests, within a very restricted area.

Desert-like vegetation occupies the plains and slopes at lower altitudes in particular. There is no overall definition of precisely what constitutes a desert (von Willert et al. 1992). Some authors point to the sparse vegetation cover as a characteristic feature. A review of the various definitions was given by McGinnies et al. (1968), but any definition used is arbitrary. In the following, the term is used to include both deserts and semi-deserts. One characteristic feature of deserts is that they are usually regarded as areas in which annual potential evapotranspiration exceeds annual precipitation.

Although most terms used to describe the vegetation of arid zones are ill-defined, dwarf shrubs, in some cases graminoids, tend to be the principal (i.e. visually dominant) components of the vegetation, as the climate is usually too harsh to support any substantial growth of trees. The system outlined by Shmida (1985) has proved practical for a broad classification of deserts, especially those of the Middle East, because it combines two attributes that to a certain extent correlate with each other, namely the amount of annual precipitation and the degree of cover by perennial woody species (Table 6.1).

However, it should be stressed that the amount of precipitation can be used only as a rough guide. Of greater relevance for vegetation development in semi-arid and arid areas is the capacity of the substrate to store water (Noy-Meir 1973).

Since the arrival of the first major wave of settlers on the islands, probably about 3,000 years ago (Naumkin 1993), or up to about 11,000 years ago as a more recent study has suggested (Cerny et al. 2009), and continuing up until the present day, the native vegetation has been subjected to various human influences, leading in some cases to pronounced changes in species composition. In former times, the people of Socotra eked out a meagre existence based on traditional subsistence farming. Complex local land management systems controlling the access of livestock to the limited natural resources during the dry season largely prevented the over-exploitation of the rangelands that is now a characteristic feature of huge tracts of the mainland Arabian Peninsula, thus ensuring their sustainable use. However, the lifestyle of the local inhabitants has begun to change enormously in recent years as a result of increasing exposure to external influences, and this is having profound negative effects on the natural landscape, as outlined in Chap. 7.

Today, much of the landscape of Socotra, in excess of 80% of the land, including some of the least productive areas, is used for grazing by domestic livestock, mainly goats, sheep, cattle, camels and donkeys. These animals have probably been responsible for the most profound effects on the natural vegetation. Furthermore, wood has been collected from wherever possible, but especially in the vicinity of human settlements, and this activity too has undoubtedly greatly modified the original vegetation cover. Settlements and plantations (such as date palm groves and small gardens), where the most radical changes to the native vegetation have occurred, account for less than 2% of the land area. The remaining area, less than 18% of the island, is covered by semi-natural shrubby vegetation types, some of which undoubtedly bear a strong resemblance to the natural vegetation before the arrival of man.

As a considerable proportion of the native flora is endemic, most of the natural plant communities are unique to Socotra, although physiognomically, they are very similar to vegetation units of comparable habitats in the adjacent mainland areas of Africa and Arabia. Only relatively few neophytes have so far managed to gain a foothold in the archipelago, and these are mostly pantropical annual weeds (see Sect. 7.3.8). Those that have been successful in establishing themselves do not form distinct non-indigenous communities, unlike the situation in some other parts of the world, such as Hawaii, where up to 66% of plants species are introduced (Stone et al. 1992).

Especially at lower altitudes, vegetation cover on the islands is often patchy, displaying the open, contracted vegetation pattern ("mode contracté"; Monod 1954) typical of arid ecosystems, with more dense plant growth restricted mainly to wadi beds, runnels and other sites that receive an additional input of water in the form of run-on. Le Houérou (2005) noted that contracted vegetation is a specific feature of true deserts with a mean annual rainfall usually below 100 mm, and is characteristic of desert pavements and gravel plains.

6.2 Broad Vegetation Classification

Vegetation classification can be problematic in arid areas due to the low, patchy vegetation cover (frequently less than 25% at lower and middle altitudes on Socotra) and the paucity of perennial plants that can be used as potential characteristic species.

Those perennials that are present can display a fairly wide range of tolerances regarding their edaphic conditions, so that they often occur in different vegetation units (e.g. *Adenium obesum*, *Buxanthus pedicellatus*, *Croton socotranus* and *Dracaena cinnabari*). Furthermore, taxonomic confusion still surrounds many species throughout the region, although much work is being undertaken to rectify the situation (e.g. Miller and Morris 2004), and knowledge on the distribution of many species is still far from complete. It is generally not feasible to use annuals for vegetation classification in arid areas for two principal reasons. (1) In dry years, which have a tendency to occur in regular cycles, many annuals can be largely, if not totally, absent. With respect to Socotra, given that most phytosociological studies have been carried out by scientists visiting the islands, rather than ones permanently based there, this can lead to strongly biased results. (2) As with some perennials, annuals often appear quite indifferent to the prevailing environmental conditions, unless these are too extreme. Only detailed studies or substantial experience over years can lead to the detection of subtle differences among species regarding their environmental preferences.

In order to facilitate a clearer understanding of the vegetation to a broader audience, the general, non-committal term "community" is used to refer to vegetation units, rather than the specialised terminology of the Zürich–Montpellier school (also known as the Braun-Blanquet floristic association system; see Müller-Dombois and Ellenberg 1974), which has a wide following in Central Europe. In any case, the relatively few studies and poor understanding of specific vegetation units on Socotra and in the adjacent regions do not allow this hierarchical system to be applied with any authority at present. Where sound information is forthcoming, specific phytosociological units are mentioned briefly in the text. Mies (1999c) undertook a phytosociological study of the *Croton socotranus* stands and some other vegetation types on the plains of Socotra, and Kürschner et al. (2006a) have recently carried out a study of the forests and woodlands in the north of the island. In both cases, however, classification is in part problematic, because, for example, some of the character species used to denote certain plant associations also occur in other vegetation types not examined by the authors. As observed by Miller and Morris (2004), vegetation description is not straightforward on Socotra, because although there is a fairly clear altitudinal zonation, a mosaic of floristically closely related vegetation units often exists, which is usually better characterised physiognomically. The practical approach adopted in this chapter is – where possible – to describe vegetation units that are readily identifiable by characteristic dominant species or combinations of dominants and co-dominants in homogenous stands, which are also physiognomically quite distinct. At the same time, a pragmatic approach is required to avoid describing all possible combinations of species. The naming of vegetation units is based primarily on the visually most dominant life-form present, with the following hierarchy being adopted for the naming of communities: trees (woody plants taller than at least 3 m) assume priority over all other species, followed by shrubs (woody plants between 0.5 and 3–5 m tall), dwarf shrubs (up to 0.5 m tall), graminoids and perennial herbs. In a number of cases, we have refrained from naming plant communities more specifically because of the lack of detailed knowledge.

In accordance with the UNESCO Structural-Ecological Formation System, as described in Müller-Dombois and Ellenberg (1974), vegetation units can be assigned to one of seven formation classes. In respect of the formations characterised by woody plants, and which are the most relevant to many parts of Socotra, "forest" (Formation Class I) is used to denote communities dominated by trees with a closed canopy, whereas "woodland" (Formation Class II) describes open communities in which trees play a predominant role, covering at least 30% of the surface. "Shrublands" (Formation Class III) is a term used to refer to communities dominated by shrubs, while "(dwarf) scrub" (Formation Class IV) is a rather general term to describe dwarf shrub communities that may or may not be interspersed with individual shrubs or trees. "Terrestrial herbaceous communities" (Formation Class V) comprise grasslands and forb-rich steppes. "Scarcely vegetated areas" (Formation Class VI) include the chasmophytic vegetation of rock faces, the sparse vegetation of scree and sand accumulations. "Aquatic plant formations" (Formation Class VII) are also represented on Socotra, albeit locally and of small areal extent.

Most plant communities of Socotra can be broadly assigned to one of five altitudinal belts (shown in simplified form in Figs. 6.2 and 6.3) that also integrate key environmental parameters:

- coastal zone (0–200 m asl);
- low-elevation slopes (200–400 m asl);
- medium-elevation slopes (400–700 m asl);
- montane (700–1,200 m asl);
- high-montane (1,200–1,540 m asl). Exposed montane cliffs are also accommodated here.

Within these belts, three general moisture regimes can be distinguished:

- arid (less than the equivalent of 200 mm rainfall per year);
- semi-arid (between 200 and 400 mm rainfall per year);
- mesic (between 400 mm and ca. 600 mm rainfall per year, but as much as 1,000 mm effective precipitation).

As indicated in Sect. 2.3, these moisture regime categories can be somewhat misleading because, although rainfall is important, it constitutes only one component of the total precipitation in some regions of the island. For instance, in the montane belt, fog and dewfall substantially increase the effective amount of plant-available water. At lower altitudes, run-off water may accumulate in distinct microsites where it can be utilised by plants. These additional inputs, which are often difficult to quantify precisely, nonetheless play a significant ecological role.

Taking elevation, moisture regime and basic vegetation characteristics into consideration, Mies (2001) recognised five distinct ecological zones on Socotra.

1. Arid zone
   High temperatures in combination with very low (<200 mm per year) and sporadic rainfall; no fog or dew. Elevated soil salinity is a characteristic feature of some habitats. This can be due to excessive evaporation or the influence of sea-borne

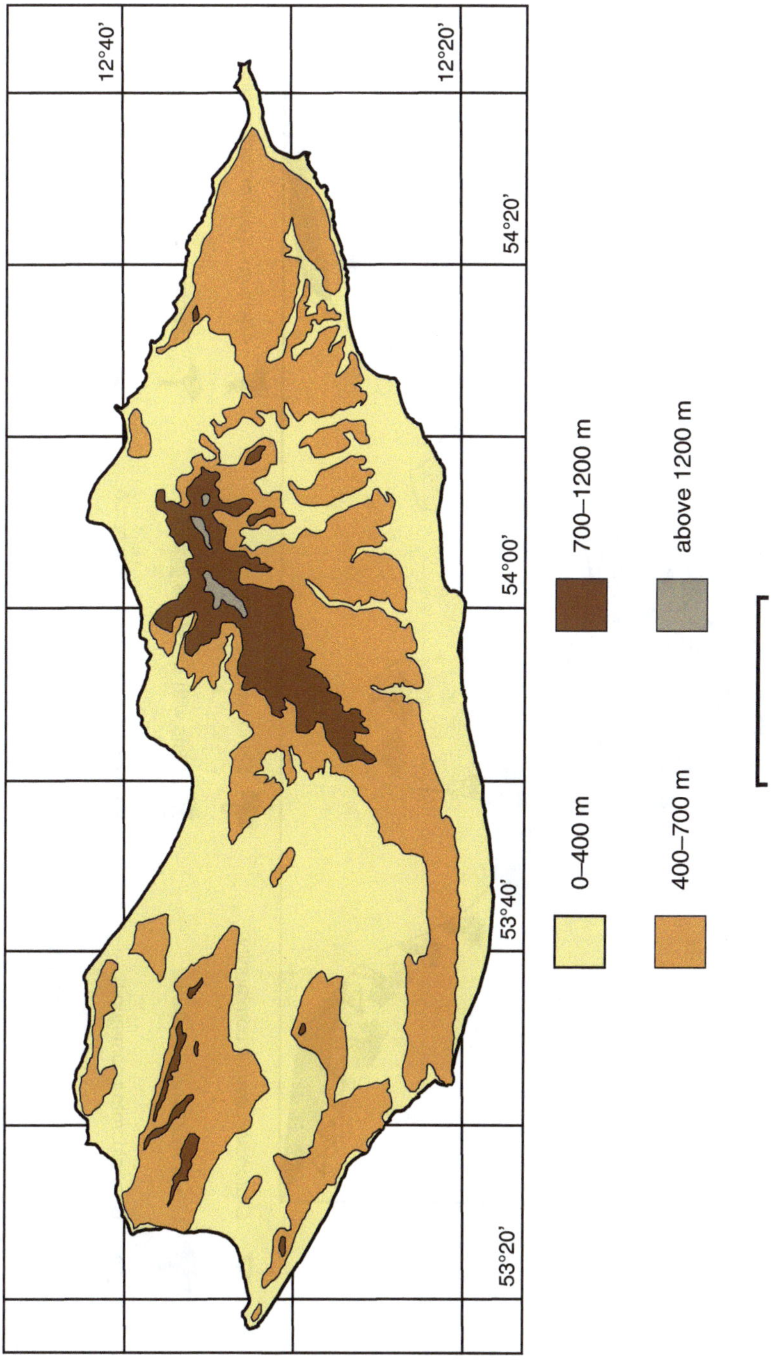

Fig. 6.2 Ecologically relevant elevation belts on Socotra

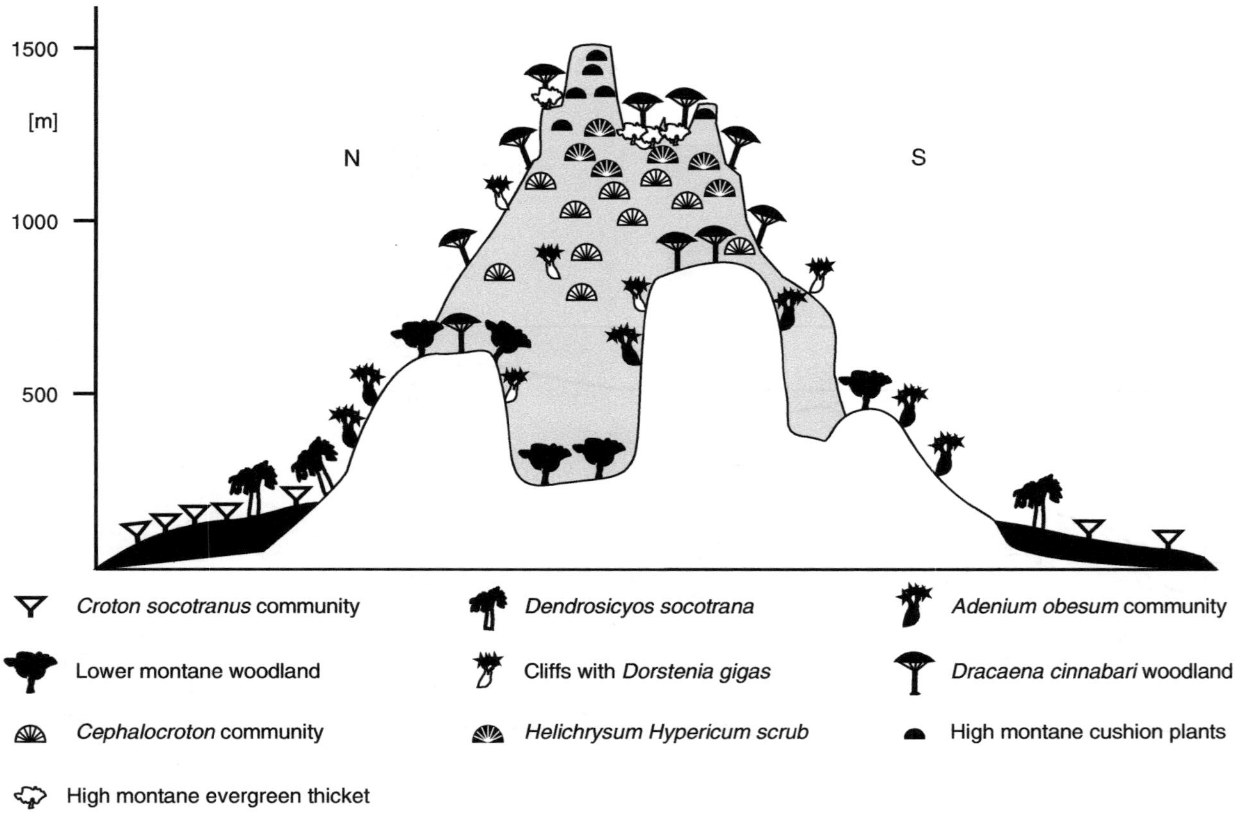

Fig. 6.3 Cross-section of Socotra with characteristic vegetation types. *Black* shading = plains, *white* = limestone, *grey shading* = granite of the Hagghier mountains

Fig. 6.4 The arid zone is characteristic of the coastal lowlands and low-elevation slopes. Here with *Croton socotranus* shrubland and the widespread endemic emergent *Euphorbia arbuscula*. Near B'ir Haarso, east of Qalansiyah, December 2008

salt. The arid zone is characteristic of the coastal lowlands and low-elevation slopes (Fig. 6.4). Vegetation cover is typically sparse, but can be remarkably high locally, and is dominated by shrubs (in particular *Croton socotranus*), dwarf shrubs and graminoids. Halophytes frequently play an important role in this zone, and succulents may be a conspicuous feature of some communities.

2. Semi-arid zone
   High temperatures with low (200–400 mm per year) seasonal rain, falling mainly between December and January; little dewfall, and restricted to the rainy period. Typically found on the lower rocky slopes and escarpments, up to 400 m asl. (Fig. 6.5). Xerophytes and succulents dominate in the scrub vegetation, with giant succulents interspersed. The Socotra Aden rose (*Adenium obesum*) is locally highly characteristic.
3. Semi-arid montane zone
   Moderately high temperatures with low seasonal rainfall (200–400 mm per year) during the winter; dewfall is a regular occurrence throughout the winter. Found on the mid-elevation slopes, including the foothills of the Hagghier range at 400–700 (–800) m. Xerophytic scrub predominates, but thickets and woodland are developed locally under favourable conditions (Fig. 6.6). Species requiring a regular input of moisture are indicative of the more favourable sites.
4. Mesic montane zone
   Occasional rainfall in two seasons: (1) from November to January, and (2) a shorter one from April to May; heavy cloud formation brought on by north-easterly

Fig. 6.5 Semi-arid zone

Fig. 6.6 Semi-arid montane zone with dense scrub

Fig. 6.7 Dense woodland in the lower mesic montane zone

trade winds and during the monsoon period; heavy dewfall and fog; temperatures moderate during the day, but dropping noticeably at night. Typical of the montane cloud belt of the Hagghier mountains and highland plateaus in the east and west of Socotra, at about 700–1,200 m (Fig. 6.7). Dense shrubland as well as evergreen woodland and forest are the characteristic vegetation types. The regular occurrence of dew and fog facilitates the luxuriant growth of mosses and lichens on trees and rocks. The highest peaks of Abdalkuri extend to this climatic zone, but the vegetation there remains more typical of an arid climate.

5. Mesic montane–sub-Afroalpine zone
   Occasional rainfall as described under (4); heavy dewfall, but less fog and precipitation; very high rates of insolation; temperatures strongly diurnal, moderately warm during the day, but falling considerably at night. Strong winds. Characteristic of the peaks of the Hagghier mountains (1,200–1,540 m). Typical of this belt (Fig. 6.8) are dwarf shrubs and cushion plants. Some of these species may also occur at lower altitudes on exposed slopes. Conversely, dense shrubland and woodland extend their range into this zone locally, mainly in sheltered locations.

Kürschner et al. (2006a) have to a large extent followed these zones in their phytosociological treatment of the forests and woodlands of northern Socotra.

Natural plant communities that have not experienced substantial disturbance for prolonged periods of time may have reached an end-point in the process of vegetation succession, corresponding to what is frequently referred to as the "climax vegetation". The concept of climax vegetation, at least of the monoclimax, is a somewhat controversial topic from an academic point of view (e.g. Tansley 1935; Whittaker

Fig. 6.8 Dwarf scrub in the upper mesic montane–sub-Afroalpine zone

1953; Sprugel 1991), a problem that is dealt with briefly in Sect. 6.5, but the concept is often useful in practical field studies. Climax plant communities can be conveniently assigned to one of three broad categories.

1. Zonal vegetation (= climatic climax vegetation)
   Zonal vegetation occurs on substrates that are not generally regarded as extreme, and it is developed primarily in response to macroclimatic influences. In areas undisturbed by humans, zonal plant communities typically occupy large tracts of the landscape. The five ecological zones mentioned above can be used to delimit zonal vegetation types. At lower altitudes on Socotra, therefore, communities dominated by dwarf scrub, shrubs and possibly grasslands will prevail, whereas the zonal vegetation in the mesic montane zone is dense shrubland or forest.
2. Azonal vegetation (= edaphic climax vegetation)
   In contrast, azonal vegetation comprises plant communities whose occurrence reflects the overriding influence of extreme substrate conditions. Such communities are frequently found under different macroclimatic conditions, and they can vary in size from small patches to large expanses. Habitats with substrates that are often wet, such as springs, inland marshy patches, wadis with high groundwater levels, the banks of streams, frequently support a highly characteristic vegetation. The substrate of inland depressions, intertidal marshland and other coastal habitats is invariably saline to some degree. This exerts a major selective pressure on plant species able to thrive there. Apart from wet azonal habitats, ones characterised by high salinity also occur in dry situations, sometimes well away from the coast. Rock crevices are another type of habitat that may support a highly specialised azonal vegetation.

3. Extrazonal vegetation
   Extrazonal vegetation units are ones characteristic of an adjacent macroclimatic region interspersed in the zonal vegetation. Their occurrence can be explained by the local prevalence of climatic conditions similar to where they would be typically found. Due to the pronounced compression of macroclimatic zones governed by altitude on Socotra, different types of zonal vegetation prevail within a very small geographical area. Where, for example, forest, the zonal vegetation of higher altitudes, is developed in shaded wadis at lower elevations, it can be regarded there as extrazonal vegetation. It can exist only due to the specific local climatic conditions, which sharply contrast with the overall macroclimatic characteristics of that lower altitude.

A concise overview of the vegetation of the islands is presented here, restricting itself to the major plant communities. Furthermore, only those vegetation units that are relevant to the present-day conditions are discussed, although a separate section explores the potential natural vegetation of the islands (Sect. 6.5). In this overview, the main plant communities are described according to altitudinal belt in which they occur, beginning at sea-level.

6.3 Vegetation of the Different Altitudinal Belts

6.3.1 Coastal Vegetation Types

The topography of the coastline of Socotra varies considerably, ranging from steep cliffs that fall precipitously to the sea, to flat, wide expanses that are well vegetated, locally at least. A striking feature of the vegetation of coastlines with low relief, especially on more saline substrates, is that the individual stands tend to be species-poor or even monospecific, i.e. one single species can dominate large patches (for example *Arthrocnemum macrostachyum*). Abrupt shifts in key abiotic factors are reflected in corresponding changes in the vegetation cover. The main abiotic factors controlling community composition in coastal habitats are salinity of the water and/or substrate, frequency and extent of inundation, and the water-holding capacity of the substrate (Deil 2000). As a consequence, a characteristic zonation of the vegetation occurs in coastal areas, with communities, predominantly azonal, broadly arranged in distinct bands, often running parallel to the coastline. Small-scale substrate heterogeneity within these broad belts is reflected by corresponding vegetation mosaics. In some areas, the typical sequence of vegetation bands may be arranged in a different order due to the local site-specific conditions. Plants growing on saline substrates in the coastal region are invariably halophytes. However, on deep sandy substrates, such as coralline sand sheets, the influence of salinity is substantially reduced, and these sites often show a preponderance of glycophytes.

If the full complement of coastal plant communities typical of the region were present in any one single locality, the characteristic sequence of vegetation bands would be (1) sea-grass beds (subtidal); (2) mangroves (a halophytic woodland formation below

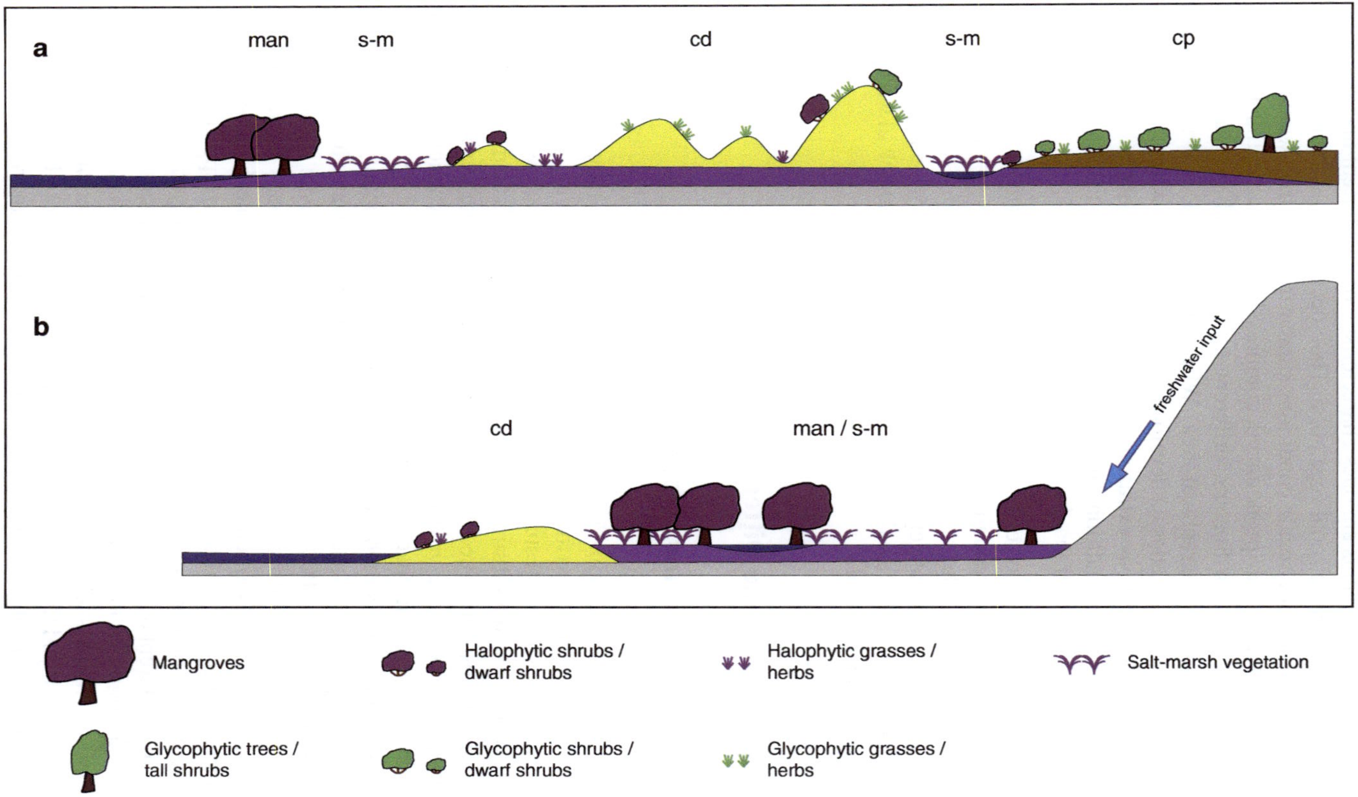

Fig. 6.9 Schematic diagram showing (**a**) the typical toposequence of coastal vegetation with mangroves in the wider region and (**b**) the toposequence found on Socotra, in particular at Niyt. *Abbreviations*: *man* mangroves, *cd* coastal dunes, *cp* coastal plains, *s-m* salt-marsh

the high-tide mark); and (3) salt-marsh (dominated to a large extent by dwarf shrubs, most commonly chenopods, and regularly inundated at high tide) or virtually barren sabkha. Depending on the locality and relief, this would be followed by (4) beach (sandy or rocky, with or without one to several barrier berms or sand bars); (5) a narrow to broad band of raised coastal sand sheets or low dunes; and (6) salt-marsh in the characteristic saline depression behind the dunes (Fig. 6.9). Further inland, true terrestrial ecosystems would follow, in particular coastal plains, in which smaller saline depressions might be interspersed locally. On Socotra, all these vegetation bands are present, but not in any single location, and often not in the typical sequence outlined above, at least not any more. For instance, the coastal waters off Socotra rapidly increase in depth, so that typical salt-marsh vegetation and coastal sabkha, which are dependent on occasional flooding at high tide, are developed only very locally on the seaward side of coastal dunes, as has been described from southern mainland Yemen by Kürschner et al. (1998), but are more commonly found in depressions behind the immediate coastline or in lagoons. However, in general terms, the zonation of coastal vegetation occurring on Socotra shows a broad similarity to that described from the Red Sea by Chapman (1974) and Deil (1998), from southern Yemen (Kürschner et al. 1998) and from the Arabian Gulf (Deil 2000; Brown et al. 2008).

With respect to salt-marsh and coastal sabkha in particular, excessive rates of evaporation invariably lead to a highly saline habitat type, and once the water has disappeared from the surface layers, a crust of crystalline salt often develops. Groundwater drawn upwards by capillary forces and the hygroscopic properties of the marine salts prevent the muddy substrate from drying out completely, and it therefore remains moist below the surface. Anaerobic soil conditions develop as a result of the high moisture content and the presence of a surface soil crust, both of which restrict the free passage of air. Hypersaline conditions occur in areas that are only occasionally inundated and that dry out for long periods. In such situations, vegetation cover is usually very sparse or absent. However, these areas were until fairly recently of some economic importance, because the salt was used to preserve fish that was exported to Zanzibar (Doe 1992) and even the Far East. More regular flooding prevents the build-up of salts in the surface layers, and as a result, dense vegetation can develop.

As emphasised by Chapman (1974), the process of succession in coastal plant communities is frequently halted and set back by regular incisive perturbations. Own observations on Socotra indicate that occasional "catastrophic" events, such as heavy storms in combination with high tides, prevent any long-term build-up of silt in the mangroves, and even remove dunes from the coastline. Furthermore, such events lead to the creation of new habitats, in particular saline depressions. The vegetation of these habitats is therefore characterised by typical proclimax communities. As part of their studies on vegetation succession on the Krakatau Islands (Indonesia), a group of islands remaining after the eruption of Krakatau in 1883 on which the vegetation was entirely destroyed, and where a new volcanic island (Anak Krakatau) emerged in 1932, Whittaker et al. (1989) found that the current floristic composition of the coastal habitats was established at a very early stage of colonisation. Furthermore, it has remained little altered since, despite the highly dynamic nature of these habitats, with the species colonising newly created sites as older ones became disrupted or destroyed by tidal action.

6.3.1.1 Sea-Grass Beds

Sea-grasses require high amounts of sunlight, and they are therefore found predominantly in shallow waters. Phillips et al. (2004) report that in the coastal waters of Abu Dhabi Emirate, sea-grass beds grow to a maximum depth of 15 m. Three main factors control their occurrence: (1) water depth; (2) the physical nature of the bottom substrate; and (3) the extent of water movement by wave action or tidal effects (Sheppard et al. 1992). As the plants are highly sensitive to exposure, extreme low tides may damage them. In the Socotra Archipelago, sea-grass beds are rather local, and restricted to the shallow coastal waters around the main island, especially on the leeward side or in sheltered bays where there is relatively little water movement. Klaus and Turner (2004) have recently given an overview of the marine biotopes of Socotra, which includes sea-grass beds. Vast subtidal banks are developed off the coast at Nuged and Qalansiyah. At Qalansiyah, the sea-grass beds consist mainly of *Halodule uninervis*, *Halophila ovalis* and *Thalassia hemprichii*, with marine algae also playing an important role (Mies 2001), and Klaus and Turner (2004) mention too the occurrence of *Thalassodendron ciliatum* there, which was found to be associated with *Porites* corals. Three species of Cymodoceaceae, *Cymodocea rotundata*, *C. serrulata* and *Halodule uninervis*, were reported from Socotra (Miller and Morris 2004). In a comparison of the marine plant communities (macroalgae and sea-grasses) of Socotra and Masirah Island (Oman), Schils and Coppejans (2003) recorded five sea-grass species from their sampling sites on Socotra: the Cymodoceaceae *Halodule uninervis* and *Thalassodendron ciliatum*, as well as the Hydrocharitaceae *Halophila decipiens*, *H. ovalis* and *Thalassia hemprichii*. Klaus and Turner (2004) list six species for Socotra: those noted by Schils and Coppejans (2003), with the exception of *H. decipiens*, but in addition, *Cymodocea rotundata* and *C. serrulata*. It seems, therefore, that at least seven sea-grass species occur in the coastal waters off Socotra.

Most, if not all, of the species known from Socotra are perennials. Generative reproduction is widespread, but floral structure is very much simplified, especially with regard to the perianth. Most species are dioecious, although *Halophila decipiens* is monoecious, and they possess unique structural adaptations to facilitate submerged pollination (hydrophily). All species mentioned above are also capable of vegetative reproduction, and because their stolons spread in lines along the sea-bed, the plants often grow in rows, like stitches produced by a sewing-machine.

The often rather uniform sea-grass beds belong to the phytosociological class *Zosteretea* (Zohary 1973). As in many other parts of the world, little research has been carried out on the marine phanerogamic vegetation, despite its enormous ecological importance (Basson et al. 1977).

A number of aquatic species also occur in brackish lagoons and creeks on Socotra, the most widespread being *Ruppia maritima*. This Ruppiaceae is found in several locations on the northern coast, for instance in the vicinity of Qalansiyah. Zohary (1973) places such stands in the vegetation class *Ruppietea*, which he himself admits is barely distinguishable from the *Zosteretea*.

Fig. 6.10 Barrier berm of coralline sand on the central Nuged coast. December 2008

6.3.1.2 Sandy Beaches, Coastal Sand Sheets and Dunes

Sandy beaches are widespread in many coastal areas of Socotra, but are generally best developed in association with the adjacent plains, in both the north and south of the island. Beaches that are frequently inundated by the sea are invariably devoid of vegetation cover, but just above the high-tide mark, individual plants begin appearing, and a characteristic, recurring sequence of plant assemblages can be observed with increasing distance from the sea. This toposequence reflects corresponding changes in key environmental conditions (see Fig. 6.9).

Immediately above the high-tide mark, a berm of coralline sand is often present, rising from just a few centimetres to several metres above the surrounding area and running parallel to the coastline (Fig. 6.10). At the base of this so-called "drift-line", a band of debris consisting of algae, seaweed, sea-grasses and other marine organisms may be washed up. The berm itself is inundated only during extreme tidal events, and is usually colonised by scattered perennials and various annuals (often chenopods). Perhaps the most widespread species in such situations is the endemic dwarf shrub *Limonium sokotranum*, and in some areas it can form dense stands (Fig. 6.11). Very locally, the halophytic tussock grasses *Urochondra setulosa* and *Halopyrum mucronatum*, as well as the small shrub *Atriplex griffithii* ssp. *stocksii* (Fig. 6.12), are characteristic species. They profit from the elevated nutrient status of the substrate, especially nitrogen, phosphate and sulphate, but high salinity levels restrict the number of other plant species able to colonise these sites.

Fig. 6.11 Barrier berm colonised by dense growth of *Limonium sokotranum*. North coast, west of Ghubbah, December 2008

Fig. 6.12 *Atriplex griffithii* ssp. *stocksii*. Ra's Shu'ab. March 1997

Fig. 6.13 Barrier berm composed of pebbles and small stones on the central Nuged coast

In a few locations, such as on the central Nuged coast, the sand barrier berm is replaced by one composed entirely of pebbles and small stones. This berm is up to 3 m in height in places, and lacks any vegetation cover (Fig. 6.13).

On the landward side of the barrier berm, extensive coralline sand sheets and even coastal dune fields may be developed locally, often as a continuation of the barrier berm. The latter are particularly characteristic of the southern Nuged coastline, where individual dunes may reach 12 m in height (Fig. 6.14). Despite their proximity to the sea, it appears as if the influence of salinity on the coralline sand dunes is much reduced. This is probably due to the fact that rainwater percolating through the sand forms a typical Ghyben–Herzberg lens in the subsurface layers. The "coastal white sand associations", a term coined by Vesey-Fitzgerald (1957) to describe the vegetation of the predominantly calcareous sands running parallel to much of the western and southern Arabian Gulf coastline, and that have been described in more detail from various parts of Arabia by Kürschner (1986), Mandaville (1990), Deil and Müller-Hohenstein (1996) and Brown et al. (2007, 2008), also appear to be barely affected by salinity, as indicated by the high density of glycophytes growing there.

Apart from the widespread *Limonium sokotranum*, the tall, shrubby chenopod *Suaeda monoica* is locally common on low coastal dunes in both the north and south of Socotra, with individuals often exceeding 3 m in height. On higher dunes, *Tamarix nilotica* plays an important role, especially in the south (Fig. 6.15). These shrubs gradually serve to stabilise the sand. Some of the dunes are possibly true nebkhas, i.e. the tree itself, as an obstacle to wind-blown sand, has initiated the formation of

Fig. 6.14 Coralline dune fields on the southern Nuged coast. December 2008

Fig. 6.15 *Tamarix nilotica* on coralline dunes. Nuged coast, December 2008

the dune, and has continued to grow faster than the sand accumulates. Whereas the halophytic *Tamarix* presumably remains in contact with the underlying saline substrate through an extensive rooting system, glycophytes or weak halophytes such as

Fig. 6.16 *Heliotropium* cf. *socotranum*. Nuged coast, December 2008

Heliotropium spp. (e.g. the yellow-flowering *H.* cf. *socotranum* – Fig. 6.16), *Polycarpaea* cf. *spicata* and *Zygophyllum simplex* are typical of the higher parts of the dunes, which also act as reservoirs for rainwater. In the south of the island, *Xylocalyx* cf. *asper* was found to be fairly widespread on coastal sand sheets and dunes. Towards the base of the dunes, halophytic species are present, including the grasses *Aeluropus lagopoides*, *Sporobolus iocladus* and *S. minutus*. *Aeluropus* is often abundant in depressions between the dunes, where it is frequently accompanied by *Limonium sokotranum* (Fig. 6.17). *Sporobolus minutus*, listed by Miller and Morris (2004) for Abdalkuri only in the archipelago, was found to be fairly common in depressions on coastal dunes in the south of Socotra. Noteworthy also is the occurrence of *Lagenantha cycloptera* on dunes near Diham on the north coast, as well as on the central south coast. Miller and Morris (2004) did not record this chenopod for Socotra itself, but mentioned that it was locally common on coastal dunes on the adjacent islands of Abdalkuri and Darsah. Kilian and Hein (2006) also reported the occurrence of this species on the north-west coast of Socotra. In the far west of the island, two other chenopods, *Cornulaca ehrenbergii* and *Sevada schimperi*, are found on coastal sand dunes, where they are locally common.

Cyperus conglomeratus, a robust, tussock-forming perennial sedge that is characteristic of coastal dunes in southern Yemen (Kürschner et al. 1998) and has a wide distribution on inland dunes and sand sheets elsewhere in Arabia, appears to be very scarce and local in such situations on Socotra. The same applies to *Panicum turgidum*, a perennial grass that is a highly typical species of the coastal white sand communities of Arabia (Kürschner 1998) and the Red Sea coastal plains of eastern Africa (Kassas 1957; Woldewahid et al. 2007), and was apparently once abundant, at least locally, on Socotra (see Sect. 7.3.1).

Fig. 6.17 *Limonium sokotranum* is often common in depressions between coastal dunes, accompanied by *Aeluropus lagopoides*. Nuged coast, December 2008

Popov (1957) observed that the small tree *Calotropis procera* was more common on the coastal dunes of the Nuged plain than elsewhere on the island. However, this species, which in many parts of its range is usually regarded as an indicator of disturbance and overgrazing, was noticed only sporadically in such habitats, but was found to be abundant in parts of urban Hadibuh, and it is now also becoming established alongside stretches of the newly constructed main roads, especially at lower altitudes (see Sect. 7.3.8).

Micro-nebkhas or phytogenic hillocks, small asymmetric mounds of loose to fairly compact aeolian sand with a steep windward side, but gently tapering off in the lee of the plant, may occasionally form around the base of dwarf shrubs and perennial grasses in open habitats. It has been suggested that such structures, which serve to promote habitat heterogeneity, may be instrumental in maintaining local species diversity in coastal and arid ecosystems, especially of the annuals in degraded environments (Brown and Porembski 1998; El-Bana et al. 2007). Brown and Porembski (1998) attributed higher densities and diversity of annuals on micro-nebkas in northern Arabia primarily to the role of these structures as moisture reservoirs, particularly as a number of authors stress the importance of water availability for species diversity in desert ecosystems (Danin 1978; Olsvig-Whittaker et al. 1983; Kutiel and Danin 1987). No studies of this phenomenon, which is widespread throughout arid regions of the world, have been conducted to date on Socotra.

By gradually stabilising the coastal dunes, plants serve to protect the coastline from erosion to a certain extent. Furthermore, the dune systems are of local importance in that, by retaining large amounts of freshwater deep below the soil surface, they

Fig. 6.18 Date palm plantation on coastal sands. Young trees are protected from livestock. The plantation is being invaded by *Tamarix nilotica*. Nuged coast, December 2008

prevent seawater intrusion (Carter 1988). Date palms, which are quite sensitive to saline substrates, are therefore a common sight on low dunes in close proximity to the sea (Fig. 6.18).

6.3.1.3 Mangroves: *Avicennia marina* Community

Throughout the region, mangroves are characteristic of the intertidal zone where the sea recedes for a few hours at low tide, but which is inundated at moderately high tides. *Avicennia marina* (Fig. 6.19), an Avicenniaceae typical of the coasts of the Arabian Peninsula (Deil 1998), is the only species currently known to occur on Socotra. It tends to grow best in brackish water, and the species is therefore best developed in coastal waters where there is an occasional input of freshwater from the land. During an expedition in 1967, a second species of mangrove with wide, leathery leaves, but lacking fruits or flowers, was found on the northern coast (Lavranos, pers. communication). It was probably the Rhizophoraceae *Rhizophora mucronata*, which has also been reported from a few localities along the coasts of the Arabian Peninsula (Deil 1998), as well as from the Red Sea and the Somali coastline (Pichi-Sermolli 1957).

Local inhabitants of Socotra report that in former times, mangroves were a conspicuous feature on the seaward side of many coastal marshland areas as well as in lagoons and protected bays. This is also indicated in the maps of Wellsted (1835), Balfour (1888) and Popov (1957). Furthermore, photographs from 1967 provided by Lavranos (see Fig. 7.18) show dense mangrove growth with old trees at Gubbah on the

Fig. 6.19 *Avicennia marina* with Dr. Saad Al-Qadoomy. Shu'ab, March 1997

Fig. 6.20 Remaining old individual of *Avicennia marina*. North coast at Ra's Kharmah, west of airport, December 2008

north coast, but these stands have now disappeared. A few individual older trees still survive in *Arthrocnemum* marshland to the west of the main airport (Fig. 6.20). The widespread destruction of the mangroves probably began in the nineteenth century

Fig. 6.21 *Arthrocnemum macrostachyum* community. North coast at Diham, December 2008

and continued until fairly recently. Reasonably intact stands now remain only at Shu'ab, where they occur in a small estuary, and on the western-most Nuged coast at Niyt. In both locations, the trees do not grow in shallow waters on the seaward side of the coast, but occur in a narrow band running parallel to the coastline in depressions behind the dunes (see Fig. 6.9). Access to the stands at Niyt is arduous from the land due to the presence of steep cliffs behind the narrow strip of coastline where they grow, and they are most easily reached by boat, at least when the sea is calm.

The stands are species-poor and dominated by *Avicennia* itself. Cover values are in excess of 90%, and the trees can grow up to 5 m in height, so that this azonal vegetation type is a highly conspicuous, albeit highly localised feature of the landscape. Due to the dominance of *Avicennia*, undoubtedly also the main species before the large-scale destruction of the mangroves, the stands on Socotra can be assigned to the phytosociological class *Avicennietea marinae* (Zohary 1973).

6.3.1.4 Salt-Marsh and Sabkha Vegetation: *Arthrocnemum macrostachyum* Community

On the northern and western coasts of Socotra, for instance at Diham and along a coastal strip from Ra's Shu'ab south-eastwards, salt-marsh vegetation occurs in which the semi-succulent dwarf shrub *Arthrocnemum macrostachyum* predominates, forming dense, low stands in favourable sites (*A. macrostachyum* community – Fig. 6.21). The stands are usually located in depressions behind the immediate coastline, and are largely monospecific, with small mats of *Aeluropus lagopoides*

Fig. 6.22 *Limonium sokotranum* community on the south coast. December 2008

and other species occasionally interspersed. Vegetation cover of this azonal community is remarkably high, often in excess of 80%, contrasting strongly with the more open vegetation of the adjacent plains. Kürschner et al. (1998) described a comparable *Urochondra setulosa–Arthrocnemum macrostachyum* community from southern Yemen that is characteristic of the landward side of low coastal dunes in the transition zone to saline depressions.

The substrate of the *Arthrocnemum* community, often of a sabkha type, is subject to occasional inundation at exceptionally high tides, but is probably more regularly flooded as water rises up from below the surface at normal high tides.

In the west of Socotra, *Arthrocnemum* is typically associated with mangroves, and similar *Avicennia–Arthrocnemum* stands ("*Avicennia marina–Arthrocnemum macrostachyum* association") were mentioned briefly by Western (1989) from Abu Dhabi Island and the adjacent sections of the UAE coastline. Various authors have reported the close association between the two species along much of the Arabian Gulf coastline.

6.3.1.5 Salt-Marsh and Sabkha Vegetation: *Limonium sokotranum* Community

In some areas that appear ideal for the development of *Arthrocnemum* salt-marsh, the physiognomically similar *Limonium sokotranum* with its succulent, linear-cylindrical leaves predominates on sabkha-like substrates (Fig. 6.22). Vegetation

Fig. 6.23 Salt rings around the base of *Limonium sokotranum*. Nuged, December 2008

cover ranges from very open patches to dense stands. Elevated substrate salinity and high rates of evaporation are indicated by the formation of salt deposits in circles around the micro-nebkhas at the base of the shrubs (Fig. 6.23), as described by Brown (2001) in the *Zygophyllum qatarense* community of coastal areas of northern Kuwait. The formation of such salt rings is presumably due to the fact that micro-nebkhas represent micro-reservoirs for water, and that evaporation is most pronounced towards their margins, where the mound substrate becomes very shallow.

Accompanying species include *Cressa cretica* (Fig. 6.24), the mat-forming grass *Aeluropus lagopoides*, *Sporobolus iocladus*, *Zygophyllum qatarense* and occasionally *Arthrocnemum macrostachyum* (Fig. 6.25).

6.3.1.6 Other Salt-Marsh and Sabkha Vegetation

On sandy patches associated with coastal marsh, the chenopod *Suaeda vermiculata*, a small, straggling shrub that, depending on its stage of development, can be morphologically very variable, is locally common.

In brackish conditions, the rush *Juncus rigidus* occurs in a few areas, forming mainly monospecific stands. Furthermore, especially at the mouth of wadis where there is the occasional input of freshwater (Fig. 6.26), but also quite frequently on the landward side of the salt-marsh, the tamarisk *Tamarix nilotica* develops small, species-poor thickets.

Fig. 6.24 *Cressa cretica*, a common species on saline substrates on the south coast. December 2008

Fig. 6.25 A single individual of *Arthrocnemum macrostachyum* (central plant) in a stand of *Limonium sokotranum*. Nuged, December 2008

Fig. 6.26 Small thicket of the tamarisk *Tamarix nilotica*. North coast at Bandar Fikhah, Mumi, December 2008

6.3.2 *Vegetation of the Coastal Plains*

In the northern part of Socotra, alluvial fans radiate from the foothills and mountains to form extensive plains. They are gently sloping, and in the vicinity of the mountains consist of coarse detritus such as pebbles and rocks that overlie a gravelly alluvium or marine limestone crusts. As the slopes gradually flatten out towards the coast, the particle size of the debris decreases noticeably, and the plains are interspersed with much finer fluvial deposits. Here the main wadi systems and channels peter out, although distinct lines of vegetation indicate intermittent watercourses and other areas where water availability is higher. Coralline sand has been blown across the plains and deposited as undulating sand sheets in a few areas, perhaps most conspicuously at Ra's Haulaf on the northern coast. In several places, including at Ra's Haulaf and in the extreme north-east of Socotra, large amounts of sand have also encroached high upon the hillside (see Fig. 3.3).

The most extensive plain on Socotra, the Nuged plain, is located along the central coastal strip in the south. It extends for nearly 70 km and is up to 7 km wide. As is the case on the northern side of the island, the Nuged plain is covered predominantly by gravelly or stony material, as well as compact or loose sand. Apart from the coralline sand dunes that occur along the immediate coastline in many areas, individual dunes up to several metres in height are scattered across

Fig. 6.27 Network of drainage channels (and one vehicle track) on the northern coastal plain. December 2008

the plain locally. A number of these dunes have been partially stabilised by a sparse vegetation cover.

Wadis, originating from the higher mountain areas and occasionally carrying water, transverse both the northern and southern plains. However, whereas in the south they tend to lead directly to the coast, in the north they are quite frequently of a distinct braided type with a dense network of channels (Fig. 6.27).

It is still the case that vegetation cover is generally more dense on the northern plains than in the south, a fact attributable particularly to the higher cover of shrubs, although excessive grazing and other anthropogenic pressures, which are more pronounced in the north, are beginning to blur this distinction. The higher vegetation cover in the north is probably due to the more favourable climatic conditions there, especially higher rainfall. A conspicuous feature of the Nuged plain is that a fairly dense band of shrub vegetation is developed towards the foot of the escarpment in many areas, ending quite abruptly on the adjoining plain where scattered dwarf shrub vegetation takes over (Fig. 6.28). However, as indicated by the presence of scattered Cucumber trees (*Dendrosicyos socotrana*) on the Nuged plain (see Fig. 6.33), it is conceivable that the vegetation cover was much more dense in former times, but due to the harsh environmental conditions, even moderate grazing has had a greater relative impact than it would have had in the north, leading locally to a quite severe degradation of the vegetation (see Sect. 6.3.2.5).

Fig. 6.28 Dense band of shrub vegetation on the slopes ends abruptly at the foot of the escarpment. Nuged coastal plain, December 2008

6.3.2.1 *Croton socotranus* Community

The endemic spurge *Croton socotranus*, known locally as "metred" (Figs. 6.29 and 6.30), is the most common shrub on Socotra, and is represented in various plant communities. *Croton* shrubland is highly characteristic of the northern plains, where the fairly uniform, loose stands attain a height of about 2–3 m (*C. socotranus* community). Pockets of zonal *Croton* shrubland also occur in the south on the Nuged plain and elsewhere, especially in the broader wadis. The community is best developed on coarse sandy and gravelly soils. Furthermore, *Croton* stands are locally common on the lower foothills and even higher on the limestone plateaus. Various stages of degradation can be distinguished; one of the more extreme is described in Sect. 6.3.2.2.

The stands often appear at first sight to be almost monospecific, even resembling young tree plantations, but they usually contain other shrubs and dwarf shrubs such as the endemics *Ballochia amoena*, *Blepharis spiculifolia* (restricted to the northeast), *Indigofera pseudointricata*, *Justicia rigida* (locally very common, especially in the west) and *Placopoda virgata* (Fig. 6.31). The taxonomically problematic endemic shrub *Trichocalyx orbiculatus* (Acanthaceae) is a common associate, especially in the western part of Socotra. Other low-growing woody species include *Asparagus africanus* var. *microcarpus*, *Fagonia luntii* (local), *Grewia erythraea* and the dwarf shrub *Seddera latifolia*. Locally, the dwarf shrub *Pulicaria stephanocarpa* is common in the ground layer, for instance in Wadi Qalansiyah (Fig. 6.32).

Fig. 6.29 *Croton socotranus* community. East of Qalansiyah, December 2008

Fig. 6.30 *Croton socotranus* in flower and fruit. Nuged plain, December 2008

Fig. 6.31 *Placopoda virgata*. Homhil, December 2008

Fig. 6.32 *Pulicaria stephanocarpa*, a locally common and dominant dwarf shrub in the ground layer of the *Croton socotranus* community. Near B'ir Haarso, east of Qalansiyah, December 2008

Fig. 6.33 *Dendrosicyos socotrana* in a mixed *Croton–Jatropha* stand on the Nuged plain. December 2008

In the south-west of the island, the endemic small shrub *Hibiscus quattenensis* is fairly widespread. Emergent trees are scattered throughout the *Croton* shrubland. Undoubtedly the most striking species is the highly characteristic bottle tree *Dendrosicyos socotrana* (Fig. 6.33). On the plains, it is found mainly in the west of Socotra, but it is more common and widespread in succulent shrubland on the rocky slopes of the foothills. Furthermore, it is also a very local, but conspicuous species on the Nuged plain. Mies and Zimmer (1994) attribute the rarity of the species on the plains to human activity, as the trees are regularly used as drought fodder for domestic livestock. The Emta tree, *Euphorbia arbuscula*, is a common and widespread emergent (see Fig. 4.15). Other emergents include the myrrhs *Commiphora parvifolia* (mainly in the west) and *C. socotrana* (especially in the north), as well as *Ziziphus spina-christi*. These trees tend to be more common on the edges of shallow, dried-out wadi beds or depressions where there is an occasional subterranean water supply. Furthermore, in the very broad Wadi Qalansiyah, the frankincense *Boswellia socotrana* is a locally conspicuous emergent (Figs. 6.34 and 6.35). The form growing there is somewhat different from that found in low-elevation woodland. These and other trees are prized for a number of reasons, in particular as an important livestock fodder during extended periods of drought, but also for construction purposes. Increased exploitation is beginning to reduce the tree populations locally quite markedly. Heavy grazing during dry periods also affects many shrubs and dwarf shrubs. Some of these are well protected against serious overgrazing, such as

Fig. 6.34 A characteristic form of *Boswellia socotrana* is a locally conspicuous emergent in *Croton* shrubland east of Qalansiyah. March 1997

Fig. 6.35 Leaves of *Boswellia socotrana*. Near B'ir Haarso, east of Qalansiyah, December 2008

the spiny *Ballochia amoena*, *Campylanthus spinosus* and *Lycium sokotranum*. *Ballochia* often has its foliage protected within a twisted framework of branches with spike-like twigs.

The vine *Cissus subaphylla* is a common succulent trailing shrub in the *Croton* community. Other succulents that are occasionally encountered include *Caralluma socotrana* and *Aloe perryi*. The inconspicuous dwarf succulent stems of *Euphorbia hadramautica* often die back after prolonged drought, but regenerate from the subterranean portion of the stem after rainfall (see Fig. 4.7). In his description of the *Croton* community, Lavranos (1994) gave a vivid account of the chance discovery of this species on Socotra.

A conspicuous ground layer of grasses develops with the onset of the first rains of the season, especially on loose gravelly and sandy substrates, providing good fodder for domestic livestock. Typical species, albeit rather local in some cases, include the perennials *Bothriochloa insculpta*, *Cenchrus ciliaris*, *C. setigerus*, *Enneapogon desvauxii*, *Lepturus pulchellus* and *Panicum rigidum*, as well as the annuals *Aristida adscensionis* (widespread and often dominant, frequently occurring as a short-lived perennial), *Brachiaria ramosa*, *Cenchrus pennisetiformis*, *Chloris virgata*, *Dactyloctenium aristatum*, *Eragrostis* spp., *Melanocenchris jacquemontii* and *Tetrapogon tenellus*. It seems highly probable that overgrazing is gradually leading to the destruction of the perennial grass understorey in many places, especially of the more palatable species, less so *Aristida*, which appears to be thriving and even increasing.

A host of perennial herbs occur, some of which are short-lived, and include *Abutilon fruticosum*, *Boerhavia diffusa*, *Cleome austroarabica*, *C. brachycarpa*, *C. scaposa*, *C. socotrana*, *Corchorus erodioides*, *Hybanthus enneaspermus*, *Indigofera articulata*, *I. nephrocarpa*, *I. pseudointricata*, *Lindenbergia sokotrana*, *Oldenlandia pulvinata*, *Neuracanthus aculeatus* (restricted to the north-east), *Polycarpaea caespitosa*, *Rhinacanthus scoparius*, *Ruellia dioscoridis*, *Solanum incanum* and *Tephrosia apollinea*. Species such as *Aerva lanata*, *Citrullus colocynthis* and *Cucumis prophetarum* regenerate from thick underground tubers. A number of these species, including *Citrullus* as well as various Solanaceae and Amaranthaceae, are unpalatable or even toxic due to the presence of alkaloids in their tissues, and are generally reliable indicators of overgrazing.

Ephemeral herbs that appear after rainfall include *Aizoon canariense*, *Anagallis arvensis*, *Argemone mexicana*, *Convolvulus rhyniospermus*, *Digera muricata*, *Glossonema revoili*, *Mollugo nudicaulis*, *Pedalium murex*, *Polycarpaea balfourii*, *P. hayoides*, *Polygala erioptera*, *Senna holosericea*, *Tephrosia strigosa* and *Tripteris vaillantii*.

During extended periods of drought, the surface layers of the soils dry out completely, possibly down to depths exceeding 2 m. The density of the *Croton* vegetation is determined by the availability of water, not just in the form of direct rainfall, but also from run-on. Furthermore, local soil characteristics, in particular the water-holding capacity of the substrate, are likely to affect shrub density and vitality. As a consequence, the densest stands of *Croton* are found on the plains towards the foot of mountains or in other areas that receive run-off water. On the

Fig. 6.36 Mixed *Croton–Jatropha* stand. Ma'naifah, south of Hadibuh, December 2008

extensive plains of Wadi Thar and in the south, the individual *Croton* shrubs tend to be much more scattered. Between the shrubs, the soil surface is colonised by cyanobacteria and blue-green lichens that form characteristic biological crusts.

On limestone slopes towards the foot of the mountains, the endemic shrub *Jatropha unicostata* usually replaces *Croton* as the dominant species, in both the north and south. This is especially the case on more rocky substrates, and the stands dominated by *Jatropha* represent a distinct transition to *Adenium* shrubland characteristic of the rocky hillside slopes at lower altitudes, such as at Hoq (see Sect. 6.3.4.1). Mixed *Croton–Jatropha* stands are often found along the sides of broad wadis throughout the island (Fig. 6.36). It is also noticeable that some wadi beds with a coarse substrate support a fairly dense growth of *Jatropha* down to nearly sea-level.

On the foothills of the central mountain range, *Croton* shrubland intermingles with woodland dominated by the locally common endemic *Carphalea obovata* (Fig. 6.37), the endemic *Trichocalyx obovatus* and *Buxanthus pedicellatus*, as well as other species. In such situations, *Croton* often colonises the drier, slightly elevated sites, whereas the other species occur in depressions that have a more favourable moisture availability.

Popov (1957), who first attempted a general vegetation description of Socotra, recognised the unique nature of what he referred to as the "dominant *Croton*-short grass community". He assigned it to the "Subdesert shrub and grass" type of Pichi-Sermolli (1955), fully aware that it was absent from the East African

Fig. 6.37 Flowers of the common endemic tree *Carphalea obovata*. Wadi Ayhaft, December 2008

mainland. Pichi-Sermolli (1955) initially did not include lowland vegetation units from Socotra in his vegetation descriptions of tropical East Africa, but later (1957) assigned them to the Fruticeto subdesertico, which is highly characteristic of the Red Sea coast from Eritrea to Somalia. In this unit, *Acacia* and *Commiphora* are common trees of scrub and grasslands, and they are accompanied by species of the genera *Ziziphus*, *Maerua*, *Cadaba* and *Boscia*. However, as had already been pointed out by Popov (1957), the *Acacia*–short grass savanna so typical of the adjacent African mainland is conspicuous by its absence on Socotra. In his detailed phytosociological treatment of the East African region, Knapp (1968) ascribed the *Croton* community to xeromorphic woodlands of the classes Acacietea tortilis and the Boscio-Commiphoretea. However, given the low contribution or even lack of character species, apart from the local co-dominance of endemic myrrhs in certain shrubland vegetation units, this classification is not satisfactory.

Mies (1999c) described the Crotonetum socotrani as a typical plant association of northern Socotra, being found predominantly at lower elevations, usually below 200 m. The shrubs *Croton socotranus*, *Placopoda virgata* and *Trichocalyx orbiculatus* were regarded as character species, but Kürschner et al. (2006a) suggested restricting the association character species to *Croton* itself, as the other plants listed were also character species of higher-ranked syntaxa. None of the herbs or grasses mentioned above that are frequently found in the understorey and that form a distinct temporal synusium can be used as character species. The Crotonetum socotrani was assigned to the phytosociological alliance Crotonion socotrani (Mies 1999c; Kürschner et al. 2006a).

Fig. 6.38 Severely degraded *Croton socotranus* community dominated by *Tephrosia apollinea*. Northern plain at mouth of Wadi Dihzafaq, December 2008

6.3.2.2 *Tephrosia apollinea* Stands

Stands dominated by the short-lived perennial *Tephrosia apollinea* are widespread, particularly in the vicinity of settlements and livestock pens (Fig. 6.38). They can be regarded as a highly degraded form of the *Croton* community. Cover values of the patchy stands range from about 5% to 30%. In extreme cases, *Croton* is completely absent, and the number of species present per unit area can be restricted to fewer than five, most notably *Tephrosia apollinea*, *Aristida adscensionis*, the neophyte *Argemone mexicana* and the endemic, but fairly widespread, shrub *Ballochia amoena*. The last of these is often heavily stunted due to severe browsing. Other typical species, some neophytes, include *Aerva javanica*, *Bidens biternata*, *Senna holosericea* and *Solanum incanum*. Species composition is highly characteristic, especially the combination of *Tephrosia* and *Argemone* (Fig. 6.39). Similar stands, also growing under comparable environmental conditions, are found throughout parts of mainland Arabia and northern Africa. Initial observations show that if the grazing pressure in such highly degraded stands is reduced, it is mainly the unpalatable weedy species that first benefit in terms of increased cover, as well as *Cissus subaphylla*. Depending on the degree of anthropogenic impact, it appears that natural *Croton* regeneration may take decades to proceed, even in the complete absence of grazing (see Sect. 7.4.1).

Fig. 6.39 The widespread neophyte *Argemone mexicana*

6.3.2.3 *Acacia edgeworthii* Community

Popov (1957) gave a very brief account of the *Acacia edgeworthii* community, which appears to be restricted to a small area of coastal plain at Ra's Haulaf on the northern coast of Socotra east of Hadibuh. The gently undulating plain is covered by thin to thick deposits of wind-blown coralline sand and is interrupted by a network of shallow channels that have a more gritty to stony substrate.

A. edgeworthii is the dominant perennial, with cover values ranging from about 10% to 20%. The shrubs are up to about 1.50 m in height, but the trunk is often buried in small nebkha-like sand accumulations with only the dome-shaped crown protruding (Fig. 6.40). Regular associates include *Ballochia amoena*, *Cymbopogon jwarancusa* and *Tephrosia apollinea*. The last of these is indicative of overgrazing, and this is also underlined by the fact that the *Ballochia* shrubs are usually heavily browsed and stunted (Fig. 6.41). *C. jwarancusa* appears to be largely avoided by the goats that frequently roam the area. Other widespread species at Ra's Haulaf include *Aerva lanata*, *Arnebia hispidissima*, *Indigofera* spp., *Seddera* sp., *Sporobolus spicatus* and *Tribulus terrestris*.

Interestingly, small populations of a second species of *Acacia*, *A. sarcophylla*, are currently known from at least two locations in the north-west of Socotra, but the species does not form extensive stands. It is, however, fairly widespread on the plains of Abdalkuri.

6.3.2.4 *Pulicaria stephanocarpa* Community

Particularly in the west of Socotra, some areas of the plains, extending locally for quite a distance up the hillsides, are dominated by the endemic dwarf shrub *Pulicaria stephanocarpa* (Fig. 6.42). The stands are low-growing and rather monotonous,

Fig. 6.40 *Acacia edgeworthii*. Ra's Haulaf, December 2008

Fig. 6.41 Heavily browsed and stunted individuals of *Ballochia amoena*. Ra's Haulaf, December 2008

Fig. 6.42 The endemic dwarf shrub *Pulicaria stephanocarpa*. Near B'ir Haarso, east of Qalansiyah, December 2008

varying in cover from about 5% to 30%, locally even higher. On the coastal plains, the species often appears to be associated with somewhat saline soils, but this is not the case in other areas, especially at higher elevations on the limestone plateau. Under the canopy of the *Pulicaria* plants, organic detritus accumulates, facilitating the growth of a limited number of annuals, including the widespread *Zygophyllum simplex* and *Arnebia hispidissima*. As noted in Sect. 6.3.2.1, *Pulicaria* is a locally abundant associate in the ground layer of the *Croton* community. *Convolvulus arabicus*, a prostrate herb up to 2 m in diameter, is occasionally found in this community, as well as in *Croton* shrubland.

Mies (2001) described the distinctive, rather species-poor azonal community from somewhat saline areas of northern Socotra as a new phytosociological association (Pulicarietum stephanocarpae), with *Lagenantha cycloptera* as a further character species. Given the relative abundance of *Pulicaria* on rocky, presumably non-saline slopes at low altitudes without *Lagenantha*, the designation of the Pulicarietum stephanocarpae appears premature until further phytosociological studies have been conducted.

6.3.2.5 *Indigofera pseudointricata–Salsola spinescens* Community

Indigofera pseudointricata is a widespread and locally abundant dwarf shrub on coastal plains, in both the north and south of Socotra (Fig. 6.43). The joint occurrence with the chenopod *Salsola spinescens*, however, appears to be typical of the Nuged plain. *Salsola*, which can attain a height of up to 1.5 m in favourable conditions,

Fig. 6.43 *Indigofera pseudointricata* community. Nuged plain, December 2008

is often heavily browsed by domestic livestock, and so the community typically presents itself as rather patchy, open dwarf scrub. Other frequent associates include *Aerva javanica*, *Aizoon canariense*, *Cenchrus pennisetiformis*, *Cleome brachycarpa*, *Convolvulus hildebrandtii*, *Corchorus erodioides*, *Euphorbia schimperi*, *Fagonia paulayana* (locally), *Heliotropium* spp., *Indigofera nephrocarpa*, *Launaea crassifolia*, *L. massauensis* and *Tephrosia apollinea*. The local abundance of the last of these is a sure indicator of overgrazing (Fig. 6.44), as is possibly the encroachment of *Jatropha unicostata* in some areas. Furthermore, *Ballochia amoena* and *Lycium sokotranum* are frequently interspersed as small shrubs, but are usually quite severely stunted as a result of heavy browsing. Patches of *Ziziphus spina-christi* and other shrubs or small trees also occur, most conspicuously in the shallow channels that mark the extension of the wadis leaving the escarpment. *Lachnocapsa spathulata*, an endemic, cushion-forming dwarf shrub, is occasionally found in this community, as well as in other coastal dwarf scrub vegetation from various parts of the island. Very locally, the spinescent dwarf shrub *Launaea* sp. (possibly *L. hafunensis*) occurs on the Nuged plain, the only place on Socotra where this species has been found (Mies 2001), although it is common in similar habitats on the neighbouring islands of Abdalkuri and Samhah (Miller and Morris 2004). It is also known from Somalia and southern Arabia (Kilian 1997). *Leucas flagellifolia*, a mat-forming perennial herb, appears to be restricted to dwarf scrub on the west coast, including on the far west of the Nuged plain.

It is quite noticeable that in some areas, loose stands of *Croton* occur in the *Indigofera pseudointricata–Salsola spinescens* community, so that the distinction between the two units can become blurred, especially as various transitions exist.

Fig. 6.44 Overgrazed *Indigofera pseudointricata* community with high proportion of *Tephrosia apollinea*. Nuged plain, December 2008

The *Indigofera–Salsola* community probably represents a degradation stage of *Croton* shrubland in many areas on the Nuged plain, and the further trajectory of deterioration under heavy grazing pressure is likely to lead to the monotonous *Tephrosia apollinea* stands described above (Sect. 6.3.2.2).

6.3.2.6 *Limonium paulayanum* Community

Limonium paulayanum, an endemic dwarf shrub, is found in a number of very different plant communities, including on coralline coastal sands and inland cliffs, but locally it dominates a distinctive dwarf shrub assemblage on the coastal plains, presumably where slightly elevated soil salinity is a decisive factor (Fig. 6.45). This is also indicated by the regular presence of *Zygophyllum qatarense*. However, the community avoids highly saline substrates along the immediate coastline. The substrate is silty-gravelly with small rocks and stones strewn across the surface. *Limonium* is dominant, with *Zygophyllum* subdominant to sporadic. Total perennial cover is in the region of 10–30%. The community is remarkably species-rich, and in overall structure somewhat reminiscent of the *Zygophyllum qatarense* community found on gravelly coastal plains in Qatar, and described in brief by Brown et al. (2008). Other species recorded in the *Limonium paulayanum* community in northeastern Socotra, a combination of glycophytes and weak halophytes, include *Cleome viscosa* (Fig. 6.46), *Indigofera nephrocarpa*, *Launaea crassifolia*, *Lotus garcinii*, *Lotus* sp., *Oldenlandia pulvinata* (Fig. 6.47), *Panicum rigidum*, *Sporobolus iocla-*

Fig. 6.45 *Limonium paulayanum* community. North-east coast near Gobhil, December 2008

dus and *Trachyspermum pimpinelloides*, as well as the rare and local endemic *Dipcadi balfourii*.

6.3.2.7 Other Vegetation Types of the Plains

Very locally, on somewhat saline ground, *Indigofera oblongifolia* dominates an open to dense dwarf shrub community, forming a transition to the vegetation of the plains. Despite its local abundance, the species appears to be restricted to a limited number of localities (Miller and Morris 2004).

Grasslands are developed locally on the Nuged plain, where the principal dominant species appears to be the endemic perennial *Panicum rigidum*. Due to heavy grazing, the determination of other grass species can be problematic.

Miller and Morris (2001) reported that the evergreen shrub *Salvadora persica*, which occurs sporadically on coastal plains, forms a belt of dense shrubland on the western part of the Nuged plain.

6.3.3 Vegetation of the Wadis at Lower and Middle Altitudes

Following heavy rain, some of the wadis become severely flooded, and water can flow for considerable distances. Wadis with strong currents are invariably more deeply incised (Fig. 6.48). The larger wadis of the Hagghier mountains carry water

Fig. 6.46 *Cleome viscosa*. North-east coast near Gobhil, December 2008

Fig. 6.47 *Oldenlandia pulvinata*. North-east coast near Gobhil, December 2008

throughout the year, along certain stretches at least. Some of them, such as Wadi Ayhaft, Wadi Dihzafaq and Wadi Thruba on the north coast, and Wadi Farho, Wadi Ireh and Wadi Mahfirin, draining to the Nuged plain in the south, are exploited to provide water for small settlements. Water reaching the lower wadi regions serves

Fig. 6.48 Mouth of a wadi in the south of the island

to recharge the groundwater in the alluvial fans. Only after exceptionally heavy rains does surface water reach the sea. Due to the risk of occasional flooding, no settlements have been established in the wadis. A major advantage for people living in the immediate vicinity of these wadis is that they have permanent access to groundwater. This is used to irrigate quite extensive date palm groves (*Phoenix dactylifera* – Fig. 6.49), and in former times to cultivate finger millet (*Eleusine coracana*). Fields and terraces were built on slightly elevated ground immediately adjacent to major wadis to protect the crops from flooding. The millet fields could be surprisingly small, frequently as little as a few square metres. At higher altitudes, these fields are now used to grow bananas and more exotic vegetables. Date palm groves tend to be very small in inland valleys, but are distinctly larger (up to several hectares) on coastal alluvium, where good quality groundwater is in more plentiful supply. Typically, the farmers will divert flood water through a series of channels to the fields. Despite increasing anthropogenic pressure, natural vegetation types, albeit sometimes much altered, are still present in many wadis.

6.3.3.1 Characteristic Species and Assemblages of the Wadis

Dry wadi beds on the plains that have subterranean watercourses support a noticeably denser vegetation than that of the surrounding areas, and thus exhibit a distinct, contracted vegetation type. Dry sandy beds are dotted with small shrubby individuals

Fig. 6.49 Date palm grove on the northern coastal plain

of *Aerva lanata*, *Indigofera coerulea*, *Senna socotrana* and other plants, as well as patches of the grass *Cenchrus ciliaris*. Small stands of *Ziziphus spina-christi* occur towards the mouth of many wadis and fringe the margins of the larger wadi beds, especially in the south. These trees provide shade for livestock, but they are also occasionally developed as low-growing shrubs.

The substrate of the lower reaches of wadi beds in the hilly areas, especially before emerging onto the plain, can be remarkably varied within small areas, ranging from sand to huge boulders (Fig. 6.50), the latter often supporting typical elements of cliff vegetation. This combination of substrate variability, pronounced heterogeneity in microclimate and microtopography, light availability and regular access to water provides a patchwork of favourable small-scale habitats that greatly promotes species diversity, especially as numerous safe sites for seedling establishment exist. Furthermore, due to occasional "catastrophic" events, i.e. severe flooding, wadi beds are highly dynamic habitats, not only in terms of their physical characteristics, but also with respect to floristic composition. Many species are decimated during torrents, but at the same time, new seeds are transported with the water and deposited in favourable microhabitats. It is interesting to note that many species are perennials, an apparent exception to the general theory (e.g. Barbour et al. 1998) that unpredictable habitats are usually colonised by annuals. This contradiction can probably be explained by the fact that several years may pass without flooding, facilitating the establishment of perennials, which may then withstand periodic short-term inundation. Many plants may also grow in more protected sites, such as on elevated banks or sheltered by boulders in the wadi bed. On the whole,

Fig. 6.50 Large boulders at the mouth of Wadi Dihzafaq. December 2008

vegetation cover is conspicuously low in the wadi beds, despite the large number of species involved. Endemics such as *Ochradenus socotranus* and *Aerva microphylla* are occasionally found growing among boulders. The succulent *Caralluma socotrana* is common on gravels and pebbles in wadi beds, especially at lower altitudes on limestone, but is easily overlooked when not in flower. Other typical lowland species include *Jatropha unicostata* (often dominant) and *Adenium obesum* ssp. *socotranum*, the latter usually growing on top of boulders or in other exposed locations. In the west, the endemic shrub *Metaporana obtusa* is locally dominant, both in association with rocky wadi beds and in adjacent *Croton–Jatropha* shrubland.

In hilly areas that have a more favourable microclimate and water supply, the wadis are often fringed by a fragmentary to dense gallery forest (Fig. 6.51), in which the figs *Ficus vasta* (Fig. 6.52) and *F. cordata* ssp. *salicifolia*, as well as *Ziziphus spina-christi*, feature prominently. This riparian forest can grade into fairly dense woodland vegetation in shaded areas, even at lower elevations (locally down to 150 m in northern wadis). Characteristic shrubs or small trees include *Lannea transulta* (Fig. 6.53), *Buxanthus pedicellatus*, *Carphalea obovata* and *Tamarindus indica* (tamarind). The last of these, a tree up to 20 m in height, was probably introduced by early settlers to the island. *Acacia pennivenia* and several species of *Boswellia*, in particular *B. elongata* and *B. socotrana*, are occasional associates in undisturbed locations. Various shrubs and small trees probably originate from forests at higher altitudes where they have their main areas of growth,

Fig. 6.51 Gallery forest in Wadi Ayhaft. December 2008

Fig. 6.52 *Ficus vasta*. Muqadrihon pass, south of Hadibuh, December 2008

Fig. 6.53 *Lannea transulta.* March 1997

and their seeds and propagules are occasionally washed down. Associates characteristic of moist highland forests are *Crotalaria socotrana*, *Dichrostachys dehiscens*, *Hypericum balfourii* and *Rhus thyrsiflora*. *Cephalocroton socotranus* and *Croton sulcifructus* are typical members of highland vegetation types that are occasionally found in woodland associated with wadis at lower elevations. *Pergularia tomentosa* and the endemics *Cissus paniculata* and *Vincetoxicum linifolium* are climbers that frequently cover shrubs in such shaded woodlands. *Pergularia* and *Vincetoxicum* are also widespread in cultivated areas, where they grow on hedges and wooden fences (Fig. 6.54).

6.3.4 Vegetation of Lower-Altitude Rocky Slopes

Rocky slopes, exposed rock faces and boulder fields are widespread throughout many parts of Socotra, including in coastal areas where the underlying bedrock is limestone. Headlands and limestone cliffs located directly on the coast are referred to as "Ra's" in Arabic, and many such examples can be found on Socotra, including Ra's Hebak, Ra's Bedu, Ra's Shu'ab, Ra's Qatanahan and Ra's Mumi. One of the most striking and impressive features of steep rocky slopes is the distinctive xerophytic vegetation with giant caudex succulents, and

Fig. 6.54 *Pergularia tomentosa* growing on fence. Near Hadibuh, December 2008

this is perhaps best developed in various locations along the northern coast (Fig. 6.55).

High rates of insolation, strong winds, low rainfall and the rocky substrate are some of the principal environmental factors with which the natural vegetation has to contend. Furthermore, the often loose nature of the substrate, ranging from small rocks to huge boulders up to 50 m in size, results in occasional landslides. Rainfall is frequently less than 200 mm p.a., but large fluctuations from one year to the next are commonplace. In extreme cases, rainfall is close to zero, whereas in some years it may be in excess of 300 mm. Despite the maritime influence, with relative humidity up to 60%, the climate of the low-altitude rocky slopes is distinctly arid, especially as dewfall and mists do not occur. The dry season lasts for most of the year, and rainfall is received almost exclusively during the period of the north-east monsoon. However, the island is occasionally hit by cyclones, which can occur at any time of the year, and these are accompanied by torrential rainfall.

In spite of the adverse environmental conditions, the flora is remarkably rich, with the large succulents, in particular *Adenium obesum*, the most conspicuous species. Dwarf shrubs predominate in the ground layer, but there is also a substantial proportion of ferns, geophytes and annuals. The large number of endemic taxa found in such habitats often matches that of the vegetation at higher elevations. Many species profit from the shade afforded by large rocks, but spiny dwarf shrubs are equally at home in well-exposed sites. Frequently, cover is fairly sparse, but a

Fig. 6.55 The distinctive xerophytic vegetation at Ra's Hebak. December 2008

large number of species contribute to the characteristic vegetation. Another important factor benefiting the flora is the relatively low intensity of grazing, as livestock are often unable or unwilling to cross extensive rocky areas.

6.3.4.1 *Adenium obesum* Community

The giant Aden rose of Socotra, *Adenium obesum* ssp. *socotranum*, is the characteristic plant of rocky slopes, forming extensive patches of succulent shrubland in favourable locations. Although various different forms of *Adenium* shrubland exist, the most common and regular associate in the shrub layer is the endemic *Jatropha unicostata* (Fig. 6.56). This shrub is perhaps one of the most common species on Socotra, but, like *Adenium*, is surprisingly absent from the adjacent islands. The close association between the two species is probably due to their apparent preference for, or ability to cope with, coarse rocky substrates. Numerous regeneration niches exist in such substrates, offering the seedlings protection from trampling and affording them some degree of shade. The seeds of both species germinate readily after rainfall – our own initial laboratory studies show that seeds of *Adenium* germinate within one day of wetting. Furthermore, it is important to point out that both species are highly toxic to domestic livestock and as such are rarely browsed.

Depending on the local environmental conditions, species composition of the *Adenium* shrubland varies markedly, and numerous transitions exist from *Croton*

Fig. 6.56 *Adenium* shrubland with *Jatropha unicostata*. Ra's Hebak, December 2008

shrubland, on the one hand, to patchy woodland on limestone plateaus as well as dense forest at higher elevations and shaded wadis, on the other. In extreme cases, *Jatropha* is absolutely dominant (Fig. 6.57), and *Adenium* may be virtually absent. The reverse situation also occurs, with *Jatropha* playing very much a subordinate role to *Adenium*.

Some of the best examples of succulent shrubland in which *Adenium* features prominently are to be found on the steep slopes at Ra's Hebak, to the west of Hadibuh on the north coast (see Fig. 6.55). The stands extend onto rocky patches of the flat coastal plain, sometimes very close to the immediate shoreline. The coarse rocky substrate at Ra's Hebak is interspersed by huge boulders and rocky exposures. *Jatropha* often forms the matrix of the community, although locally, *Croton* shrubland up to 1.5 m in height and attaining cover values up to 75% plays a predominant role, presumably on finer substrates. On account of its massively swollen stems and local abundance, *Adenium* is one of the most prominent elements of the vegetation. Other conspicuous associates include the succulent tree *Dendrosicyos socotrana*, which can superficially resemble *Adenium*, *Euphorbia socotrana* and the emergent *Sterculia africana* (Fig. 6.58), with its massive trunk. *Sterculia* is only occasionally interspersed, but towers above the shrub layer and is therefore a striking feature of the community (Fig. 6.59). In the vicinity of Hadibuh, these trees are favourite roosting sites of the Egyptian vulture.

Apart from the species mentioned above, a host of others occur in the shrub and ground layer. A rather gangling form of *Maerua angolensis* var. *socotrana* (Fig. 6.60) is scattered throughout the stands. *Commiphora* spp., *Cordia obovata*,

Fig. 6.57 Dominant *Jatropha unicostata* in *Adenium* shrubland. Near Homhil, December 2008

Fig. 6.58 The emergent *Sterculia africana* in *Adenium* shrubland. Near Di-Rohan, south of Diham, December 2008

Fig. 6.59 *Adenium* community with *Sterculia africana* in Wadi Ayhaft. December 2008

Fig. 6.60 The gangling form of *Maerua angolensis* var. *socotrana*. Near Di-Rohan, south of Diham, December 2008

Fig. 6.61 The endemic *Hibiscus noli-tangere*. Wadi Dizyaf (Wadi Shi'faar), December 2008

Lannea transulta and *Trichocalyx obovatus* are other species found in the shrub and small tree layer. The endemic *Boswellia popoviana*, a rock-dwelling frankincense that displays considerable morphological variability, can occasionally be observed growing on vertical rock faces. Along the north-facing coastal limestone escarpment below Hamaderoh, the very rare and local endemic small tree *Hibiscus noli-tangere* (Fig. 6.61) occurs. Figure 6.62 shows the typical habitat of this highly endangered species there. Further to the west, in the vicinity of Ra's Hebak, the rare endemic small tree *Hibiscus malacophyllus* is found in similar situations. The only known station of *Hibiscus socotranus*, originally discovered by John Lavranos and Alan Radcliffe-Smith in 1967, is located on the limestone escarpment along the north-west coast near Qalansiyah. *Hibiscus diriffan*, a small endemic tree, is much more widely distributed, occurring on limestone slopes and cliffs down to nearly sea-level in the south (Fig. 6.63), extending to the highest granite peaks further north.

Despite the overall arid nature of the rocky slopes, a number of temporarily moist habitats are occasionally interspersed, such as shady overhangs and seepage areas. Local pastoralists have often protected small rock pools with slabs to ensure a limited supply of water during periods of drought. In such sites, domestic livestock is usually able to find small quantities of lush fodder on the slopes throughout the dry season. Common species include *Barleria tetracantha*, *Cleome* spp., *Fagonia luntii*, *Heliotropium* spp., *Launaea massauensis*, *Lavandula nimmoi*, *Pergularia tomentosa*, *Withania riebeckii* and various others. Some of these species can be very harmful to livestock due to the toxic compounds they contain, but goats appear to be less affected by them.

Fig. 6.62 Typical habitat of *Hibiscus noli-tangere*. Wadi Dizyaf (Wadi Shi'faar), December 2008

Fig. 6.63 The fairly widespread *Hibiscus diriffan*. Dixam plateau, December 2008

Apart from species-rich stands, which thrive mainly on the exposed slopes of the north coast where they intergrade with woodland and forest communities in partially shaded wadis at lower altitudes, a variety of other easily recognisable recurring assemblages occur in which *Adenium* features prominently. For instance,

Fig. 6.64 Relatively species-poor, dense stands of *Adenium* on the coastal plain at Hoq. December 2008

at several locations along the north coast (e.g. Hoq; Fig. 6.64), relatively species-poor, dense stands of *Adenium* colonise the rocky substrate of coastal plains and at the foot of the hills. The main associate in the shrub layer is *Jatropha unicostata*, although locally *Croton* plays an important role. Other frequent plants include *Cissus subaphylla*, *Ballochia amoena* and various weedy species. The stands are heavily grazed, as witnessed by the stunted appearance of the more palatable species.

Kürschner et al. (2006a, b) carried out a detailed phytosociological study of the *Adenium* community, which they refer to as the "Adenio sokotrani-Sterculietum socotranae", an association belonging to the alliance Crotonion socotrani. They regard *Adenium obesum* and *Sterculia africana* var. *socotrana* as the primary character species. Apart from the typical subassociation, which encompasses the highly species-rich stands on the north coast described above, these authors also delimit two further subassociations, based on the presence of the two frankincense species *Boswellia elongata* and *B. socotrana*, respectively. To what extent this further differentiation is justified remains subject to more detailed studies, as the stands could well represent degraded lower montane woodland (see below), in part at least. For instance, at Homhil, where according to Miller and Morris (2001) the most extensive stands of *B. elongata* occur (see Fig. 7.15, Sect. 7.3.2), it is quite clear that the entire area has been subject to massive degradation, allowing "increasers" such as *Adenium* to invade and become fairly common.

Fig. 6.65 *Aloe perryi*. Ra's Hebak, March 1997

6.3.4.2 Dwarf Stem Succulent Communities

On rocky substrates that dry out rapidly after rainfall, dwarf stem succulents often predominate. These plants root in rock fissures and on shallow soils where larger species cannot gain a foothold. Abundant are *Aloe perryi* (Fig. 6.65), *Cissus subaphylla*, *Euphorbia schimperi* (including the very similar and possibly conspecific *E. oblanceolata*), *E. spiralis*, *Kalanchoe robusta* (Fig. 6.66) and *Sarcostemma socotranum*.

These succulents grow in readily accessible places because they are generally less palatable to livestock, protected by their numerous spines, irritating or toxic sap and other features. Such small succulent communities can be accommodated in the phytosociological class Kleinio-Carallumetea (Deil and Al-Gifri 1998), and they show distinct affinities to communities of the alliance Euphorbion adenensis from the southern Yemeni highlands (Deil 1991), as well as to corresponding communities from Eritrea to the Red Sea. With the exception of the rather small *Euphorbia spiralis*, densely branched spiny or cactoid species of *Euphorbia* are absent from Socotra, in contrast to adjacent regions of the Horn of Africa and southern Arabia. On the western island of Abdalkuri, the very rare spineless endemic *E. abdelkuri* is found (Fig. 6.67), a species resembling *E. canariensis* of the Canary Islands in habit and size (although with more rounded stems, yellow milk sap and lacking the small spines of the latter species), and also occurring in similar habitats to *E. canariensis*. Both taxa belong to the section Diacanthium. On Abdalkuri, the total population of *E. abdelkuri* comprises less than 300 individuals, and occurs on Jebel Saleh where

Fig. 6.66 *Kalanchoe robusta*

it is associated with *E. schimperi*, and remarkably, *E. balsamifera*, a species with a pronounced disjunct distribution (see Sect. 4.2.4.1). Why this last species should be restricted to Abdalkuri in the Socotra Archipelago is unclear.

6.3.5 Vegetation of the Cliffs and Crevices

Cliffs, defined as rock surfaces over 2 m in height and sloping at an angle of more than 60°, are widespread on Socotra (Fig. 6.68). Globally, they represent one of the least disturbed habitat types and therefore often contain extensive areas of natural vegetation. However, around the world relatively few studies have been conducted on cliff vegetation (Larson et al. 2000). Apart from the obvious difficulties of accessibility on steep, high cliffs, cliff vegetation is inherently problematic to describe and classify because of the discontinuous nature of cliff habitats that incorporate a wide range of different microhabitats. Further complications arise because cliffs support a very diverse flora that is in part habitat-specific, but is also sampled from the surrounding areas. In fact it has been shown from other parts of the world that cliffs harbour some of the most floristically diverse communities, as for instance in Ireland, where a large proportion of the country's flora has been recorded on them

Fig. 6.67 Habitat of the very rare endemic *Euphorbia abdelkuri* on Abdalkuri (Photograph by Wolfgang Wranik 1999)

Fig. 6.68 Cliffs are a prominent feature on Socotra. Muqadrihon pass, south of Hadibuh, December 2008

(Webb 1977). Cliff habitats therefore contribute substantially to biodiversity, more than their actual surface coverage would suggest (Larson et al. 2000). This undoubtedly also applies to Socotra, where steep cliffs that are largely inaccessible to goats and other livestock hold a very diverse flora. Species occurring on cliffs are in part restricted to them, but some may appear to be confined to them due to heavy browsing in more accessible habitats. Cliffs therefore represent important refugia for many Socotran plant species.

Although vertical rock faces receive less direct rainfall than flat terrain or slopes, there seems to be a fairly constant supply of underground water on many cliff faces. Water either slowly percolates through the porous rock, or passes quite rapidly through the cavernous system of karstic limestone. North-facing cliffs are shaded throughout most of the day and as a result are often well vegetated. Due to the inaccessibility of the terrain, little information on the specific habitat requirements of the plant species concerned is forthcoming. It is likely that depth of the crevice or ledge and the extent to which detritus has accumulated are key factors determining colonisation, depending on the rooting profile of the plant species concerned. Once a plant has become established in a favourable microsite, competition would make it difficult for other species to gain a foothold (site pre-emption). Chamaephytes are the dominant life-form, followed by hemicryptophytes. It appears that annuals and geophytes are poorly represented on rock faces that are exposed to the sun for lengthy periods, but they can be widespread on shaded cliffs.

Apart from the regular stochastic occurrence of more widespread species on cliffs, the description of specific vegetation units is further complicated by the presence of species with highly contrasting environmental demands. Not only do calcicole and calcifuge species sometimes grow in close association with each other, but some species also display a marked habitat plasticity. For instance, the two *Limonium* species occurring on Socotra, both endemics, grow predominantly in coastal saline habitats, but they also occur on limestone cliffs inland, well away from the maritime influence, as described below. Interestingly, this phenomenon is by no means restricted to Socotra, but has been described from elsewhere. For example, in Scotland, maritime halophytes such as *Armeria maritima* and *Plantago maritima* are also known to grow on inland cliffs (Birks 1973). The description of the cliff vegetation of Socotra should therefore be regarded as being preliminary at this stage. If vegetation cover on the cliffs exceeds about 10%, then the vegetation can usually be better assigned to one of the units of the plains, slopes or plateaus.

6.3.5.1 Characteristic Species and Assemblages of the Cliffs

A highly conspicuous species of limestone cliffs is the endemic bottle tree *Dorstenia gigas* (Fig. 6.69), which is able to colonise crevices and holes in the rock that contain very little soil. It is also widespread on the granite cliffs of the Hagghier mountains. Cliffs and huge boulders possibly represent the primary natural habitat of this succulent, and at lower altitudes, it is restricted to such rocky habitats that remain largely inaccessible to goats. However, light grazing by these animals probably does

Fig. 6.69 The endemic bottle tree *Dorstenia gigas* is highly characteristic of cliffs. Muqadrihon pass, south of Hadibuh, December 2008

not pose any serious threat to the individuals, as huge old specimens up to 2 m in height can still be seen on gentle slopes in the more remote areas of the Hagghier mountains. Numerous scars and secondary shoots on these individuals probably indicate very occasional grazing, or perhaps mechanical damage from loose rocks. The fairly dense, partially hanging shrub *Periploca visciformis* is a common associate, often forming large patches on many cliffs (Figs. 6.70 and 6.71).

Mies (2001) described the limestone cliff vegetation characterised by *Dorstenia gigas* as a new association (Dorstenietum gigantis), with *Periploca visciformis* and *Boswellia popoviana* (see Figs. 6.76–6.78) as further character species of this phytosociological unit. Many of the accompanying species are opportunists that colonise suitable microsites on cliff ledges and in crevices, and so at present, this association should be treated with a certain amount of caution. In addition, further studies are required to examine to what extent species composition on granite is similar to that on limestone.

Another member of the genus, *Dorstenia socotrana*, was recently described from seaward-facing limestone cliffs and large boulders at Hamaderoh in north-eastern Socotra (Miller 1996). Being a perennial herb, it is morphologically very different from *D. gigas*, and ecologically is rather similar to *D. foetida* of the adjacent mainland areas.

Several of the eight species of *Boswellia* known to occur on Socotra, all endemics, are restricted to limestone cliffs and huge boulders, mainly below about 600 m. These species, *B. bullata* (Fig. 6.72), *B. dioscorides* (Fig. 6.73), *B. nana*

Fig. 6.70 *Periploca visciformis*, a common species on cliffs. Wadi Ayhaft, December 2008

Fig. 6.71 Flower of *Periploca visciformis*. Wadi Ayhaft, December 2008

(Figs. 6.74 and 6.75), *B. popoviana* (Figs. 6.76, 6.77 and 6.78) and a so-far undescribed species "A" (see Miller and Morris 2004), possess swollen holdfasts that securely attach them to even vertical cliffs in the vicinity of the crevice in which they germinated.

Perhaps the most widespread of these species, occurring mainly in the north, but also locally in the south and west of the island, is *B. popoviana*, a small tree up to

Fig. 6.72 *Boswellia bullata*. Ra's Hebak, December 2008

Fig. 6.73 Flowers of *Boswellia dioscorides*. February 1999

5 m in height. The leaves are morphologically quite variable, but are usually undivided and serrate, and up to 15 cm in length (see Fig. 6.78). The tree is locally common on cliffs and occasionally between huge limestone boulders. On the limestone cliffs and pavement at Hamaderoh, a closely related frankincense, *B. nana*, occurs.

Fig. 6.74 *Boswellia nana*. Hamaderoh, January 1998

Fig. 6.75 Flowers of *Boswellia nana*. Hamaderoh (Photograph by John Lavranos, 1967)

This species is generally much smaller than *B. popoviana* and usually has a different leaf shape (see Fig. 6.74). In contrast to the opinion of Thulin and Al-Gifri (1998), who regarded it as conspecific with *B. popoviana*, *B. nana* should probably best be treated as a separate taxon, as also suggested by Mies et al. (2000) and Miller and Morris (2004). This dwarf tree species is often hidden within deep clefts in the

Fig. 6.76 The cliff-dwelling *Boswellia popoviana*. Ra's Hebak, December 2008

Fig. 6.77 Flowers of *Boswellia popoviana*. Ra's Hebak, July 2002

Fig. 6.78 Leaves of *Boswellia popoviana*. Ra's Hebak, January 1998

limestone rock, and for most of the year when it is leafless, it remains virtually invisible on account of its white and usually lichen-encrusted bark. *B. dioscorides* is locally abundant on limestone cliffs in the south and east of the island, whereas *Boswellia* sp. "A" appears to be restricted to limestone cliffs in a very small area of central Socotra (Rokeb di Firmihin), where it is locally common.

A common and widespread species on limestone cliffs and large boulders at lower altitudes, usually below about 300 m, is the trailing evergreen caper *Capparis cartilaginea*, a species that produces large, white, showy flowers and conspicuous red fruits. The leathery leaves are ovate to round, and quite large, and the shrubs can cover extensive patches in suitable locations. The species occurs throughout the island. In contrast, the liana *Cocculus pendulus*, a fairly widespread species in the adjacent mainland areas, is rare and local on Socotra, growing on both trees and cliffs. On the drier, but partially shaded, cliffs in the south, *Capparis* is accompanied by *Aerva microphylla*, *Angkalanthus oligophylla* (in one wadi at least – Fig. 6.79), *Cryptolepis socotrana* (see Figs. 5.25 and 5.26), *Ficus cordata*, the cliff-dwelling form of *Limonium paulayanum* (Fig. 6.80), otherwise known from coastal communities, and *Tragia balfouriana*, the last of these growing in larger rock crevices with some soil development. Like *Limonium paulayanum*, *L. sokotranum* thrives on cliffs in both the north and south of the island, often well away from the direct influence of the sea, very much in contrast with its more usual salt-marsh and coastal white-sand habitat.

Two very local endemic small shrubs, found primarily on cliffs in the north of the island at lower altitudes, are *Psilotrichum aphyllum* and *Tephrosia socotrana*. In contrast *Poskea socotrana* (Fig. 6.81) is a small cliff-growing shrub that is

Fig. 6.79 The extremely rare *Angkalanthus oligophylla* in flower in a wadi in the south of Socotra. December 2008

Fig. 6.80 The cliff-dwelling form of *Limonium paulayanum*. Wadi Ireh, December 2008

Fig. 6.81 *Poskea socotrana*. Wadi Ayhaft, February 1999

fairly widespread, growing over a broad altitudinal range. *Leucas spiculifolia* is a fairly common woody herb on cliffs at lower elevations throughout most parts of the island.

At higher altitudes, the umbellifer *Rughidia milleri* is a locally common endemic plant of vertical limestone cliffs. The species was initially thought to be restricted to the Rewged plateau (ca. 600–750 m), located in the north of the island (Miller and Morris 2004). In the meantime, we have recorded it from similar habitats on the Dixam plateau further to the south (Fig. 6.82), at about 620 m. The rare small tree *Turraea socotrana* is scattered throughout mainly the western parts of the island on limestone cliffs at medium elevations, and is an example of a species that in addition occurs on granite in the Hagghier mountains. In contrast, the small tree *Tamridaea capsulifera* (Fig. 6.83) is fairly widespread, especially in the north, occurring over a broad altitudinal range, also on both limestone and granite.

The endemic umbellifer *Oreofraga morrisiana* is a perennial woody-based herb growing on limestone and granite cliffs above ca. 500 m, mainly in areas strongly affected by mists and drizzle. *Begonia socotrana* (Fig. 6.84) also prefers shaded, damp locations, mainly on wooded, north-facing limestone and granite cliffs at medium to high elevations. On Samhah, it is replaced by the morphologically similar, extremely local *B. samhaensis* (Hughes and Miller 2002). *Hypericum balfourii* is a species that occurs primarily on granite at higher altitudes, but is also locally

Fig. 6.82 The umbellifer *Rughidia milleri*. Dixam plateau, December 2008

Fig. 6.83 The endemic *Tamridaea capsulifera* is fairly widespread on rocky slopes. Dixam plateau, December 2008

Fig. 6.84 The endemic *Begonia socotrana*

present on limestone cliffs, such as on the Dixam plateau at 950 m (Fig. 6.85), and according to Miller and Morris (2004), even lower in the north. Two species of *Dyerophytum*, both endemics, are typically found on cliffs. Whereas *D. socotranum* is characteristic of limestone cliffs at lower altitudes, *D. pendulum* occurs primarily on partially shaded cliffs above ca. 600 m, and mainly on granite.

Several species are currently known only from maritime cliffs in the far west of the island, often where atmospheric moisture is locally intercepted in an otherwise predominantly arid area. These include two endemic woody herbs, *Nesocrambe socotrana* and *Pulicaria dioscorides*, as well as the more widespread *Camptoloma lyperiifolium*. Furthermore, the endemic succulent *Aloe squarrosa* is largely restricted to mid-elevation limestone cliffs above Qalansiyah.

In shaded situations on ledges and in crevices, a host of other species, some endemics, others common weeds, occur. More regular herbs, some woody-based, include *Ageratum conyzoides*, *Bidens pilosa*, *Corchorus erodioides*, *Euphorbia* spp., *Exacum affine*, *Hybanthus enneaspermus*, *Hypoestes pubescens* (Fig. 6.86), *Lindenbergia sokotrana*, *Oldenlandia bicornuta*, *Peristrophe paniculata*, *Portulaca quadrifida*, *Reichardia tingitana* and *Tripteris vaillantii*, as well as various graminoids. Common poikilohydric ferns such as *Actiniopteris semiflabellata* and *Adiantum*

Fig. 6.85 *Hypericum balfourii* occurs locally on limestone cliffs. Dixam plateau, December 2008

Fig. 6.86 *Hypoestes pubescens*. Dixam plateau, December 2008

balfourii are typical resurrection plants, capable of surviving extended dry periods in a desiccated state.

From seepages on limestone cliffs overlooking the Nuged plain, Thulin (1996) described the very rare and local endemic dwarf shrub *Erythroxylon socotranum*, the first representative of the Erythroxylaceae to be found in south-west Asia.

Fig. 6.87 Seepage areas are occasionally found on limestone cliffs. Nuged plain, December 2008

Growing in similar situations is another extremely rare and local dwarf shrub, *Scaevola socotraensis*.

Occasionally damp overhangs fed by water seepage are found under huge limestone blocks or in large caves (Fig. 6.87). A species-poor vascular plant community dominated by *Adiantum capillus-veneris* (Maidenhair fern) is frequently encountered in such situations. Accompanying plant species are usually rare, but liverworts can be plentiful. Deil (1991, 1996) places such hydrophilous communities with Maidenhair fern in the vegetation class Adiantetea capilli-veneris.

At high altitudes on granite, moist crevices are colonised by a variety of ferns and herbs. Where sufficient soil has accumulated, the latter include *Peperomia blanda* and *Peperomia tetraphylla*, apart from the already mentioned *Begonia socotrana*. Moist cliff ledges provide an important habitat for *Euphorbia socotrana*, allowing this tree to ascend to the higher peaks on the island.

As already noted, cliffs represent important refugia for a number of plant species that are failing to regenerate elsewhere on the island, and one of the most conspicuous of these, occurring from medium to the high elevations, is the enigmatic dragon's blood tree (*Dracaena cinnabari*), dealt with in Sect. 6.3.6.1 below.

6.3.6 Vegetation of the Mid-Elevation Plateaus and Gentle Slopes

Plateaus between 400 and 700 m and composed predominantly of Palaeocene to Eocene limestones, in part also dating back to the Cretaceous, occupy a large portion of Socotra (Fig. 6.88). In many areas, these limestones exhibit distinct

Fig. 6.88 Limestone plateau at Dixam. December 2008

karst features, and are riddled with sink holes and fissures of various sizes. Limestone pavement, a natural karst landform, refers to near horizontal surfaces that are irregularly incised and furrowed, and often dissected by deeper, more regular fissures ("grykes") that correspond to naturally occurring joints within the rock (see Fig. 6.92). The coarse, stony to rocky substrate supports a sparse to fairly dense vegetation cover dominated physiognomically by drought-deciduous shrubs and dwarf shrubs, in which small trees are occasionally interspersed (Fig. 6.89). In some areas, significant perennial vegetation cover is restricted to patches in favourable locations such as gullies, ravines or even fissures in the limestone. A highly distinctive feature of the plateaus in the eastern part of Socotra is the open stands of dragon's blood tree (*Dracaena cinnabari* – Fig. 6.90). After rainfall, the accompanying shrubs develop fresh foliage, and ephemeral herbs and grasses germinate in depressions, crevices and other microsites. There is no sharp boundary delimiting the vegetation of the mid-elevation plateaus from that of the lowland communities. However, the absence of *Euphorbia arbuscula* and the appearance of *Dracaena* above 400 m can be used as a broad indicator of the mid-elevation plateaus.

6.3.6.1 *Buxanthus pedicellatus–Dracaena cinnabari* Woodland of the Mid-Elevation Limestone Plateaus

Due to their extremely variable floristic composition and physiognomy, the conspicuous semi-evergreen *Dracaena cinnabari* woodlands are inherently difficult to accommodate within a vegetation classification system, as outlined in more detail

Fig. 6.89 The substrate on the limestone plateaus can be very rocky

Fig. 6.90 The dragon's blood tree is a distinctive feature of the plateaus in the eastern part of Socotra. Dixam plateau, December 2008

below. However, a frequent associate that occupies a similar elevational niche is the small evergreen shrub *Buxanthus pedicellatus*, and it is therefore convenient to refer at least provisionally to a "*Buxanthus–Dracaena* community". This community is largely restricted to eastern Socotra, as the endemic *D. cinnabari* is virtually absent from most of the western portion of the island, apart from isolated occurrences in the south-west. *Dracaena* is also a conspicuous element in various plant communities, including as an emergent in damp woodland at higher altitudes on granite, such as in the *Leucas hagghierensis–Pittosporum viridiflorum* community (see Sect. 6.3.9.1) at the highest point on the island. However, by far the most impressive stands of *Dracaena* are found on rocky limestone slopes and cliffs between 400 and 700 m, where the tree is part of a spectacular, species-rich woodland community. The general distribution of *Dracaena* on Socotra reflects the extent of areas that are affected by the monsoon mists. The virtual absence of the species in most of the west of the island appears to be attributable to the insignificance of the monsoon influence there. Speculation abounds that the apparent decrease in precipitation in recent years, and ascribed to anthropogenic changes in local weather patterns, may be affecting the natural regeneration capability of *Dracaena*, a topic discussed in more detail in Sect. 7.5.1.

Canopy cover of *Dracaena* can reach up to 40% locally, with trees attaining a height of between 5 and 8 m. The tree is a typical emergent, far overtopping the woody layer below in which various species may be present. *Buxanthus pedicellatus* is widespread throughout many parts of the island, but has its main occurrence above 500 m. In favourable conditions, its cover values may reach 40%, too, but the species can also be conspicuous by its near absence in other locations. Apart from *Dracaena*, other trees such as *Boswellia* sp. (similar in appearance to *B. ameero* and locally abundant, such as at Rokeb di Firmihin), *Carphalea obovata*, *Commiphora planifrons* (Fig. 6.91), *Ficus cordata* and *Sterculia africana* are occasionally interspersed as emergents. *Ficus* is locally common in very narrow fissures on otherwise virtually barren limestone pavement, where it protrudes above the surface as a low shrub (Fig. 6.92).

Buxanthus–Dracaena woodland on the mid-elevation slopes is typically accompanied by various shrubs or straggling trees, and depending on the site-specific conditions, these include *Adenium obesum*, *Croton socotranus*, *Croton sulcifructus* (locally dominant), *Jatropha unicostata*, *Lycium sokotranum* and *Rhus thyrsiflora*. The endemic small tree *Punica protopunica* is locally common in some areas, especially at higher elevations where moister conditions are more prevalent. According to Lavranos (pers. comm.), this species is also abundant along the limestone ridges to the east of Hamaderoh in north-eastern Socotra. The small tree *Euphorbia socotrana* is locally common on the mid-elevation slopes, extending into the thickets and forest towards the highest peaks of the island as a typical emergent.

The unpalatable dwarf shrub *Euryops arabicus* is locally abundant in central parts of the island, and may well be on the increase as a result of overgrazing. Species such as *Hypericum scopulorum* and *Rhus thyrsiflora* are indicative of stands in sheltered localities on the plateaus at higher elevations.

Fig. 6.91 *Commiphora planifrons*. July 2002

Fig. 6.92 *Ficus cordata* in a gryke. Dixam plateau, December 2008

At lower elevations, species most commonly associated with the plains are very much in evidence, most notably *Ballochia amoena* (Fig. 6.93), *Croton socotranus*, *Jatropha unicostata* and *Placopoda virgata*, with *Buxanthus* becoming rare or largely absent. The local abundance of *Adenium obesum* in the understorey, often accompanied by *Jatropha*, indicates the transition to the drier stands of the *Adenium* community (see Sect. 6.3.4.1), in which *Dracaena* is frequently absent.

Fig. 6.93 *Ballochia amoena*. Dixam plateau, December 2008

An isolated occurrence of *Dracaena* has been found in the far south-west of Socotra along the summit of Qatariyah, overlooking the Nuged plain (Miller and Morris 2001). This exposed locality catches low cloud and the monsoon mists, and so offers environmental conditions amenable to the development of fairly dense woodland in an otherwise sparely vegetated area. The few *Dracaena* trees at Qatariyah are found in woodland dominated by *Buxanthus pedicellatus* and accompanied by various shrubs and small trees, such as *Allophylus rubifolius*, *Carphalea obovata*, *Commiphora ornifolia*, *Commiphora planifrons*, *Cryptolepis intricata*, *Euclea divinorum*, *Euphorbia socotrana*, *Ficus cordata* and *Hibiscus quattenensis*. In somewhat more open stands, where *Dracaena* is lacking, the perennial grass layer is well developed, with *Heteropogon contortus* attaining very high cover values, locally in excess of 50%.

Even on the mid-elevation slopes, there is a conspicuous presence of lichens on trees and shrubs locally, underlining the influence of the monsoon mists, and species include *Enterographa* spp., *Heterodermia* spp. (Fig. 6.94), *Opegrapha dracaenarum*, *Physcia dimidiata* and *Ramalina* spp. These become more common with increasing moisture input from the monsoons.

Despite the visually prominent stands of *Dracaena*, it is worth noting that to date there has been no attempt to classify them phytosociologically. Integration of the *Buxanthus–Dracaena* community into the Braun-Blanquet floristic association system is problematic for a number of reasons, but most importantly the following. (1) The two chief species occur with varying degrees of frequency in other communities, especially at higher altitudes, on both limestone and granite. The vertical

Fig. 6.94 *Heterodermia leucomelos*. Dixam plateau, February 1999

compression of habitats, enabling these and other species to occur in quite different stands within a short distance on Socotra, provides strong arguments for those who rebut the notion of plant communities as discrete, recognisable units. From an academic standpoint, *Dracaena* can be viewed as a particularly clear example of a species that occupies different segments along an environmental gradient, often independent of the distribution of other species, very much in line with the "continuum view" of Gleason (1926). (2) Stands with most of the major species, but lacking *Dracaena*, are widespread on the limestone plateaus, and more intensive vegetation studies are required to distinguish recurring units. However, this undertaking will be severely hampered by the massive overgrazing that has led to the degradation of large areas, which is also greatly restricting the natural regeneration of many key species, including *Dracaena* itself.

Nonetheless, from a practical, but highly subjective point of view, the designation of the *Buxanthus–Dracaena* community is a useful preliminary construction to delimit the remarkable *Dracaena* stands of the mid-elevation slopes, i.e. those at the drier extreme of the environmental gradient from the woodland and thickets of the moister high mountain peaks.

The *Buxanthus–Dracaena* community does not appear to be characterised by any particular species assemblage in the ground layer, and species composition can vary substantially from one area to the next, in part determined by the capacity of the substrate to store water. In the following, therefore, a general summary is given of the characteristic species of the ground layer on the limestone plateau, irrespective of the predominant shrubs or trees.

Fig. 6.95 *Cyanixia socotrana*. Near Di-Rohan, south of Diham, December 2008

The grass layer, a mixture of perennials and annuals, is locally well developed, especially between rocky substrates or in small depressions and fissures in the limestone. Typical species include *Aristida adscensionis*, *Cenchrus ciliaris*, *Digitaria* sp., *Eragrostis* spp. (including *E. papposa*) and *Heteropogon contortus*. A large variety of herbs and small woody species can be frequently encountered. Some are widespread, others rather local, but sometimes common within a very restricted area, and include *Aerva lanata*, *Cyanixia socotrana* (Fig. 6.95), *Barleria tetracantha*, *Cometes abyssinica* (locally common on the limestone plateau in eastern Socotra), *Commicarpus heimerlii*, *C. simonyi*, *Convolvulus sarmentosus* (typical of exposed places, such as overgrazed vegetation on the limestone plateau; Fig. 6.96), *Corchorus erodioides*, *Crotalaria strigulosa* (Fig. 6.97), *Dicliptera effusa*, *Haya obovata*, *Helichrysum gracilipes*, *Indigofera nephrocarpa*, *Ipomoea obscura* (a climbing annual in shrubland on limestone and granite, and also grows in crevices on the limestone pavement; Fig. 6.98), *Lavandula nimmoi*, *Macledium canum*, *Melhania muricata*, *Oldenlandia bicornuta*, *O. pulvinata*, *Oxalis corniculata*, *Priva socotrana*, *Plantago amplexicaulis*, *Pulicaria diversifolia*, *Rhinacanthus scoparius* (Fig. 6.99), *Ruellia dioscoridis* (Fig. 6.100), *R. patula*, *Senna holosericea*, *Talinum portulacifolium*, *Tephrosia odorata* (Fig. 6.101), *Trichodesma laxiflorum*, *Tripteris vaillantii* and many others. The shrub *Zygocarpum coeruleum* is locally common, but is often developed as a small prostrate plant, closely adpressed to the ground (Fig. 6.102).

Various succulents are locally common on the limestone plateau, some occurring also in the *Buxanthus–Dracaena* woodland, although the smaller species are easily overlooked when not in flower. *Caralluma socotrana* is widespread and common, as is *Aloe perryi*. In contrast, *A. jawiyon*, a recently described species similar to *A. perryi*, appears to be restricted to limestone pavement at medium elevations in

Fig. 6.96 *Convolvulus sarmentosus*. Homhil, December 2008

Fig. 6.97 *Crotalaria strigulosa*. Dixam plateau, December 2008

Fig. 6.98 *Ipomoea obscura* in a gryke. Dixam plateau, December 2008

Fig. 6.99 *Rhinacanthus scoparius*. Dixam plateau, December 2008

the west of the island. *Kalanchoe robusta, K. rotundifolia* and the trailing succulent *Cissus hamaderohensis* are also characteristic of this altitudinal belt. *Echidnopsis socotrana*, which grows in pockets of soil in the shade of boulders, is the most widespread of the various *Echidnopsis* species on Socotra. The rather rare *E. insularis* thrives in similar situations. *Socotrella dolichocnema*, a small succulent that is very

Fig. 6.100 *Ruellia dioscoridis*. Homhil, December 2008

Fig. 6.101 *Tephrosia odorata*. Dixam plateau, December 2008

inconspicuous when not in flower, was noted by Miller and Morris (2004) as being very rare and restricted to limestone pavement in the far west, but has since been found to be more widespread in the west of the island (Řičánek and Hanáček 2006). Our own observations suggest that the species is in fact fairly widespread throughout the island. The endemic succulent composite *Kleinia scottii* (see Figs. 4.28 and 4.29) is locally common within its very restricted range at higher altitudes on the limestone plateau.

Fig. 6.102 *Zygocarpum coeruleum*. Hamaderoh, February 1999

Several species of *Cyperus* are common in temporary damp places, including in shallow depressions in which water accumulates after rain, and *C. bulbosus* is one of the first species to emerge after rain. The fern *Ophioglossum polyphyllum* grows in similar situations on limestone plateaus, and is also found on shaded cliff ledges.

6.3.6.2 Other Vegetation Types

As indicated above, some of the accompanying species in the *Buxanthus–Dracaena* woodland of the mid-elevation slopes can themselves predominate locally, often without *Dracaena* (as throughout most of the west of the island), and sometimes with low participation of *Buxanthus*. This is particularly obvious with species such as *Croton sulcifructus* and *Rhus thyrsiflora*, both species that can be dominant at higher altitudes receiving more rainfall. Furthermore, the endemic small shrub *Cryptolepis intricata*, occasionally interspersed in *Buxanthus–Dracaena* woodland, can be abundant and dominate the vegetation locally. The same applies to the endemic evergreen shrub *Gnidia socotrana*, a species that is locally abundant in various vegetation types in the northern half of Socotra, with cover values up to about 50%. The high abundance of this species may in part be due to the fact that it is largely avoided by goats. In exposed localities, it is often found on northern slopes or summit areas that receive enhanced moisture input (e.g. at Hamaderoh). It can be accompanied by a host of other species, and woody associates include *Acacia pennivenia*, *Acridocarpus socotranus* (Fig. 6.103), *Anisotes diversifolius*, *Ballochia rotundifolia*, *Boswellia nana*, *Carphalea obovata*, *Commiphora socotrana*, *Cryptolepis intricata*, *Jatropha unicostata*, *Lannea transulta*, *Rhus thyrsiflora*,

Fig. 6.103 *Acridocarpus socotranus*. Jebel Rewged, April 1993

Vernonia cockburniana and *Ziziphus spina-christi*. For instance, on the limestone plateau at Rewged (ca. 700 m), immediately to the south of Wadi Ayhaft, the main emergent in *Gnidia* scrub is *Commiphora socotrana*, with other trees and shrubs occasionally intermingled. The rare occurrence of *Dracaena cinnabari* indicates the affiliation to *Buxanthus–Dracaena* woodland, in which *Gnidia* largely replaces the physiognomically similar *Buxanthus*. In some areas on the limestone plateaus, where pedogenesis has taken place to a certain degree, there are fairly extensive stands of the shrub *Lycium sokotranum*. This spiny species is usually only lightly browsed by domestic livestock.

Quite extensive open stands of *Croton* occur locally on the limestone plateau in various parts of the island. The precise taxonomic status of some of the highly variable forms is unclear. On the Dixam plateau, Miller and Morris (2004) treat the dominant *Croton* species that has a conspicuous white bark as *C. socotranus*, as it is otherwise morphologically indistinguishable from lowland forms. On the rugged Hamaderoh plateau, a highly distinctive *Croton* taxon was initially referred to as *Croton pachyclados* by Radcliffe-Smith (see Radcliffe-Smith 1973). He regarded it as a local endemic, but later relegated it to a variety of *C. socotranus* (Radcliffe-Smith 1989). The population of this taxon is very small, and Miller and Morris (2004) also treat it as another variant of the widespread and highly variable *C. socotranus*. However, due to its striking appearance and differences in floral structure (Figs. 6.104 and 6.105), we feel that more detailed studies are required that may yet warrant its status as a distinct species.

An almost unique type of dense grassland in which the endemic tussock grass *Andropogon bentii* predominates was first described by Miller and Morris (2001)

Fig. 6.104 *Croton pachyclados*. Hamaderoh, March 1997

Fig. 6.105 Flowers of *Croton pachyclados*. Hamaderoh, March 1997

from Jebel Ma'alah, about 5.5 km to the south-south-east of Qalansiyah. It occupies only a small patch on the undulating limestone plateau at about 600 m. One of the visually most prominent associates is *Buxanthus pedicellatus*. This type of grassland is presumably dependent on fairly deep soils. In adjacent sites, *Andropogon* is

Fig. 6.106 The rare endemic *Wellstedia socotrana*. Jebel Ma'alah (Photograph by F. Boltz, December 2003)

frequently accompanied by other woody species, including *Jatropha unicostata* and *Pulicaria stephanocarpa*.

In the more arid areas of the south and south-west of Socotra, the vegetation of the limestone plateaus consists of very open dwarf shrubland with a high proportion of cushion- or rosette-forming woody-based herbs. Common species include *Campylanthus spinosus*, *Convolvulus sarmentosus*, *Cryptolepis intricata*, *Distephanus qazmi*, *Leucas flagellifolia* (at lower altitudes), *Lotus mollis*, *Lycium sokotranum* and *Oldenlandia pulvinata*. The rare and local prostrate dwarf shrub *Wellstedia socotrana*, a genus with a remarkably disjunct distribution (see Sect. 4.2.2), is found in such vegetation on a limestone plateau at about 700 m in the central part of southern Socotra. This rather nondescript species, initially believed to be restricted to the south (see Miller and Morris 2004), has also been collected by one of us from Jebel Ma'alah in the far west (Fig. 6.106), and from the Hamaderoh plateau in the north of the island (see Kilian and Hein 2006). It is therefore probably much more widespread than previously assumed.

6.3.6.3 Lower Montane Woodland

With increasing elevation and enhanced moisture availability, the succulent shrubland described in Sect. 6.3.4 gradually or abruptly gives way to woodland and even forest on the slopes of northern Socotra. The typical boundary occurs around 500 m,

Fig. 6.107 *Boswellia ameero*. Muqadrihon pass, south of Hadibuh, February 1999

but it can be considerably lower in heavily shaded wadis where a gradual transition from riparian forest, recognised by the high proportion of figs, to lower montane woodland is apparent, as outlined in Sect. 6.3.3.1. With cover values between 40% and 70% in the tree layer, this woodland can be quite dense, and it is locally developed as forest. Tree height ranges from 4 to 8 m. The often well-developed shrub layer can make the stands virtually impenetrable, and this inaccessibility is frequently reinforced by the steep slopes and coarse nature of the substrate, with huge boulders interspersed. Both the tree and shrub layer are usually species-rich, and depending on the location, different species or species combinations can predominate. In many cases, the low-growing tree *Rhus thyrsiflora* plays an important role, being locally dominant, and the same applies to *Allophylus rubifolius*. Some of the more regular associates in the tree and shrub layer include *Acridocarpus socotranus*, *Anisotes diversifolius*, *Boswellia ameero* (Figs. 6.107 and 6.108), *B. elongata* (Fig. 6.109), *B. socotrana*, *Buxanthus pedicellatus*, *Clerodendrum leucophloeum* (Fig. 6.110), *Commiphora ornifolia* (Fig. 6.111), *C. socotrana*, *Croton sarcocarpus*, *Gnidia socotrana*, *Grewia turbinata*, *Maerua angolensis*, *Paracalyx balfourii* and *Ruellia insignis* (Fig. 6.112). The endemic *Dirachma socotrana* (see Fig. 4.9), a

Fig. 6.108 Flowers of *Boswellia ameero*. April 1993

Fig. 6.109 *Boswellia elongata*. Muqadrihon pass, south of Hadibuh, December 2008

small, rather nondescript tree and easily overlooked when not in flower, is locally common within its restricted range in this woodland, occurring from about 300 to 1,000 m. *Dracaena cinnabari* is also occasionally present, especially in more inaccessible exposed sites such as on cliff ledges and amongst large boulders

Fig. 6.110 *Clerodendrum leucophloeum*. Muqadrihon pass, south of Hadibuh, December 2008

Fig. 6.111 *Commiphora ornifolia*. Wadi Ayhaft, December 2008

(see Fig. 6.68), and our observations suggest that it is regenerating fairly well there. Climbers are a characteristic feature of lower montane woodland, and typical representatives include *Cissus paniculata*, *Cryptolepis volubilis*, the annual *Ipomoea nil* (Fig. 6.113) and *Tragia balfouriana*. In the ground layer, a variety of species occurs, including *Adiantum balfourii*, *Bidens biternata* (usually in disturbed areas),

Fig. 6.112 *Ruellia insignis*. Muqadrihon pass, south of Hadibuh, March 1997

Fig. 6.113 The annual *Ipomoea nil*. Muqadrihon pass, south of Hadibuh, December 2008

Campanula balfourii (Fig. 6.114), *Commelina forskalaei*, *Craterostigma pumilum* (Fig. 6.115), *Cystostemon socotranus*, *Dicliptera effusa*, the orchid *Habenaria socotrana*, *Hibiscus sidaeformis*, *Indigofera marmorata* (Fig. 6.116), *Ipomoea sinensis* (Fig. 6.117), *Oplismenus burmannii* and *Withania riebeckii* (Fig. 6.118).

Fig. 6.114 *Campanula balfourii*. Dixam, January 1998

Fig. 6.115 *Craterostigma pumilum*. Wadi Ayhaft, December 2008

Fig. 6.116 *Indigofera marmorata*. Muqadrihon pass, south of Hadibuh, December 2008

Fig. 6.117 *Ipomoea sinensis*. Wadi Ayhaft, December 2008

Kürschner et al. (2006a) have recently described a new woodland plant association (Ruellio insignis–Boswellietum ameero) from northern Socotra based on the character species *Ruellia insignis* and *Boswellia ameero*. The stands of this community are typically developed on steep slopes above 500 m, extending up to about 850 m. Good examples are to be found in the upper reaches of Wadi Ayhaft and Wadi Dihzafaq. However, *B. ameero* is possibly not particularly suitable as a character species in the Braun-Blanquet floristic association system, as it is also found

Fig. 6.118 *Withania riebeckii*. Wadi Ayhaft, December 2008

in more open woodland types (Fig. 6.119), and even along parts of the arid southern escarpement. Mies et al. (2000) regarded it as a common species above 400 m in the Hagghier mountains, and the altitudinal vicariant of *B. elongata*. Kürschner et al. (2006a) assigned the Ruellio insignis–Boswellietum ameero to the phytosociological alliance Crotonion socotrani, underlining its floristic affinities to lowland plant communities on the island.

6.3.7 Vegetation of the Montane and High-Montane Cloud Zone

Above 700 m, temperatures drop markedly during the late afternoon and evening, leading to the regular formation of clouds. Dewfall is also a common occurrence, and both this and intercepted mist are probably a major source of moisture for plants in this altitudinal belt, as outlined in Sect. 2.3. This is underlined by the luxuriant growth of ferns, bryophytes and lichens. In fact the input of moisture is so high that epiphyllic bryophytes and lichens are able to thrive in suitable locations, organisms otherwise typical of tropical rainforests (Mies et al. 1994). The northern side of the Hagghier mountains receives the largest amount of moisture, in terms of precipitation as well as mist and dewfall, and this contributes to the formation of a fairly thick (several centimetres) humus layer under the more dense vegetation cover of the cloud zone.

The natural vegetation is characterised primarily by dense shrubland, ca. 2–3 m in height, locally with scattered emergents. The proportion of evergreen species

Fig. 6.119 *Boswellia ameero*. Dixam plateau, December 2008.

(ones with either malacophyllous or sclerophyllous leaves) rises with increasing altitude and moisture availability (Popov 1957). This author assigned the shrubland vegetation to the "Evergreen Scrub (Types 3 and 4)" of Pichi-Sermolli (1955), and belonging to the semi-arid zone. White (1983) described it as "East African Evergreen and Semi-evergreen Bushland and Thicket". The affinities with East African vegetation are evident in the occurrence of species such as *Buxanthus pedicellatus* and *Dodonaea viscosa*, as well as various vicariant taxa belonging to a number of genera.

Due to the lack of detailed studies, it is possible to outline only broad vegetation types, although recently, Kürschner et al. (2006a) included some of the predominant vegetation types in their phytosociological classification of the forests and woodlands of northern Socotra.

6.3.7.1 *Cephalocroton socotranus* Community

On the basis of initial field studies from the Adho Dimele area (Fig. 6.120), Mies (2001) referred to the dense thickets occurring on granite slopes between roughly 700 and 1,000 m, and dominated by *Cephalocroton socotranus* (Figs. 6.121 and

Fig. 6.120 Adho Dimele, high in the Hagghier mountains. March 1996

Fig. 6.121 *Cephalocroton socotranus*. Muqadrihon pass, south of Hadibuh, December 2008

Fig. 6.122 Male inflorescence of *Cephalocroton socotranus*. Muqadrihon pass, south of Hadibuh, December 2008

6.122) with scattered emergents, provisionally as the *Cephalocroton socotranus* community. Furthermore, this author underlined the necessity of carrying out urgent vegetation studies due to rapid deforestation and erosion. Miller and Morris (2004) referred to this vegetation type as submontane (semi-deciduous) shrubland. Recently, Kürschner et al. (2006a) have examined this "dwarf forest" community in more detail, describing it as a new phytosociological association ("Trichodesmo scottii–Cephalocrotonetum socotrani"), referred to in the following simply as the *Cephalocroton socotranus* community. It is well developed in several areas, in particular in the upper reaches of Wadi Dihzafaq, Adho Dimele and around Tweleq. The main shrub layer of this semi-evergreen low woodland or forest rarely exceeds 3–5 m, and apart from the character species *Cephalocroton socotranus*, other regular associates in the tree and shrub layer are *Buxanthus pedicellatus* (restricted mainly to stands at lower elevations), *Carphalea obovata*, *Cocculus balfourii*, *Croton sulcifructus*, *Graderia fruticosa* (Fig. 6.123), *Hibiscus scottii*, *Hypericum balfourii*, *Rhus thyrsiflora* (locally co-dominant – Fig. 6.124) and *Trichodesma scottii* (Fig. 6.125). Cover of the shrub/small tree layer may reach over 90%, but in other places, the community has more the character of an open woodland, with tree cover as little as 30%.

Conspicuous are the scattered emergents of *Dracaena cinnabari* in some places, but other, taller tree species are also occasionally interspersed, such as various Burseraceae, *Euphorbia socotrana*, represented at these higher elevations by the recently described subspecies *E. socotrana* ssp. *purpurea* (Kilian and Hein 2006), *Lannea transulta* (at lower elevations) and *Sterculia africana*. Wild orange trees

Fig. 6.123 *Graderia fruticosa*. Adho Dimele, January 1998

Fig. 6.124 *Rhus thyrsiflora*. Adho Dimele, January 1998

(*Citrus aurantium*) are found in the dense thickets locally, but Doe (1992) suggested that these were introduced by Portuguese visitors to the island from South-East Asia.

Lianas are well represented by *Cryptolepis volubilis*, *Dioscorea lanata* and *Tragia balfouriana*. A number of species are found in the ground layer of dwarf shrubs, grasses, herbs and ferns, some of the more widespread being *Adiantum balfourii*, *Ageratum conyzoides*, *Cystostemon socotranus*, *Euryops arabicus*, *Hypericum*

Fig. 6.125 *Trichodesma scottii*. Adho Dimele, January 1994

scopulorum (locally abundant), *Hypoestes pubescens*, *Oplismenus compositus* and various other perennial grass species.

A further notable feature of this woodland is a remarkable epiphytic bryophyte community described in detail by Kürschner (2004a), and named after *Lejeunia rhodesiae* and the endemic *Sematophyllum socotrense* ("Lejeunio rhodesiae–Sematophylletum socotrensis"). Other bryophyte species indicative of moister site conditions include *Hyophila involuta* and *Macrocoma abyssinica*, whereas *Brachymenium nepalense* and *Frullania socotrana* are characteristic of locally occurring drier situations.

6.3.8 Other Montane Vegetation Types

A closely related assemblage to the *Cephalocroton socotranus* community, and thriving locally in similar habitats in the south and east of the Hagghier mountains, is one dominated by the endemic shrub *Croton sulcifructus*. It too is accompanied by *Rhus thyrsiflora* and *Allophylus rubifolius*.

Grazing and the clearing of woodland have greatly modified the natural vegetation in the cloud zone, so that a mosaic of grasslands and evergreen thicket is now found in many areas. Popov (1957) listed *Arthraxon lancifolius*, *Hyparrhenia hirta* and *Themeda triandra* (possibly wrongly recorded as *T. quadrivalvis* by Popov) as

Fig. 6.126 *Aerva revoluta*. Ascent to Adho Dimele, March 1996

characteristic species of the higher altitude grasslands. The perennial *Heteropogon contortus* is also very common, and *Arthraxon micans* is locally dominant. It is quite remarkable that the endemic rosette-forming herb *Helichrysum balfourii* forms extensive dense carpets at higher altitudes, somewhat resembling grasslands in some areas. These plants appear to be very effective at preventing soil erosion. A number of other herbs with a more widespread distribution, including some neophytes that were probably fairly recently introduced from Africa, Asia or Europe, are typically found in woodland clearings, usually at higher altitudes. Examples are given by *Anagallis arvensis*, *Capsella bursa-pastoris*, *Medicago polymorpha*, *Melilotus indica*, *Misopates orontium*, *Plantago ovata*, *Sida acuta* (locally abundant), *Veronica polita*, *Waltheria indica* and *Zornia glochidiata*.

Where woodland is absent on granite slopes, either naturally or more commonly due to anthropogenic causes, a low shrub formation with a high cover of cushion-like representatives and reminiscent of tropical highland vegetation is developed. The stands can be quite species-rich, and two of the more characteristic shrubs are *Hypericum scopulorum* and *Hypericum tortuosum*. Other regular small shrubs or dwarf shrubs include *Aerva revoluta* (Fig. 6.126), *Allophylus rubifolius*, *Euryops arabicus* and *Gnidia socotrana* (Fig. 6.127). The Caryophyllaceae *Gymnocarpos bracteatus* is a rare dwarf shrub found locally on rocky granite slopes mainly in the west of the Hagghier range. The *Hypericum* stands, which need to be examined in more detail, represent a low-elevation form of the *Helichrysum–Hypericum* dwarf shrub communities described below.

Fig. 6.127 *Gnidia socotrana*. Ascent to Adho Dimele, March 1996

6.3.9 Vegetation of the High-Montane Granite Peaks

With the exception of the recently described *Leucas hagghierensis–Pittosporum viridiflorum* community, the high montane thickets gradually peter out by about 1,200 m (Figs. 6.128 and 6.129). Low nocturnal temperatures throughout the year and the poor water-holding capacity of the granite substrate are probably the chief environmental factors restricting their growth. They give way to vegetation types dominated by dwarf shrubs and cushion plants. The distinctive morphology of these plants serves to reduce transpirational water-loss and to protect the underside of the leaves from overheating during the day. Depending on the local conditions, vegetation cover varies from scattered individual plants to dense, knee-high dwarf scrub, with the latter occupying the more protected sites. Few detailed vegetation studies have been carried out at this altitude. One conspicuous aspect is that a large proportion of the species appears to be endemic to Socotra.

6.3.9.1 *Leucas hagghierensis–Pittosporum viridiflorum* Community

A remarkable type of evergreen woodland was recently described by Kürschner et al. (2006a) from the highest parts of the central Hagghier mountains between 1,300 and 1,540 m, although it was mentioned in broad terms by Miller and Morris (2001, 2004) as "dense montane evergreen woodland and thicket". The main tree

Fig. 6.128 Sparse vegetation cover high in the Hagghier mountains. View towards Ghubbah on the north coast from the Fieri peaks. February 1999

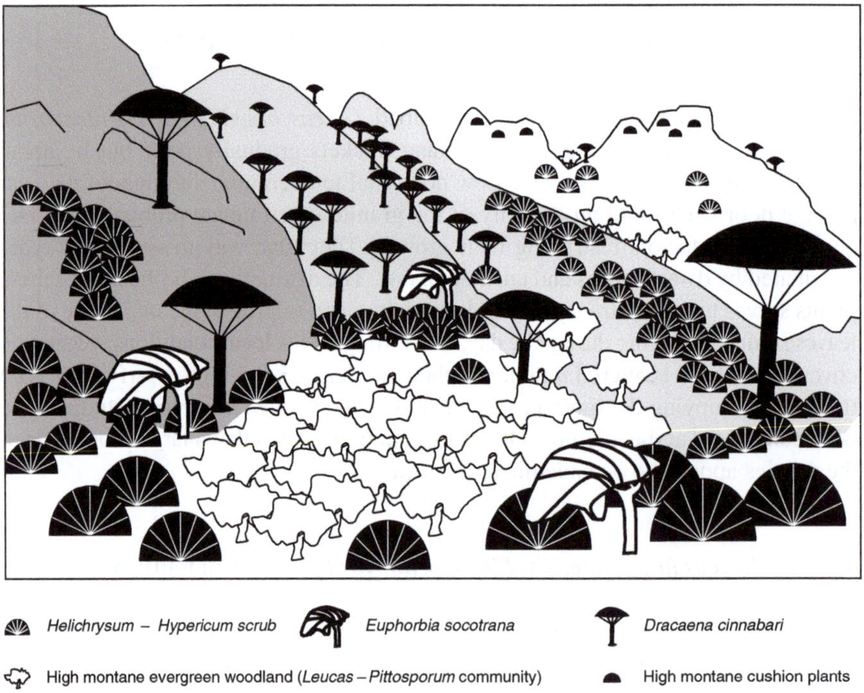

Fig. 6.129 Schematic representation of vegetation in the high-montane zone

Fig. 6.130 Thickets with emergent dragon's blood trees. View towards Skent from the Fieri peaks. February 1999

layer, which typically does not exceed 6 m in height, is rather open, with cover values generally below 60% (Fig. 6.130). However, the species-rich shrub layer is dense, and individuals of *Dracaena cinnabari* overtop the stands as typical emergents. According to Kürschner et al. (2006a), the characteristic tree and shrub species of this woodland or thicket are *Clerodendrum galeatum*, *Helichrysum nimmoanum*, *Hypericum fieriense*, *Leucas hagghierensis*, *Olea capensis* ssp. *macrocarpa* (= *O. hochstetteri*, a new species for the island mentioned by Kilian and Hein (2006)) and *Pittosporum viridiflorum*. Other tree, shrub and dwarf shrub species that regularly occur include *Ballochia rotundifolia*, *Carissa spinarum*, *Coelocarpum haggierensis* (Fig. 6.131), *Croton sulcifructus*, *Euclea divinorum*, *Hypericum balfourii*, *H. scopulorum*, *Rhus thyrsiflora* and *Spiniluma discolor*. *Buxanthus pedicellatus*, characteristic of the medium-elevation slopes, and extending to about 11,00 m, appears to be absent at the highest altitudes on the island, as do a number of other shrubs and small trees such as *Gnidia socotrana* and *Lannea transulta*. The high cover of *Croton sulcifructus* indicates the transition from stands dominated by this species at lower altitudes, so that the *Leucas–Pittosporum viridiflorum* community could be regarded as a species-rich, high-altitudinal form of *C. sulcifructus* scrub. *Trichodesma microcalyx* is a fairly frequent herb in the high elevation thickets. The very delicate, prostrate *Parietaria debilis* is restricted to the moist thickets at the higher elevations, as is the minute epiphytic *Utricularia striatula*, a carnivorous species with broader tropical distribution from West Africa to SE Asia.

Fig. 6.131 *Coelocarpum haggierensis*. Fieri plateau, February 1999

The proportion of endemics at the higher altitudes probably accounts for about 50% of the species present, a remarkably high percentage.

6.3.9.2 *Helichrysum–Hypericum* Dwarf Shrub Communities

Scrub communities dominated by various species of *Helichrysum* and *Hypericum* are the predominant vegetation type at the highest altitudes on Socotra (Fig. 6.132), although some species are also frequently found on granite slopes at lower elevations. *Dracaena cinnabari* occurs as a locally common or even abundant emergent in some areas. The endemic cushion plant *Helichrysum rosulatum* (Fig. 6.133) is perhaps one of the most characteristic species at higher altitudes, typically growing on cliffs and steep slopes, and it is accompanied by the endemics *H. aciculare* and *Hypericum scopulorum*, both species that also occur at much lower altitudes. *H. aciculare* locally replaces *H. rosulatum* as the dominant species, although in other sites it can be noticeably rare. Other species generally restricted to this zone, although not necessarily exclusively so, include the dwarf shrubs *Helichrysum nimmoanum*, *Hemicrambe fruticosa* (Fig. 6.134), *Nirarathamnos asarifolius* (Fig. 6.135) and *Thamnosma socotrana* (Fig. 6.136), and the annual *Dichrocephala chrysanthemifolia*. The last of these, a quite widespread species in southern Arabia, has only rarely been found on Socotra. *Pulicaria lanata* (Fig. 6.137) is a locally common woody herb in shrubland around the highest granite pinnacles, and *Justicia alexandri*, a recently described shrub species (see Miller and Morris 2004), is apparently restricted to such situations.

Fig. 6.132 Scrub communities dominated by *Helichrysum* and *Hypericum* are characteristic of the higher altitudes. April 1993

Fig. 6.133 The endemic cushion plant *Helichrysum rosulatum*. West of Adho Dimele, March 1996

Species that cover a wider altitudinal range, occurring in rocky communities from about 600 m to the summit, include the endemic herbs *Begonia socotrana* (see Fig. 6.84) and *Plectranthus socotranus* (Fig. 6.138), as well as the endemic dwarf shrub *Pluchea obovata*.

Fig. 6.134 *Hemicrambe fruticosa*. Fieri peaks, February 1999

Fig. 6.135 *Nirarathamnos asarifolius*. Fieri peaks (Photograph by John Lavranos, 1967)

Fig. 6.136 *Thamnosma socotrana*. Fieri peaks, February 1999

Fig. 6.137 *Pulicaria lanata*. Fieri peaks, February 1999

The cosmopolitan fern *Pteridium aquilinum* grows in large patches on some of the more gentle granite slopes at the highest elevations. According to Miller and Morris (2004), it is restricted to the western part of the Hagghier range, and is for instance common in the immediate vicinity of Jebel Skent, but our observations show

Fig. 6.138 *Plectranthus socotranus*. Peak west of Adho Dimele, March 1996

it also occurs in the central Hagghier region. However, it may well be absent from the eastern part of the range, as this area does not exceed much above 1,200 m.

A characteristic community in which the endemic dwarf shrub *Helichrysum nimmoanum* plays a dominant role forms dense, low stands at the base of granite pinnacles and in other sheltered locations. Other frequent species include *Carphalea obovata*, *Croton* cf. *socotranus*, *Euryops arabicus*, *Hypericum scopulorum* and *Rhus thyrsiflora*, with *Heteropogon contortus* a prominent feature of the grass layer. The annual *Dichrocephala chrysanthemifolia* was registered in these stands. *Dracaena cinnabari* is occasionally interspersed, easily overtopping the low scrub layer.

Geophytes are fairly well represented in this altitudinal belt. They are usually encountered in moist depressions where deeper soils have accumulated, and where quite dense scrub is developed. Apart from species such as *Cyanixia socotrana*, *Chlorophytum graptophyllum*, *Dioscorea lanata*, *Dipcadi* cf. *balfourii*, *Eureiandra balfourii* and *Habenaria socotrana*, the rare hyacinth *Drimia porphyrostachys* was recorded during fieldwork, possibly on one of the few occasions since it was first discovered by Balfour in 1880.

Increased grazing pressure by goats at higher altitudes is indicated by the fact that the dwarf shrubs *Nirarathamnos asarifolius* and *Hemicrambe fruticosa* are now mainly restricted to inaccessible cliffs. The populations of other slow-growing rare species are likely to suffer considerably if the grazing pressure is not reduced.

Fig. 6.139 *Eleocharis geniculatus*. Hamaderoh, February 1999

6.3.10 Vegetation of Wet Flushes and Other Semi-Aquatic Habitats

The vegetation of wet flushes and the margins of streams and pools, which occur at all altitudes on Socotra, is characterised by a limited number of species, although it is likely that various others may well have been overlooked due to identification problems. No specific studies have been conducted on the vegetation of these habitats, and so the following account is restricted to a brief overview of some of the main species.

Juncus socotranus can form dense stands by permanent watercourses at medium elevations, and the cosmopolitan sedge *Cladium mariscus* is a common perennial in similar situations. At higher elevations (above 1,200 m), the rush *Juncus bufonius*, a widespread annual in many parts of Europe, including in coastal salt-marshes, thrives in wet flushes.

Eleocharis caduca has so far been found in damp wadis and pools in shaded locations at medium elevations. *E. geniculatus* was thought to be restricted to similar habitats close to sea-level (Miller and Morris 2004), but we have also found it at medium elevations (Fig. 6.139). Several species of *Fimbristylis* occur in wet flushes and by streams, the most common probably being *F. cymosa*. The annual grass-like sedge *Fuirena ciliaris* occurs in similar habitats. *Phyla nodiflora* is a common plant of wet areas, as are *Eclipta prostrata*, often associated with *Phyla*, and *Bacopa monnieri* (Fig. 6.140). The last of these occasionally forms dense mats on the surface of

Fig. 6.140 *Bacopa monnieri*. Homhil, December 2008

Fig. 6.141 *Ceratopteris cornuta*. Wadi Manifoh, March 1997

slow-flowing streams. *Echinochloa colona*, a widespread grass throughout many tropical and subtropical regions of the world, is commonly found in the vicinity of damp ground, as is *Paspalum vaginatum*. *Ludwigia palustris* appears to be fairly widespread in marshy ground by springs at all altitudes, although it is easily overlooked. Whereas *Ammannia baccifera*, *Ceratopteris cornuta* (Fig. 6.141) and

Lythrum hyssopifolium typically grow in mud by permanent freshwater at lower altitudes, the pantropical umbellifer *Centella asiatica* is found by seepages predominantly at higher elevations, especially on granite.

6.4 Vegetation Map of Socotra

Up-to-date maps of natural resources, including vegetation, are an important tool for effective decision-making in land-use planning, ecosystem management and conservation, and for academic purposes. A detailed map of the current vegetation of Socotra based primarily on floristic, rather than structural, attributes is not feasible at present due to the lack of adequate information.

One of the first attempts to map the vegetation of Socotra, at least on a very coarse scale, was produced by Gywnne (1968). A more refined map, comprising six main structural units, was presented by Mies (2001), who stressed its preliminary nature. It distinguished the following units: (1) halophytic vegetation/mangroves; (2) lowland scrub dominated mainly by *Croton socotranus* and/or *Jatropha unicostata*; (3) *Adenium* community; (4) mesic-montane thicket and riparian woodland; (5) montane scrub; (6) high-montane thickets and grasslands of the Hagghier mountains above 1,200 m. Furthermore, the map indicates the rough distribution of *Dracaena cinnabari* on the island. A broadly similar map based on seven main structural vegetation types was included in Miller and Morris (2004). A useful feature of this map is that accompanying panoramic photographs show a typical example of each major vegetation type. Král and Pavliš (2006) have recently published a land-cover map of Socotra containing a total of 21 classes, many of which are based on vegetation attributes. Vegetation classes in these maps are distinguished according to structural criteria, i.e. dominant life-form and vegetation structure, rather than species. As demonstrated by these authors, remote sensing is an invaluable tool in coarse vegetation mapping, not least in that it can provide a synoptic overview of large areas very quickly. Although the land-cover map by Král and Pavliš (2006) possibly provides the best overview of the vegetation structure of Socotra currently available, it is important to be aware of the limitations of remote-sensing technology, especially in arid environments with sparse vegetation cover, or with vegetation that may be largely dormant at certain times of the year or even for several years in succession. Some of these limitations were highlighted by the authors themselves, and relate to problems in differentiating between various structural vegetation types, but there are a number of other issues. For instance, in very sandy areas, it seems that vegetation cover is greatly underestimated. The authors, although noting that *Acacia edgeworthii* is present, assign the vegetation at Ra's Haulaf to the class "Bare soil". Our recent observations show that over large areas at Ra's Haulaf, perennial vegetation cover is as high as 20% (Fig. 6.142). In respect of discriminating among dominant plant species in the vegetation, this is often virtually impossible, even with high-resolution satellite imagery and when vegetation cover is high. For detailed vegetation studies in arid areas, we have found that high-resolution aerial photographs remain the best

Fig. 6.142 Fairly dense vegetation cover, dominated by *Acacia edgeworthii*. Ra's Haulaf, December 2008

option, even though their use may be feasible for relatively small areas only. Figure 6.143 shows a generalised vegetation map of Socotra based on the studies mentioned above and our own more recent observations.

6.5 Potential Natural Vegetation

As noted in Sect. 6.2, the concept of potential natural vegetation, i.e. the vegetation that would prevail in a given habitat if all anthropogenic influences on the site and in its immediate surroundings were to stop and the terminal stage of vegetation development were immediately reached (Tüxen 1956), may be somewhat controversial from an academic viewpoint, but can be of great practical benefit in conservation and land management. For instance, with ecosystem restoration, a knowledge of the natural climax vegetation may serve as a useful reference point on which to base remediation efforts. In this context, an understanding of vegetation dynamics and possible successional pathways is also desirable. However, Zerbe (1998) has expanded upon some of the pitfalls of relying too heavily on the potential natural vegetation for conservation purposes.

Identifying the natural climax vegetation of Socotra is problematic for a variety of reasons, including the following three. (1) Whereas in the past it was generally

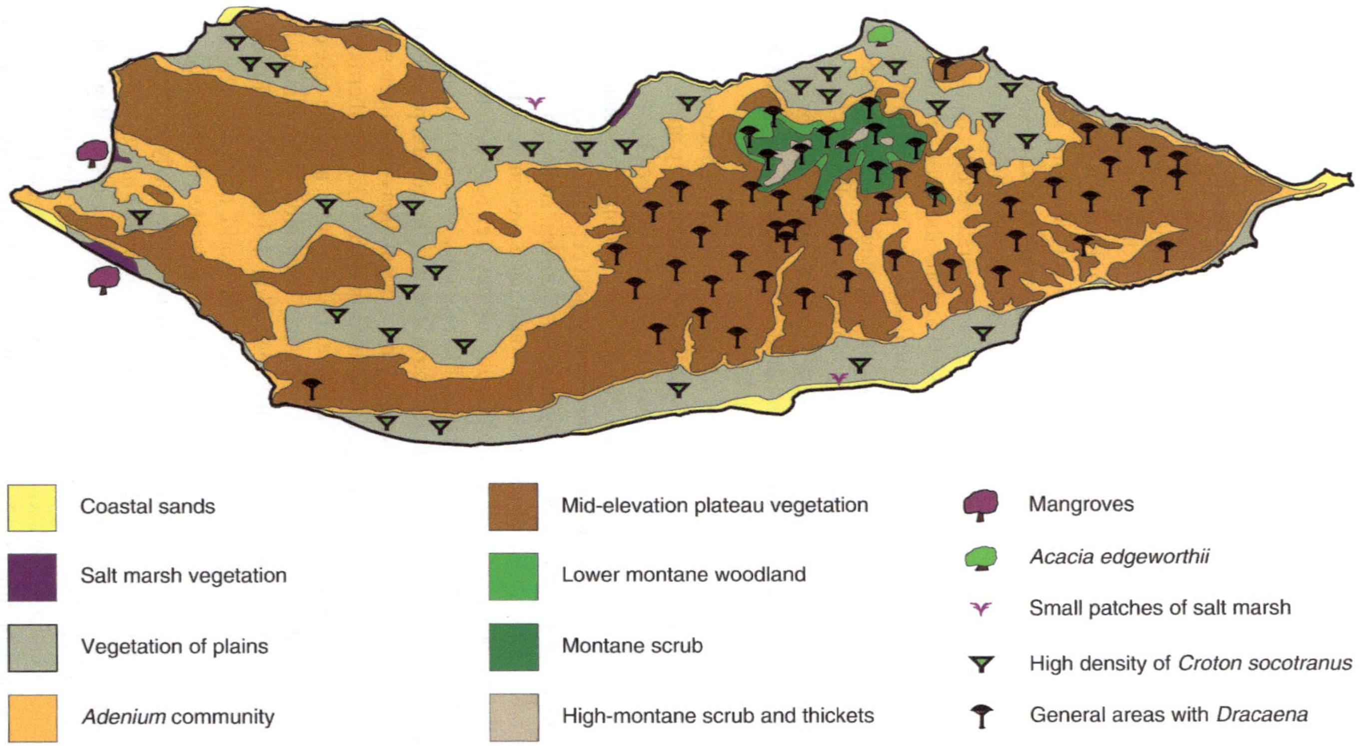

Fig. 6.143 Generalised vegetation map of Socotra. Coastal sands include localised patches of sabkha-like substrate and coastal dune fields. Vegetation of the plains includes predominantly the *Croton* community (mainly in the north) and *Indigofera–Salsola* community (in the south), but also the other minor communities mentioned in the main text. Mid-elevation plateau vegetation encompasses a broad variety of different vegetation types that cannot be mapped out accurately, including nearly barren land, annual and perennial grasslands, dwarf shrub and shrub communities, the *Buxanthus–Dracaena* community and other types of open woodland. Montane scrub and high-montane vegetation cover the various scrub and woodland communities, as well as cushion plant vegetation and anthropogenic grasslands

assumed that climax stages would be reasonably stable for centuries, it is clear that on Socotra climate patterns can shift quite rapidly, and this will obviously influence natural climax vegetation over a much shorter time-frame. For instance, if it is assumed that the climate is drying out rather quickly on the island, woodland will become much less common at lower altitudes where it still occurs today. As elaborated in Sects. 7.3.2 and 7.3.6, excessive anthropogenic pressures are likely to amplify strongly even very subtle changes to a more arid climate. More generally, because vegetation is by its very nature dynamic, the concept of potential natural vegetation has been criticised as merely representing a snapshot of a moving target. (2) It can be reasonably assumed that human influence on Socotra is likely to have been most intense in the more accessible areas. Even low-intensity grazing by domestic livestock over centuries can greatly modify vegetation structure. Attempting to reconstruct an accurate picture of the natural vegetation without human influence is therefore substantially complicated, especially as no accurate records on the condition of the vegetation at lower altitudes exists before the very recent past. Retrogressive succession in the sense of human-mediated ecosystem deterioration, i.e. degradation towards a simpler, more depauparate community, is now widespread at lower and middle altitudes. It can proceed very rapidly, as underlined in Sect. 7.3. (3) As mentioned in Sect. 7.3.1, if native megaherbivores were once present on Socotra, they would have helped to shape vegetation communities quite noticeably, in terms of both species composition and abundance.

It is clear that mangroves were much more widely distributed on Socotra in former times and that the loss of this intertidal woodland can be directly attributed to human destruction. Many other types of coastal vegetation, especially the more halophilic ones, possibly correspond largely to the natural climax vegetation, as they are usually of lower economic value. The same almost certainly does not apply to non-saline inland communities, especially at lower altitudes. Although it can be expected that some form of *Croton* shrubland would have predominated on the plains, as it still does today, "natural" *Croton* shrubland may well have differed from the current stands in several important respects. For example, it is easy to conclude that a probable major difference from natural *Croton* stands is the general paucity of economically important tree and shrub species, whose populations have been seriously reduced by excessive wood-cutting, as outlined in Sect. 7.3.3. Furthermore, where precipitation is reasonably high, i.e. above ca. 120–150 mm, as is the case on the north coast, perennial grasses would have possibly been more predominant in the ground layer, creating a more savanna-like vegetation. It has been well documented from semi-arid and arid environments around the world, quite dramatically for instance in South Africa (Acocks 1953), that even moderate grazing causes a decline in perennial grass cover, leading to the increase of small woody species (see Sect. 7.3.1). In addition, perennial grasslands in environments with strongly seasonal rainfall have been transformed to annual grasslands as a result of grazing (Westoby et al. 1989), and this seems to be the case on Socotra. However, studies from the USA have indicated that when grazing has ceased, perennial grass dominance has not always been re-established (Biswell 1956; Heady 1958), although observations from Kuwait (mean annual rainfall ca. 120 mm) show that in some highly degraded desert areas, perennial grasslands dominated by *Stipagrostis plumosa* with *Centropodia forsskaolii* can become fairly rapidly established once protected from grazing (Brown 2009). In the more arid south of Socotra, the relative proportion of perennial grasses was possibly always lower at the expense of dwarf shrubs, and vegetation cover would have been much more patchy than in the north.

The vegetation of the limestone plateaus has possibly suffered much more over large areas than that of the plains, because of the shallow, more fragile substrate, which is more readily eroded. It is conceivable that woodland communities would have been more widespread, not only in sheltered locations with deeper accumulations of soil, but also in more exposed areas. This is indicated by the presence of scattered older trees, especially *Dracaena*, which is apparently a very long-lived species. Adolt and Pavlis (2004) estimate the age of the oldest trees on Socotra to be in the region of 550 years, and the scattered individuals on the Dixam plateau to be about 350. It is also highly likely that perennial grasslands would have been quite extensive, but as with the situation in lowland areas, grazing has led to the conversion of vegetation dominated mainly by annual grasses. According to Naumkin (1993), the traditional use of these areas was for emergency grazing by herds from the coastal plains during periods of prolonged drought.

In forested areas at higher elevations, natural grasslands will probably exist only as a result of "catastrophic" events, such as landslides on mountain slopes, and will therefore be highly localised. Such impacts will abruptly revert the vegetation to an earlier successional stage, i.e. to one dominated by grasses. As these events are likely to occur with some degree of regularity and in different locations, they can be regarded as important drivers of allogenic cyclic succession. Grasslands are now more widespread due to anthropogenic degradation of woodlands and forests, and are maintained by grazing. Subsequent successional stages are characterised by scrub vegetation in which *Hypericum* and *Helichrysum* predominate.

As already outlined in Sect. 6.3.5, inaccessible cliffs, which abound on Socotra, can be expected to support mainly natural vegetation types. Where larger patches occur, for instance on larger cliff ledges or on remote plateaus, the vegetation there could provide a reliable indication as to the composition and structure of vegetation that previously existed over large tracts in the more accessible areas.

Chapter 7
Environmental Management

Abstract This chapter sets out some of the most pressing environmental problems affecting Socotra, and attempts to offer long-term solutions to some of these major challenges. "Desertification" is generally perceived to be a serious threat in the countries of the Arabian Peninsula, and many parts of Socotra at lower to mid-altitudes are affected by this menace. The drivers are the same as in many other parts of the world. Rapid population expansion, of both humans and livestock, and poverty are two of the most important of these. In former times, complex local land management systems controlling the access of livestock to the limited natural resources during the dry season largely prevented the over-exploitation of the rangelands. However, with changes in the lifestyle of the local inhabitants and the abandonment of traditional practices, the challenge now is to develop land management systems that are relevant to the present-day situation. Implementation and enforcement of the existing conservation zoning plan would be a substantial achievement. Even so, given the severity of degradation, habitat restoration will have to play an important role, and a clear concept of what is to be restored has to be developed, bearing in mind that vegetation regeneration can be a slow process in arid and semi-arid environments. Finally, the outstanding importance of islands such as Socotra for the conservation of global plant diversity is reiterated.

7.1 General Introduction

Many countries of the Arabian Peninsula have experienced a dramatic increase in wealth over recent decades, which has enabled them to exploit their natural resources, including water, at an unprecedented level (Brown et al. 2006). This in turn has had severe consequences for the rangelands, which occupy a substantial part of the area, accelerating the demise of traditional grazing systems that have maintained the productivity of the land for centuries, and negatively affecting biodiversity and biomass production. Coastal ecosystems have also suffered immensely from extensive infrastructure and agricultural development (Brown et al. 2008). Although Socotra

G. Brown and B.A. Mies, *Vegetation Ecology of Socotra*, Plant and Vegetation 7,
DOI 10.1007/978-94-007-4141-6_7, © Springer Science+Business Media Dordrecht 2012

belongs to a very poor region of the Arabian Peninsula, a variety of factors, including uncontrolled development, rapid population growth and the collapse of traditional land-use practices, have led to a marked deterioration of the landscape since the early 1990s. There is a generally held view that up until about 1994, which corresponds to a substantial increase in immigration from mainland Yemen (Mies 2001), the vegetation of Socotra was to a large extent relatively untouched, although this picture may be somewhat distorted, as outlined in Sect. 7.2.

This chapter sets out some of the most pressing environmental problems affecting Socotra, and attempts to offer long-term solutions to some of these major challenges, without jeopardising the desire of the local inhabitants for improved material wealth. In this pursuit of a better standard of living, it can only be hoped that the same mistakes will not be committed as elsewhere on the Arabian mainland and throughout many other parts of the world, often for short-term economic gain, but invariably with enormous negative long-term environmental consequences. In the longer term, it is more cost-effective to adopt a cautious approach to development than to have to initiate expensive environmental remediation measures at a later stage, which may or may not meet with the desired level of success. This applies particularly to Socotra, whose chief assets are its unique ecosystems, which, if carefully managed, will drive ecotourism and provide a major source of income. As underlined by the Millennium Ecosystem Assessment (2005), the creation of a "culture of prevention" can go a long way towards protecting drylands from the onset of desertification or its continuation. This approach requires a change in governments' and people's attitudes through improved incentives. The establishment of the Socotra Archipelago Conservation and Development Programme (SCDP) to coordinate sustainable development in the archipelago should be viewed as a positive initiative. The recently completed zoning plan, presented in brief below, should serve as a useful guide for sustainable development if implemented and enforced. However, due to the ever-increasing pressure on the lands, it should be stated that the initial signs do not appear overly positive, and general acceptance in the local population for conservation measures seems half-hearted at best. A recent assessment of the past and present human impacts on the biodiversity of Socotra also foresees a rather bleak outlook for conservation prospects in the archipelago (van Damme and Banfield 2011).

7.2 Habitat Degradation and Desertification

Although up until fairly recently Socotra had experienced virtually no development, and in the words of Alexander and Miller (1995) this lack of development meant that "Socotra is much as it would have been in prehistoric times", these comments should not be misinterpreted as indicating that the vegetation was in a pristine condition. It is highly unlikely that humans and their introduced livestock have had no impact on the natural vegetation of Socotra over the ca. 3,000–11,000 years since their arrival. As underlined by Le Houérou (2000), Mediterranean-type vegetation and

Fig. 7.1 Woodland clearing with montane meadow high in the Hagghier mountains. Near Adho Dimele, July 2002

ecosystems have undergone intense processes of degradation for decades, centuries or even millennia under prolonged pressure from humans and domestic livestock. It is the case that the inhabitants of Socotra developed complex systems to prevent over-exploitation of many – but not all – species. Despite these measures, it is inconceivable that livestock on the plains will not have modified the original vegetation cover to some extent, however subtly. At higher altitudes, the clearing of woodland and forest to create grazing land has probably taken place for as long as humans have inhabited the island (Fig. 7.1). From other islands around the world (e.g. Hawaii, the Canaries), it has been well established that anthropogenic plant extinctions occurred hundreds of years ago before the advent of scientific documentation (Whittaker and Fernández-Palacios 2007). An obvious example from Socotra where direct human impact has led to the massive decline of a keystone species and an entire ecosystem is given by the demise of the mangroves, which have been largely eradicated due to excessive wood-cutting (see Sect. 7.3.3).

However, with respect to most of the terrestrial ecosystems on Socotra, it would be fair to assess the effect of human intervention over the centuries as being relatively benign, especially compared to that on other islands such as St Helena (see Cronk 1989). As a very broad generalisation, it is probably the case that before the recent rapid deterioration of the landscape, the degree of human impact was inversely correlated with altitude, and on the lowland plains, where most of the island's inhabitants reside, the magnitude of impact declined with increasing remoteness from human settlements. Following the substantial environmental degradation of recent years, relatively pristine vegetation types are probably now found only in small parts of the island. Examples include some halophytic communities along the immediate coastline, for instance dominated by *Arthrocnemum macrostachyum* (an exception to the more general assertion that lowland vegetation types have been the most seriously impacted), the vegetation of steep cliffs, and some forest stands on inaccessible sites towards the highest elevations of the island.

"Desertification" is generally perceived to be a major threat in the countries of the Arabian Peninsula (Thalen 1979; Khalaf 1989; Barth 1999; Peacock et al. 2003; Brown et al. 2006), and many parts of Socotra at lower to mid-altitudes are affected by this phenomenon. The definition of desertification adopted by the United Nations Convention to Combat Desertification, and used by Middleton and Thomas (1997) in the *World Atlas of Desertification*, is "land degradation in arid, semi-arid and dry sub-humid areas resulting from various factors, including climatic variations and human activities". The basic indicator of desertification used in the *Atlas* is soil or land degradation. The term "desertification" is controversial (Verstraete 1986), and is especially confusing in regions where desert-like conditions already prevail. However, given that vegetation degradation can also be regarded as a useful indicator of desertification, it seems more appropriate to adopt the concept of Dregne (1986), who listed the key features of this menace as: (1) the decline in vegetation productivity; (2) the loss of biodiversity; and (3) loss of soil fertility and increase in aeolian activity. It is therefore possible to differentiate between vegetation degradation and soil (land) degradation, with the latter usually a direct consequence of the former. In the following, the term "ecosystem degradation" will be used to emphasise the joint impacts of vegetation and soil degradation.

The drivers of ecosystem degradation are the same on Socotra as in many other parts of the world. Rapid population expansion, of both humans and livestock, and poverty are two of the most important of these. As a biodiversity hotspot in the wider region, which has led to its designation as the "Galápagos of the Indian Ocean", Socotra has attracted substantial external funding, in part to help preserve its natural ecosystems. However, this funding has also contributed to accelerated ecosystem degradation by interfering with socio-economic factors that until very recently served to maintain a balance between the natural productivity of the land and the size of the livestock herds.

7.3 Direct Causes of Ecosystem Degradation

As complex entities consisting of multiple species and individuals, ecosystems display highly variable responses to different types of disturbance. Large disturbances may sometimes have no perceivable impact on ecosystem processes. Conversely, seemingly modest perturbations may be amplified to have quite significant effects (Brown et al. 1997). Furthermore, it is important to realise the synergistic effect of different impacts. Severe overgrazing is usually cited as the main cause of "desertification" on the Arabian mainland (Thalen 1979; Khalaf 1989; Le Houérou

Fig. 7.2 Overgrazed coastal plain. Wadi Manifoh near Hadibuh, March 1997

1996; Schuster 1998; Barth 1999), and this also applies to Socotra. In addition, the direct destruction of woody perennials for construction materials and fuel plays a major role, at least locally (Mies 2001). The local clearance of woodland and forest probably has a long history, so that it is now impossible to gauge the former extent of these vegetation types, and the negative effects of habitat fragmentation. Indiscriminate off-road driving is on the rise as the material wealth of a section of the population of Socotra increases, especially on the northern plains, but it has not yet reached anywhere near the same catastrophic proportions as in many places on the Arabian mainland. Overzealous infrastructure development, funded in part by international agencies, is also taking its toll on some highly sensitive habitats, leading again to substantial habitat fragmentation. Van Damme and Banfield (2011) have recently discussed this last aspect in some detail.

7.3.1 *Effects of Overgrazing*

The current nature of vegetation in many parts of the Socotra Archipelago has been determined to a large extent by traditional grazing regimes, and more recently by serious overgrazing (Fig. 7.2). Although it is possible that indigenous megaherbivores, which could have had a similar effect on the native vegetation as low-intensity grazing by domestic livestock, were never present in the archipelago due to its geographical isolation, their occurrence cannot be ruled out entirely. Von Heuglin (1861, cited from Wranik 1999) mentioned that gazelles occurred in the interior of Socotra itself, although this apparently remains the sole reference to large mammals. Indirect, albeit very weak, evidence for the former presence of native megaherbivores is forthcoming in that several endemics or near-endemics are quite tolerant of substantial levels of grazing, for instance *Ballochia amoena*. The very stunted and compact forms of this species (see Fig. 6.41) are highly conspicuous signs of the enormous grazing pressure to which the landscape is now exposed, particularly in the vicinity of settlements, and yet the species continues to thrive. The spine-tipped cladodes of the regional endemic Menispermaceae *Cocculus balfourii* (also known from the far south of mainland Arabia), which forms an impenetrable scrub in heavily grazed sites at higher altitudes, prove to be very effective at deterring livestock. This species may be regarded as a typical "increaser", in the rangeland ecology jargon, which also acts as an important nurse plant (Mies 2001). Even if larger herbivores once roamed the main island, their population density would have been tightly self-regulated (e.g. through recurring drought), and they would not have exerted anywhere near the same destructive influence on the natural vegetation as is now the case with the large herds of domestic livestock.

The direct effects of grazing in arid and semi-arid ecosystems vary according to intensity and duration. Apart from effects on the morphology of plants, as described above for *Ballochia*, even moderate grazing may cause shifts in dominant plant species, with less palatable species or ones more tolerant of grazing increasing at the expense of others favoured by livestock. It is tempting to speculate on how much of the island at lower to mid-altitudes, currently characterised by scrub or dwarf shrub communities, would have been covered by perennial grasslands, such as that still found in a very small area on Ma'alah, and dominated by the endemic *Andropogon bentii* (see Sect. 6.3.6.2), an important savanna genus on the African mainland, or by *Panicum rigidum*, a fairly common species at lower altitudes. As further evidence for the once more widespread occurrence of perennial grasses, Miller and Morris (2004) observed that one of the most important fodder grasses in the region as a whole, *Panicum turgidum*, was described as "abundant" near Qalansiyah in the nineteenth century, but today is known only locally from the Nuged plain.

The conversion of perennial grasslands to shrublands under mild to moderate overgrazing has been well documented in various arid and semi-arid parts of the world (e.g. Glendening 1952; Acocks 1953; Werger 1986; Westoby et al. 1989; Schlesinger et al. 1990; Brown et al. 1997; van Auken 2000). Shrub encroachment is often accompanied by a concentration of soil resources in certain microsites within the community, with fertile patches or "islands of fertility" located around the shrubs, and embedded in a less fertile matrix (e.g. Mott and McComb 1974; Shachak et al. 1998; Brown 2003). This progressive shift to shrub-dominated ecosystems is therefore often considered to exacerbate soil erosion and lead to desertification (Schlesinger et al. 1990; Stock et al. 1999; Millennium Ecosystem Assessment 2005). Whether this applies to the dominant *Croton socotranus* shrub community on the plains of Socotra, i.e. whether *Croton* has increased in dominance as a consequence of moderate grazing over centuries, is open to question. Mies (unpublished results) observed that nutrient and humus status, as well as soil texture, were more favourable for plant growth under *Croton* shrubs than in adjacent degraded areas.

Furthermore, even the less palatable woody vegetation (compared to perennial grasses) may experience shifts in composition of dominant species in response to grazing. This has been suggested by Brown (2003) for north-eastern Arabia, where in some sites, the dominant *Rhanterium epapposum* dwarf shrub community may have been replaced under moderate overgrazing by a less palatable *Haloxylon salicornicum* community. Le Houérou (2002) described the degradation of *Artemisia herba-alba* steppe in Tunisia to a community characterised primarily by *Anabasis oropediorum*. This shift in species composition, caused by overgrazing and wood-cutting, was also associated with a massive loss of topsoil. On Socotra, our observations suggest that in the south of the island, *Jatropha unicostata* appears to be increasing on finer-grained substrates at the expense of *Croton socotranus* in some moderately grazed sites (see Fig. 6.33). The precise reasons for this increase remain unclear, but it could be due to the fact that *Jatropha* is completely avoided by livestock, whereas *Croton* is at least browsed by goats when little else is available to them.

The perennial *Tephrosia apollinea* is one of the few species to benefit from the intensive grazing pressures because it is avoided by livestock. It is therefore not surprising that it is now one of the dominant species in areas worst afflicted by overgrazing, forming fairly monotonous stands in some sites (see Sect. 6.3.2.2).

The following example illustrates the trajectory of degradation on the Hadibuh plain over a 30-year period, from 1967 to 1998. This deterioration of the landscape can be attributed directly to an increase in grazing pressure brought about by a rising human and livestock population. The plain encompasses an area of 42 km^2, roughly 1% of the island. The study area proper is located in a stand of *Croton socotranus* on the Hadibuh plain from roughly sea-level up to the 200 m contour line. In 1998, the plain was home to ca. 3,000 inhabitants (Lavranos, pers. communication), with most of them living in Hadibuh itself. Livestock numbers in 1998 were estimated to be ca. 3,200 goats and 800 sheep. A vegetation map of the study area as it was in 1967 was inferred from photographs taken from the higher peaks of the Hagghier mountains by J.J. Lavranos (who kindly made them available to us – Fig. 7.3). In 1998, a more detailed vegetation map, also based on field studies, was made of the same area. In 1967, the village of Hadibuh covered an area of 0.30 km^2. By 1999, it had increased more than eightfold to 2.54 km^2. Over this 30-year period, there was a substantial reduction in the cover of scrub vegetation. *Croton* was most affected (reduced from ca. 24 to 15 km^2); the area occupied by *Jatropha unicostata* appeared to have remained constant. This is not surprising, as *Jatropha* occupies the slightly higher elevations with a more rocky substrate, or channels choked with coarse substrate that has been washed down from higher elevations. Such rocky areas are by their very nature more resilient to degradation. As a broad generalisation, typical features of the vegetation that were observed with increasing distance from Hadibuh (from about 50 m to 2.5 km away) were: (1) an increase in total vegetation cover from about 10% to 60%; (2) a substantial increase in vegetation productivity; (3) vegetation primarily characterised by small perennials, in particular *Tephrosia apollinea*, giving way to *Croton* scrub or *Croton-Jatropha* scrub; (4) a gradual increase in species number per 100 m^2 from as low as 2 to about 20.

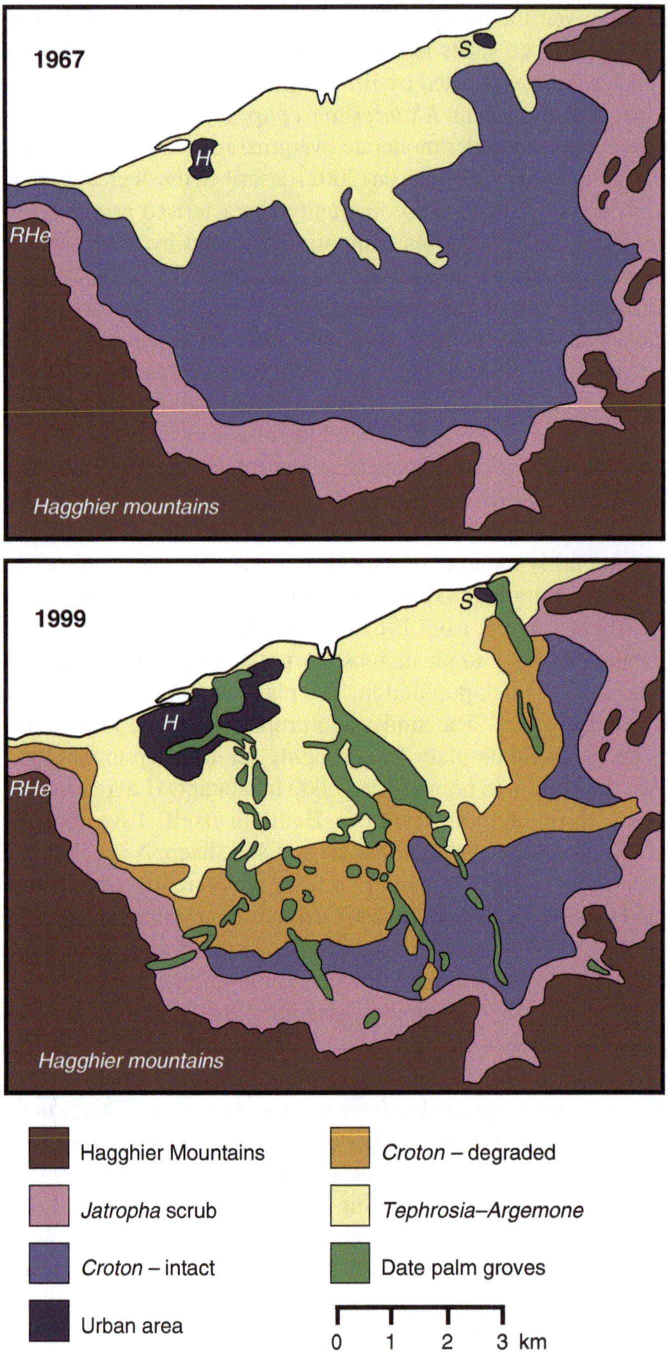

Fig. 7.3 Simplified vegetation map of the Hadibuh plain in 1967 (based on photographs provided by J.J. Lavranos taken from the adjacent mountains) and in 1999, highlighting the demise of intact *Croton* vegetation. The extent of date palm plantations is not shown in the 1967 map. Abbreviations: *H* Hadibuh, *RHe* Ra's Hebak, *S* Suq (Shiq)

Fig. 7.4 Young individual of *Dendrosicyos socotrana* growing between rocks. Wadi Dihzafaq, December 2008

The toxic succulent *Adenium obesum* is another species whose populations have undoubtedly increased as a direct result of massive overgrazing. This is in marked contrast to the situation of another prominent succulent, the endemic *Dendrosicyos socotrana*. Seedlings of this species are now observed only in microsites inaccessible to livestock, such as in pockets of soil between rocks or sheltered by other plants, especially spiny species (Fig. 7.4).

Serious overgrazing can lead to the large-scale eradication of perennial vegetation, as has occurred over wide expanses of northern Saudi Arabia and Kuwait (Brown, pers. observation). On Socotra, this is unlikely to occur at lower altitudes due to the massive increase in the unpalatable *Tephrosia apollinea*, but at higher altitudes, where *Tephrosia* is absent, there has been a catastrophic reduction in vegetation cover, with large tracts of land now virtually devoid of perennials (Fig. 7.5). The stark contrast between grazed open land and ungrazed exclosures at Homhil is clearly shown in Wranik (1999), and is also visible in Figs. 7.6 and 7.7, taken on the Dixam plateau. Brown et al. (2006) highlighted the exacerbation of desertification processes resulting from the gradual shift from nomadic to sedentary forms of pastoralism in the countries of the Arabian Peninsula as a direct consequence of increased water and fodder availability. In a similar vein, Scholte and de Geest (2010) hypothesise that it is not the absolute grazing pressure that leads to the lack of regeneration of certain plant species on Socotra, but increasingly reduced spatial and temporal grazing dynamics. The demise of traditional grazing systems on Socotra can therefore be viewed as a key cause of vegetation degradation.

Fig. 7.5 Barren landscape on the Dixam plateau. December 2008

Fig. 7.6 Exclosure to protect a *Dracaena cinnabari* nursery on the Dixam plateau. December 2008

Fig. 7.7 Comparatively dense vegetation cover with conspicuous grass cover in the *Dracaena cinnabari* nursery on the Dixam plateau. December 2008

More generally, Grainger (1992) noted that the worst forms of soil and vegetation degradation have probably taken place on lands where access is free and unregulated, resulting in what is frequently referred to as the "tragedy of the commons", a term coined by Hardin (1968).

Large herds of livestock also contribute to soil erosion, as discussed below. Wooded areas on the more inaccessible steeper slopes are also likely to suffer as the size of the herds increases. In comparable woodlands in Dhofar, well-worn tracks of livestock are visible as a dense network around most of the trees in some areas (Fig. 7.8).

7.3.2 *Soil Erosion*

In areas where the perennial vegetation has been seriously depleted, both on the plains and at higher altitudes in woodland and forest communities, the inevitable consequence is an increased risk of soil erosion and a loss of soil fertility, due to the combined effects of rain and wind (Fig. 7.9). Maintaining vegetative cover to protect soil from wind and water erosion is a key preventive measure against ecosystem degradation (Millennium Ecosystem Assessment 2005). Properly maintained vegetative cover also prevents loss of ecosystem services during periods of drought.

Fig. 7.8 Dense network of livestock tracks around most of the trees in Dhofar woodland. Wadi Darbat, Dhofar, Oman, March 2011

Fig. 7.9 Soil erosion in the Hagghier mountains. March 1996

Soil erosion is probably most acute on Socotra during the rainy season when large amounts of precipitation can fall within a few minutes. Not only is topsoil washed away, but so too is coarser material. As the topsoil usually contains most of the nutrients, these are also lost. In this context, Shachak et al. (1998) speak of "nutrient-leakage" as a characteristic feature of degraded arid ecosystems, resulting in a loss of soil fertility. A large part of this substrate gradually ends up in wadis, which can be transformed to raging torrents during heavy rainfall, and it is eventually deposited on the plains in both the north and south of the island, or even washed out to sea. The extent of soil erosion – at least locally – can be gauged by the account given in van Damme and Banfield (2011) from the Shibehon plateau, where scientists found that in 2003, substantial amounts of soil had accumulated in the main galleries of the Ghiniba Cave, which had not been observed there just two years previously.

Soils with reduced perennial vegetation cover are also highly susceptible to rapid wind erosion. Trampling by livestock and off-road driving accelerate the process of wind erosion in that they cause a partial compaction of the surface during wet periods, but during the dry season, soil particles are loosened and removed by the wind (Webb and Stielstra 1979; Novikoff 1983). Clouds of dust created by off-road vehicles and by large herds of livestock are a common sight in desert environments, and this is particularly visible on the plains on Socotra. Pandey and Singh (1991) showed that in arid regions, intensive grazing leads to a decrease of the fine soil particles (<0.1 mm), while the coarse fraction (0.5–2.0 mm) increases.

Once land degradation has progressed, organic material is usually restricted to pockets in protected microsites, such as in the shelter of boulders or stones, in rock crevices and around remaining perennial vegetation. Substrates lacking organic material dry out more quickly, and are therefore less conducive to germination and establishment of many perennial species. Without a protective layer of perennial vegetation cover, the soil surface heats up more intensely, and becomes completely dry and hard, cracking open in places. Due to reduced infiltration, rainwater flows more quickly over the surface, but it also enters cracks where it rapidly washes away any softer material. At higher altitudes, the resulting erosion channels form smaller gullies that merge into the larger wadis (Fig. 7.10). Any water that does penetrate the ground is usually held in the surface layers, where it is exposed to rapid evaporation. Increased run-off and evaporation are the two processes that accelerate loss of water from the ecosystem, with negative consequences for seed germination and recruitment.

As a consequence, even if the factors that caused degradation are removed, full vegetation recovery can be a slow process in arid ecosystems, because soil formation proceeds very slowly, if at all. Loss of critical soil resources therefore often results in irreversible (i.e. over a period of 25 years at least) ecosystem degradation. This is also the case in the climatically more favourable higher montane areas on steep slopes. Once severely degraded, the land becomes much less productive, and it is of use to livestock only if it supports a rich annual flora, or if supplementary fodder can be brought in from outside. This pathway of degradation can unfortunately be observed in many parts of the world, and on Socotra, the demise of *Boswellia elongata* woodland at fairly low altitudes provides a clear example, as outlined in the following.

Fig. 7.10 Erosion channels in the Hagghier mountains. March 1996

Eight species of *Boswellia* have been described from Socotra, all endemic, and it is possible that some forms of these species readily distinguished in the field may turn out to be distinct taxa in their own right. Of these eight, five are typical cliff-dwelling species that are also found on huge boulders, while the other three are small, ground-dwelling trees. *Boswellia elongata*, apparently the most abundant member of the genus on Socotra (Attorre et al. 2011), belongs to the latter group, and is highly characteristic of forest in deep sheltered valleys at lower altitudes (200–300 m), such as in Wadi Ayhaft. Such stands probably represent the primary habitat of the species, especially as it is typically accompanied by a variety of other small trees. In some cases, a dense understorey of lianas, shrubs and herbs is also present (Fig. 7.11). Characteristic woody species include *Ballochia amoena*, *Buxanthus pedicellatus*, *Carphalea obovata*, *Commiphora ornifolia*, *Cordia obovata*, *Lannea transulta*, *Trichocalyx obovatus* and others, among them occasionally the emergent *Sterculia africana*. Kürschner et al. (2006a) placed these stands in a distinct subassociation (boswellietosum elongatae) of the Adenio-Sterculietum socotranae, but as argued in Sect. 6.3.4.1, this classification is somewhat debatable. *Adenium obesum*, after which the phytosociological unit is named, is primarily a succulent of coarse substrates in exposed locations, and as a typical "increaser" is able to gain a foothold on finer substrates as an indirect result of overgrazing. With the more intensive grazing in such formerly relatively remote sites, degradation of the woodlands is an all too conspicuous phenomenon. The thinning of the understorey, and the increase of grazing-resistant species such as *Tephrosia apollinea* and *Senna socotrana* in particular, point to a gradual

Fig. 7.11 *Boswellia elongata* in fairly dense woodland. Wadi Ayhaft, December 2008

deterioration of the woodlands (Figs. 7.12 and 7.13). *Adenium*, which can occur naturally in such woodlands on rocky slopes, outcrops and cliffs, is able to expand its local range to colonise finer-grained substrates. This process of deterioration is probably accompanied by a loss of soil resources due to reduced ground cover, further diminishing the prospects for natural regeneration of the larger trees. The final stage of degradation of *Boswellia elongata* woodland can be clearly seen in the raised valley at Homhil on the south-western side of the Hamaderoh escarpment (Fig. 7.14). Miller and Morris (2001) regarded these stands as relicts of a former, more widely distributed woodland. This is almost certainly the case, and although the gradual drying out of the climate may have played a minor role in preventing regeneration of the trees, which are all of the same broad age class and have reached their maximum size, we are convinced that – in accordance with the general statement of Le Houérou (1996) outlined in more detail below – the primary reasons for this degradation are anthropogenic, and far outweigh any possible effects of climate change. At Homhil, we have been unable to detect any signs of natural regeneration of the *Boswellia* trees, which now exist in a park-like landscape. More detailed studies by Attorre et al. (2011) over a wider area support our observations, suggesting a general poor degree of natural regeneration. Conversely, *Adenium* is one of a handful of species avoided by livestock that is on the increase, as evidenced by numerous young plants (Fig. 7.15). Our initial thoughts on degradation at Homhil have been recently corroborated by soil studies conducted by Pietsch and Morris (2010). It appears that enhanced soil erosion, specifically

Fig. 7.12 *Boswellia elongata* in somewhat degraded woodland. Wadi Ayhaft, December 2008

Fig. 7.13 Seriously degraded woodland. Wadi Ayhaft, December 2008

the removal of humus-rich topsoil and associated loss of soil structure, can be linked to the establishment of permanent settlements in the Homhil raised valley floor in the early 1960s, leading to a more intensive grazing regime in the area.

Fig. 7.14 Seriously degraded *Boswellia elongata* woodland. Homhil, December 2008

Fig. 7.15 *Adenium obesum* with several young plants – a typical "increaser" in degraded woodland. Homhil, December 2008

Fig. 7.16 *Croton socotranus* is regularly harvested. Near B'ir Haarso, east of Qalansiyah, December 2008

7.3.3 Wood-Collecting and Deforestation

Warren et al. (1996) observed that, particularly in the poorer arid regions of the world, to which Socotra belongs, the cutting of wood for fuel and construction purposes can have a significant impact on the vegetation. On Socotra, the main source of fuel for the majority of its inhabitants is wood. Only in the larger settlements such as Hadibuh and Qalansiyah has there been a marked increase in the use of gas in recent years. Accordingly, most large woody species tend to be more common with increasing distance from settlements. However, a few woody species close to villages are left standing for other purposes. For instance, the nutritious leaves of various members of the Acanthaceae are an important source of browse for livestock. *Croton socotranus* is one of the most widely used species, and serves a variety of purposes. The slender woody trunks of older plants make excellent timber, which is regularly harvested (Fig. 7.16). The roofs of most huts on the island are constructed from *Croton* trunks. In addition, the trunks are extensively used as fence-posts or to form the framework of a fence, which is then further protected either by the planting of spiny shrubs, or by reinforcement with other materials such as netting and metal (Fig. 7.17). *Croton* also provides an excellent renewable source of fuel in the form of firewood. Although *Croton* is on the decline in many areas, other species are faring much worse. One of the most conspicuous casualties is *Dendrosicyos socotrana*, whose populations have dwindled locally, because it is used as a source of animal

Fig. 7.17 Native species are still used extensively for fencing. Hadibuh plain, December 2008

fodder during periods of drought. The succulent branches are lopped, pulped and then fed to goats. In the absence of more palatable fodder, herdsmen on Socotra will also feed the succulent cylindrical stems of *Euphorbia arbuscula* to their goats. In contrast to *Dendrosicyos*, *Euphorbia* is occasionally propagated by cuttings specifically to be used as dry season fodder, but as the species is probably still declining, this measure does not seem to be compensating for the increased pressure on it as a result of expanding livestock populations.

The scrub and woodland vegetation of the wadis and higher latitudes plateaus was until recently fairly well protected from over-exploitation by the remoteness of many of these areas and the restricted access for motor vehicles. The local inhabitants collected wood mainly for their own personal requirements. However, with the rapid expansion of graded tracks and roads, and the more widespread use of off-road vehicles, there has been a marked deterioration of the situation. Local inhabitants are now able to supplement their income by harvesting wood and selling it elsewhere. Mies and Beyhl (1999) estimated that if the then current rate of wood-cutting continued unabated, large parts of Wadi Qalansiyah would be devoid of trees and large woody species by about 2020. Unfortunately, the experience from adjacent arid regions of Africa supports this bleak outlook.

The loss of trees is perhaps best illustrated by the demise of mangroves. Fig. 7.18 shows intact mangrove vegetation near Gubbah on the north coast in 1967. This site – along with others – has been completely destroyed. Mangroves are of enormous ecological significance for various reasons (e.g. Ellison 2000). For instance, they help to retain sediments and nutrients washed down from wadis

Fig. 7.18 Dense mangrove stands at Ghubbah in 1967, now destroyed (Photograph by John Lavranos)

and as such are highly productive habitats on which many marine species depend. On Socotra, it seems as if the destruction of the mangroves has also led to a concomitant decline of adjacent coral reefs due to smothering with sediments that the trees would have otherwise filtered out, again with negative repercussions for marine wildlife.

Less well documented, but with effects on the natural vegetation cover probably as dramatic as those of the demise of the mangroves, has been the destruction of woodland and thickets to provide grazing for domestic livestock at mid- to high altitudes (Fig. 7.19). This has probably persisted for centuries, and as a consequence, woodland and forests are now usually restricted to the steeper slopes. Clearance of the forests has led to massive soil erosion, as described above, and also to habitat fragmentation, with potentially disastrous consequences for individual species and entire vegetation types. Dwarf cattle are the predominant livestock in the Hagghier mountains, but goats are also held, possibly increasingly so. In former times, the movement of livestock to higher or lower regions of the mountains was tightly controlled, but this system appears to be breaking down, especially as more permanent settlements become established in the mountains. Despite these changes, the population at higher altitudes is possibly decreasing as people move to the plains in search of employment. However, it is unlikely that forest will recover due to the loss of substrate, and also the increase, at least locally, of feral goats that are preventing natural regeneration of the woody vegetation.

Fig. 7.19 Deforestation and erosion on slopes high in the Hagghier mountains. Near Adho Dimele, July 2002

7.3.4 Infrastructure Development

The current population of Socotra barely exceeds 50,000 inhabitants, and it will undoubtedly continue to rise over the next decades. A modern road network is therefore essential for the local inhabitants, who were until recently dependent on very simple means of transport. The first tarmac road was completed in about 2002. Up until then, access to the various parts of the island was at best by rough graded tracks. These roads will bring many benefits to the local population that people in other parts of the world take for granted (Fig. 7.20). However, in several cases their planning has left much to be desired, and some road schemes are in no way justified by the actual demand. A good example is the dual carriageway around the Hadibuh, where the amount of traffic in the area barely exceeds 60 vehicles per hour at peak periods (Fig. 7.21). Such projects have not only caused considerable habitat destruction, but have also wasted development aid that could have been put to better use. Perhaps even more insidious is the planned construction of roads in more sensitive sites, where valuable habitats with extremely rare species may be destroyed. In such cases, close cooperation with the appropriate conservation authorities is absolutely imperative so that, if necessary, compromises can be sought.

Apart from the direct impacts of roads in destroying or deteriorating habitats, including the dumping of unsightly construction waste and the disposal of rubbish

Fig. 7.20 Construction of a new road on the Nuged plain. December 2008

Fig. 7.21 The dual carriageway east of Hadibuh. December 2008

immediately next to the roads, there are indirect consequences that are often just as serious or even more so. For instance, with the new Dixam road linking the north and south of the island (Fig. 7.22), a major infrastructure project has been recently

Fig. 7.22 The new Dixam road links the north and south of the island. December 2008

completed that has opened up formerly remote areas to increased grazing, as livestock can now be readily transported to these sites.

Furthermore, such roads will inevitably encourage more tracks and off-road driving in the surrounding areas. Off-road driving has long been recognised as a major deleterious factor in arid regions, causing widespread damage to the vegetation and producing tracks on the soil surface (Adams et al. 1982). Due to the fragility of arid environments and the slow rate of recovery of many native perennials, the damage caused by even isolated incidents may be visible for many years (Lathrop 1983; Le Houérou 1996). Freely available satellite imagery clearly shows the myriad of tracks criss-crossing the Hadibuh plain (Fig. 7.23). The effects of off-road driving in accelerating soil erosion are discussed in Sect. 7.3.2.

A further consequence of tarmac roads is that, due to the considerable disturbance caused during their construction, they facilitate the colonisation of some invasive or expansive species, as described below (Sect. 7.3.8).

The construction of roads to remote areas may increase pressure on the land there and inadvertently facilitate unscrupulous developers who wish to push through ill-conceived and ecologically disastrous development projects. This has been the case in the Canary Islands, which from a social and ecological perspective are in many ways comparable to Socotra. The recent boom in waterfront development schemes on the Arabian mainland does not bode well for Socotra in the longer term. It is worth remembering that in undeveloped, poor regions, it is usually external or foreign entrepreneurs who are the primary beneficiaries of such development schemes rather than the local inhabitants.

Fig. 7.23 Off-road driving is becoming a serious problem, as witnessed by the numerous vehicle tracks in the landscape. December 2008

7.3.5 *Impacts of Drought*

It is important to realise that drought, even if it occurs several years in succession, is not the cause of "desertification". In fact ecosystem degradation can proceed rapidly even under favourable climatic situations (Le Houérou 1996, 2002). There is ample evidence to show that natural arid-land vegetation, which has evolved under the overriding need to conserve water, can withstand prolonged periods of drought with virtually no damage, provided the anthropozoic impact is light, although vegetation productivity may be extremely low in dry years. It should also be borne in mind that cycles of drought lasting for several years are a recurrent feature in the arid regions of the Middle East, and probably on Socotra. However, in combination with the serious effects of anthropogenic degradation on Socotra from overgrazing, even short-term changes in rainfall characteristics can be expected to be highly detrimental to vegetation productivity and species composition, and to affect recruitment of keystone species, as discussed below (see Sect. 7.5.1). The inherent natural resilience to drought of arid-land plant communities is therefore seriously compromised due to anthropogenic ecosystem degradation.

7.3.6 *Potential Impacts of Climate Change*

There has been some speculation on the potential impacts of climate change on species composition and productivity on Socotra, without any conclusive evidence provided

Fig. 7.24 Water is intercepted in the upper reaches of the wadi and diverted to date palm groves near the coast. Wadi Ireh, December 2008

(e.g. Attorre et al. 2007). Without wanting to belittle the magnitude of the problem, we concur with Le Houérou (1996), who, in consensus with numerous scientists who have long been engaged in fieldwork in various arid regions of the world, underlines the fact that in arid and semi-arid regions, the potential effects of climate fluctuations or change are likely to be trivial compared with the past and ongoing impacts of humans and their livestock. The stark contrast between areas that are protected from grazing and those that are not provides more than sufficient insight into the overriding causes of ecosystem degradation on Socotra (see Sect. 7.5.1, Fig. 7.32), and examples also abound from the Arabian mainland (e.g. Brown and Al-Mazrooei 2003). A further discussion of this topic is given in Sect. 7.5.1.

7.3.7 Water Extraction

In some wadis, water is extracted from permanent streams near the source and piped down to the plain for irrigation of the expanding date palm plantations (Fig. 7.24), as well as for human consumption. In some instances, a situation is developing very similar to that in the Canary Islands, where a number of barrancos that would normally have permanent watercourses are now largely dry throughout the year due to the interception of water upstream to support agriculture and to satisfy the needs of a large number of tourists. The reduction of water discharge in these wadis on Socotra will inevitably lead to changes in physical habitat features (such as water depth, decrease in area wetted by the streams), disrupt natural patterns of stream dynamics and result in the decline in the populations of some plant species (apart from seriously depleting the stream fauna). Furthermore, the reduction of water flow may also have negative impacts on the microclimate in these wadis, thus impacting the survival of other species.

7.3.8 Invasive Species

The invasion of ecosystems by non-native species is a global phenomenon, which in some cases can cause serious ecological and economic problems. The impact of such introduced species on islands can be particularly insidious if so-called "transformers" are involved, species with the potential to alter entire ecosystems (Richardson et al. 2000; Henderson et al. 2006). At present, Socotra has been colonised by only relatively few invasive plant species, and many of these are pantropical annual weeds. *Prosopis juliflora* (mesquite), which has wreaked havoc in parts of mainland Arabia, has been the subject of a targeted programme of eradication (Miller and Morris 2004), and the potential threat from this species appears to have been thwarted. Perhaps the most conspicuous invasive is the Papaveraceae *Argemone mexicana*, an annual herb native to Central America, but now a pantropical weed. This noxious plant is generally toxic to livestock, although goats in particular will eat young plants when little else is available. In accordance with the general theory of Davis et al. (2000) and outlined in Sect. 4.2.4.5, the potential for invasion by alien species on Socotra is perhaps greatest in disturbed woodland areas where unused resources are likely to be available to a much higher extent, and less in the more arid ecosystems. The fact that *Argemone* has been so successful in colonising arid habitats underlines its pernicious nature. It seems unlikely that this species will be successfully eradicated, especially as it has probably been able to increase dramatically due to overgrazing, which has had the effect of reducing resource use by the native vegetation. *Calotropis procera* (Sodom's apple) can also probably be regarded as an invasive species, even though it is native to the region as a whole. It is a characteristic species of open habitats where there is little competition from other plants, including coastal dunes. As noted in Sect. 4.2.4.5, it is now spreading along the newly constructed roads to various parts of the island where it did not formerly occur, as witnessed in December 2008 by the large number of young shrubs that were observed at regular intervals along the roads in both the north and south at lower altitudes. An increase in resources can be explained by the disturbance caused during construction of the roads and destruction of the resident vegetation, allowing *Calotropis* to invade. This species, too, is avoided by livestock, and is therefore a clear indicator of ecosystem deterioration. The effects of *Calotropis* invasion are clearly visible in Dhofar, southern Oman, where it has colonised huge tracts of the escarpment summit that were once covered in grasslands 1.5 m high and dominated by species such as *Apluda mutica*, *Themeda quadrivalvis* and others (see Figs. 7.25 and 7.26). At present, the woodland and forests at higher altitudes on Socotra do not appear to have been

Fig. 7.25 *Calotropis procera* on the highly degraded former grassland plateau. In the background, an exclosure with a predominance of grasses in the ground layer can be seen. Dhofar, Oman, March 2011

Fig. 7.26 Exclosure on the plateau with a predominance of grasses in the ground layer. Dhofar, Oman, March 2011. See corresponding situation on Socotra, as shown in Fig. 7.32

Fig. 7.27 Egyptian vultures picking through rubbish dumped in the landscape. Qalansiyah, December 2008

affected by invasive aliens, possibly because of their relative remoteness. If the area continues to be opened up by future development projects, the monitoring of undesirable species will need to be taken seriously.

Van Damme and Banfield (2011) give a more exhaustive account of the potential effects of exotic species, both flora and fauna, on the ecosystems of Socotra.

7.3.9 Rubbish

The indiscriminate dumping of rubbish in the landscape currently remains a fairly localised problem, albeit one that has the potential to become much more serious as the population expands and urbanisation increases. The problem of macro-waste is most evident in the larger villages (Fig. 7.27), and apart from any other considerations, it is an eyesore that may deter tourism. Education and public awareness programmes are vital if this problem is to be tackled at an early stage.

7.4 Regeneration of the Natural Vegetation

Rigorous protection of entire ecosystems is generally regarded as the most effective approach to *in situ* conservation. Despite the high degree of protection that the archipelago now enjoys, much of the vegetation in the lower and mid-altitudinal range has become seriously degraded. Regeneration of the vegetation is therefore required to reinstate natural ecosystem functioning, including plant–animal interactions, and also to justify the conservation status of the archipelago. In accordance with the terminology of Le Houérou (2000), regeneration of degraded arid and semi-arid ecosystems can be achieved by two main approaches: (1) restoration, which implies the regeneration of an ecosystem to its pre-degradation or pristine condition; and (2) rehabilitation, which is the reclamation of degraded land to yield an ecosystem different from the natural pre-degradation situation in terms of floristic composition. This is usually accomplished through the planting of exotic or subnative species. From an ecological perspective, rehabilitation is generally not an acceptable solution, although there are examples from various parts of the world where it has been necessary first to establish non-native vegetation in highly degraded and altered habitats that are not capable of supporting native plant species. This is not the case with Socotra, and so need not be discussed further, except to emphasise that with both restoration and rehabilitation, the removal or serious reduction of the causes that brought about degradation are crucial for a successful outcome of any remediation measures.

7.4.1 Restoration

There is an urgent need for science-based methodological concepts and scientifically tested techniques for the restoration of habitats (Bainbridge 2007; Miller and Hobbs 2007), and this also applies to Socotra, where virtually no work has been done. At the onset of any restoration activities, the key components of a given habitat to be restored need to be clearly articulated, taking into account the available resources and constraints. A clear concept of what is to be restored should therefore be developed, bearing in mind that vegetation regeneration can be a slow process in arid and semi-arid environments (Bainbridge 2007). In arid rangelands, where grazing is the main factor leading to degradation, vegetation restoration can be achieved by two main procedures: (1) exclosure, which means that the area under consideration is protected from human and livestock intrusion, usually by fencing, and allows for natural vegetation recovery; and (2) assisted revegetation, which refers to the transplanting or seeding of native species, usually keystone perennials, in order to speed up the trajectory of recovery to the desired state. Some form of protection from humans and livestock, at least on a temporary basis, is usually required to ensure success within a reasonable time-frame. Exclosures can be established as a temporary, semi-permanent or permanent measure. In the last of these cases, they can be opened to pastoralists for limited periods of time during periods of drought. On mainland Arabia, the traditional use of semi-permanent exclosures was once widespread to protect the land from overexploitation, but this "hema" system has now become largely obsolete, except in a few cases. On Socotra, a comparable land management system to promote sustainable use of the natural vegetation cover was still in widespread use up until fairly recently.

As described by Le Houérou (1986), vegetation often responds very favourably to permanent exclosure, even in arid deserts, especially in places where the substrate is deep and fairly stable. Depending in part on the aridity of the region, and providing degradation has not reached an advanced stage, exclosures usually facilitate the restoration and biological recovery of vegetation structure, composition, biomass and productivity within a time-span of 3–5 years in steppe-like ecosystems (Le Houérou 2000). A striking example of rapid recovery of the dwarf shrub vegetation on the Arabian mainland in response to exclosure has been given by Brown and Al-Mazrooei (2003).

In some cases, it has been shown that tightly controlled access and limited grazing may produce even better effects than complete long-term exclusion of megaherbivores (Le Houérou 2000), especially if recovery has proceeded beyond an ecosystem-specific threshold. Very light grazing may prevent the excessive development of individual plant species in years of high rainfall, which in subsequent years may abruptly die back due to lack of water (Le Houérou 2000). In the longer term, it may therefore be possible to develop grazing programs that enhance the productivity of both the overall landscape and the livestock, as was underlined by Krueger (1990). A full understanding of the processes limiting biomass production and diversity is therefore essential to the overall management of degraded arid ecosystems (Shachak et al. 1998).

At advanced stages of vegetation deterioration, merely removing the anthropogenic stressors by establishing exclosures that restrict the access of livestock is usually insufficient to halt or even slow the continuing processes of degradation in arid and semi-arid areas. This is particularly the case in heavily impacted sites, where not only has the vegetation been heavily degraded, but also critical soil resources have been seriously depleted (see, for example, Thalen 1979). In such instances, the seedbank of key species is also likely to have been greatly reduced or even eliminated. According to Le Houérou (1996), desertification is irreversible (i.e. the vegetation will not recover to its pristine condition even after 25 years of total protection) where the environment is drier and where the soil is more shallow. In such extreme cases, a new metastable equilibrium, characterised often by unpalatable weeds, perennial dwarf ephemeroids or even cryptogams, may set in. Several authors, including Knapp (1991) and Webb and Wilshire (1980), have found the degree of soil compaction to be a major determinant of the rate of plant recovery in semi-arid landscapes in the USA dominated by *Larrea tridentata*, with compacted soils showing the slowest recovery rates. In an arid desert ecosystem in northern Kuwait (mean annual rainfall ca. 120 mm), there was no discernible natural vegetation recovery on the most compacted substrates where the topsoil had been lost after four years of protection from livestock grazing (Brown and Al-Mazrooei 2003).

The following case study serves to illustrate the limitations of natural vegetation regeneration under persistent grazing on Socotra. In his studies of the vegetation of the airfield east of Hadibuh and in the vicinity of the village itself, Popov (1957) outlined what he considered to be the probable main pathway of recovery from barren ground to what he assumed to be the climax *Croton socotranus* shrubland community. On ground that had been initially cleared of vegetation and subsequently abandoned, he observed that *Tephrosia apollinea* and various annuals were the first colonisers, followed by perennials such as *Corchorus depressus* and *Cucumis prophetarum*. Once these species had become established, *Panicum rigidum* gained a foothold, followed by the dwarf shrub *Indigofera* sp. Popov (1957) also noted that heavy grazing by goats and sheep slowed down the succession process, and that after eight years, only a few *Croton* shrubs had become re-established. Over 40 years later, our observations suggest that *Croton* has still not managed to regain its former dominance. In fact it appears as if species composition on the airfield has remained much the same as when Popov carried out his study, and must therefore be regarded as a persistent anthropogenic assemblage that has resulted from severe grazing pressure. Various authors (e.g. Westoby et al. 1989) have suggested that rangeland ecosystems have distinct alternative stable states, and that rather than gradual changes, transitions from one state to another can be quite abrupt. Transitions to a low state (invariably of lower biomass and diversity) take place once a critical threshold level of disturbance, brought about by climate or management, has been reached. The concept of thresholds of environmental change between relatively stable states has also been described by other authors (e.g. Holling 1973; Friedel 1991; Laycock 1991). If stability is resistance to change imposed by external forces, then a system is regarded as stable if it returns to its original steady state after experiencing disturbance. An unstable system or state does not revert to its prior state, but rather crosses a threshold and assumes a new state (Laycock 1991). Friedel (1991) noted that once a threshold is crossed to a more degraded state, the former state cannot be reinstated without substantial management effort. Most importantly, transitions to higher states are the most difficult to reinstate, as indicated for north-eastern Arabia by Brown and Al-Mazrooei (2003).

To accelerate natural vegetation regeneration, a number of authors, including Vetaas (1992), have indicated the importance of microsites created by trees, bushes and dwarf shrubs for plant establishment. In highly degraded semi-arid and arid habitats the shelter afforded by these woody species is also beneficial to a number of other species, protecting them from high levels of irradiance. Furthermore, litter produced by such shrubs enriches the soil with nutrients and improves the soil-water status (Charley and West 1975; Shachak et al. 1998). Tongway and Ludwig (1996) described a successful procedure for restoring productive soil patches in semi-arid landscapes in Australia, which involved laying piles of branches of native woody plants (i.e. artificially creating microsites) in open patches and allowing soil, litter and water to accumulate there. The same technique also proved feasible in restoring vegetation patches in the same area (Ludwig and Tongway 1996). If adequate protection from grazing is afforded, similar techniques could prove to be a viable option for accelerating vegetation recovery in many degraded habitats on Socotra.

7.4.2 Native Plant Production

In some cases, regeneration of the natural vegetation can proceed quite rapidly under favourable circumstances, but more often, ecological restoration of native plant communities in desert ecosystems requires seeding or the transplantation of plants to ensure biological recovery within a reasonable time-frame. The origin of this plant material and its use in revegetation activities are of critical importance to maintaining the genetic integrity of local populations (Hufford and Mazer 2003). It is well established that habitat heterogeneity, combined with natural selection, often results in genetically distinct ecotypes within a single species. This is perhaps most evident in plants of extreme habitats, for instance, in the case of heavy metal grasslands, where metal-tolerant and non-tolerant ecotypes often exist in the immediate vicinity of each other (Antonovics et al. 1971; Simon 1978; Brown and Brinkmann 1992). Recent advances in molecular techniques, reviewed in Hufford and Mazer (2003), have underlined that this is also the case for a wide variety of plant species. In the context of island floras with a high degree of endemism and isolated populations, this aspect is of even greater importance in preserving the genetic make-up of local populations. Given that our taxonomic understanding of the flora of Socotra is still rather limited, especially regarding the status of plants below the species level, extreme caution is warranted when attempting to re-establish plants in other areas. Not only should the use of non-Socotran germplasm be banned, but also the translocation of germplasm within the island should be avoided to protect local populations of plant species that may represent distinct local ecotypes, some of which may even transpire to be separate species. For instance, if augmenting or re-establishing *Dracaena*, *Boswellia* and *Commiphora* populations is identified as a conservation priority, this should be achieved using seed or plants of the relevant local population only.

A convenient way of ensuring sufficient seed and plants from specific locations are available is to propagate the species involved in specialised nurseries. On Socotra, a small nursery experienced in the production of rare and endemic species that can be used for restoration work has long been established (Alexander and Miller 1996), and it is successfully growing numerous plants (Fig. 7.28). It is interesting to note that Mies et al. (2000) suggested that in ancient times, some *Boswellia* species, most notably *B. elongata* and *dioscorides*, were planted in nurseries by the local inhabitants because of their importance for resin collecting (Fig. 7.29).

7.5 Conservation of the Flora of Socotra

Oceanic islands are of particular importance for the conservation of global plant diversity (Caujapé-Castells et al. 2010), and although Socotra may not strictly fit the definition of an oceanic island, it can for all intents and purposes be treated as one. Islands comprise only about 5% of the Earth's terrestrial surface, but Kreft et al. (2008) have estimated that they hold roughly 25% of all known vascular plant species. At the same time, islands have been heavily affected by human impacts (Whittaker and Fernández-Palacios 2007), and of the 80 extinctions of plant species documented over the past 400 years, 50 have involved island species (Sax and Gaines 2008). As a broad generalisation, it has been suggested that in the coming decades, human pressures on islands will rise disproportionately higher than in continental areas (Kier et al. 2009). The susceptibility of island floras is due directly to small and often fragmented populations, sometimes with very restricted ranges of the

Fig. 7.28 *Dracaena cinnabari* in a small nursery specialised in the production of native plants. Near Hadibuh, December 2008

Fig. 7.29 Possible ancient cultivation of *Boswellia dioscorides*. Di-Hishal, March 1997

species involved. Socotra is of outstanding universal conservation value and belongs to the biodiversity hotspot in the Horn of Africa. The high percentage of endemic species (both flora and fauna) in the Socotra Archipelago is one of the major reasons for its designation as a Biosphere Reserve and World Heritage Site. A number of these species have been documented by only single or very few records, although it should be conceded that, due to the paucity of systematic botanical exploration and the inaccessibility of many parts of the island, the true status of some of these species may be quite different. For instance, the extremely rare and local species *Angkalanthus oligophylla*, known to science from one wadi in the south of the island containing a handful of individuals, is said by local inhabitants to be common in other locations, although they were unable to demonstrate where (Miller, pers. comm.).

Effective conservation measures require that the flora is taxonomically well understood and that the size and distribution of populations can be reliably assessed. With the publication of the *Ethnoflora of the Soqotra Archipelago* (Miller and Morris 2004), a solid foundation has been established to address the first aspect, and it is hoped that this will stimulate more detailed studies, particularly dealing with the second point. Molecular studies, such as those described in Chap. 4, can help shed light on the status of complex taxa, and should be seen as a useful additional tool that is available to conservation biologists and taxonomists, but such studies cannot replace painstaking fieldwork, which remains the most important and necessary approach to evaluating many aspects of the island's flora relevant to its preservation.

In their general review, Caujapé-Castells et al. (2010) summarised by stating that habitat destruction has been substantial and overall has probably been the most important factor in the declines of plant populations on oceanic islands in the past. This statement applies without reservation to Socotra, if serious vegetation degradation is equated to habitat destruction. It should be borne in mind that there is typically a time-lag between the reduction in habitat area and the eventual disappearance of the remnant populations, and this lag represents what is often referred to as the "extinction debt" (Tilman et al. 1994; Triantis et al. 2010). In other words, it can take several generations for the full impact of habitat destruction and fragmentation to be reflected in the number of extinctions.

Conservation efforts that focus on individual plant species are often likely to be of limited success if the prime objective is to ensure that the species survives in the wild. This is perhaps even more relevant to island floras with a high degree of endemism, as these species may, for instance, have co-evolved with endemic pollinators. It is vital that the community as a whole be protected or restored to ensure that complex species interactions are preserved or reinstated. For instance, it could be argued that the planting of saplings of *Boswellia elongata* at Homhil should ensure that the population is preserved *in situ*, but if, as is the case, overgrazing continues unabated, the essence of the community and many of the interactions that characterise the natural ecosystem will be lost. In some ways, therefore, the Homhil stands currently bear more of a resemblance to a plantation in a botanic garden, designed to protect one species only, rather than to preserve the natural plant community.

In a limited number of cases, the propagation of species in botanic gardens, preferably in several locations around the world, may be the only effective approach to ensure that some species do not become completely extinct, not just in the wild. This may also be necessary to protect some species, especially some of the more exotic succulents, from unscrupulous collectors, as well as from poorly conceived and ill-fated restoration efforts.

It is not only wild plant species that are becoming increasingly rare. Many traditional crops are now being replaced by imports such as maize, rice and wheat. A good example is that of finger millet (*Eleusine coracana*), which was once the major source of starch for the local inhabitants (Bazara'a et al. 1991). The loss of such local cultivars represents a serious depletion of genetic diversity. These cultivars, which have been grown for thousands of years, may well be resistant to a whole variety of pests and diseases, and could be of enormous benefit in plant breeding in the future.

7.5.1 The Conservation of **Dracaena cinnabari**

As a flagship species that is a highly conspicuous element of parts of the landscape of Socotra, specifically between about 300 and 1,480 m, the dragon's blood tree, *Dracaena cinnabari*, has been a focus of conservation efforts and research activities on Socotra in recent years. Several detailed studies have been conducted to assess the potential impacts of various environmental factors, in particular climate change, on the long-term survivability of *Dracaena* populations (e.g. Attorre et al. 2007; Habrova et al. 2009), because it had already been noted by various authors (e.g. Beyhl 1996b; Mies 2001; Adolt and Pavlis 2004; Miller and Morris 2004) that regeneration in many natural populations is very low. Miller and Morris (2004) cautiously ruled out grazing as a prime cause for the lack of recruitment, except where the local populations were already under stress from presumed aridity. It is clear that the gradual decline of *Dracaena* is a result of a general aridification of the island (and the wider region as a whole) that has been taking place over a very long time-scale. As mentioned in Sect. 4.2.4.3, *Dracaena* woodlands on Socotra are relicts of the Mio-Pliocene thermo-sclerophyllous southern Tethys flora (i.e. Laurasian subtropical flora), and have therefore presumably been in gradual decline for the past several million years. However, our own superficial observations, for instance that the trees regenerate mainly in inaccessible locations or under spiny shrubs (see Fig. 7.31), indicate that a more recent acceleration of this decrease, including in some of the wetter regions of the island, can almost certainly be attributed to overgrazing, either directly or indirectly. This is in part corroborated by the results of a recent study by Habrova et al. (2009), who examined changes in population size on Jebel Skent over a 100-year period. These authors managed to relocate the approximate site ("Jebel Dried") from where Franz Kossmat had taken a photograph of Jebel Skent in 1899 (and later published in von Wettstein 1906) during an Austrian-led expedition to the island. Numerous individual *Dracaena* trees were

Fig. 7.30 *Dracaena cinnabari* is primarily found in areas highly affected by the monsoon clouds and mists. Dixam plateau, December 2008

clearly discernible in this image. In 2004, new images of Jebel Skent were taken from roughly the location used by Kossmat, allowing for a direct comparison after almost exactly 100 years. Despite minor technical problems with the comparison (due to slightly different locations from where the two sets of photographs were taken and the less favourable lighting conditions in 2004), it is clear that there has been a substantial reduction in the population size of *Dracaena* on Jebel Skent. The authors estimated that the decrease in the number of trees may exceed 40%. Significantly, it appears that regeneration of the species is severely hampered, as indicated by the low number of younger trees that were visible in 2004. If conservation of the natural *Dracaena* stands is the stated aim, detailed studies are required to address key questions regarding the autecology and population dynamics of the species, such as seed viability (including inter-annual variability of viability), germination characteristics, establishment success and recruitment frequency necessary to maintain current population levels. Some studies have been recently completed, but while awaiting the outcome of others, active intervention measures such as the transplanting of nursery-grown plants would seem highly advisable, as is currently being undertaken by the local Environment Protection Agency on the Dixam plateau.

Several authors, including Mies (2001) and Miller and Morris (2004), have remarked that the best stands of *Dracaena* are found in the areas most affected by the monsoon clouds and mists (Fig. 7.30), especially at higher elevations (as on Jebel Skent), and that, with the exception of various small populations, sometimes consisting of an individual tree, the species is largely restricted to the Hagghier mountains and

Fig. 7.31 Young individual of *Dracaena cinnabari* growing under *Cocculus balfourii*. West of Dimele ridge, March 1996

Hamaderoh on the north-east coast. More recently Attorre et al. (2007) have shown the current distribution to correlate with three main factors: (1) moisture index (ratio of mean annual precipitation and potential evapotranspiration); (2) mean annual temperature; and (3) slope. According to these authors, *Dracaena* occurs in areas with a mean annual precipitation ranging from 207 to 569 mm. It seems the authors did not incorporate cloud or mist occurrence in their model, but as these factors are likely to be an inherent feature of the variables they measured, they can be omitted without affecting the overall conclusions. According to this model, *Dracaena cinnabari* occupies only about 5% of its potential habitat on Socotra. The authors also suggested that increased aridity (due mainly to climate change predictions) will reduce the potential habitat of the species substantially by 2080, by up to 50%. Only three areas, Skent, Deidho and Dixam, were shown by environmental modelling to allow natural populations of *Dracaena* to thrive on Socotra, assuming grazing can be restricted, which is currently not the case for Dixam at least. Mies (2001) underlined the importance of nurse plants for the successful recruitment of *Dracaena* in areas exposed to grazing, such as on the Dixam plateau (Fig. 7.31), and Miller and Morris (2004) noted that regeneration is problematic in relict populations that lack a dense understorey of shrubs. The effects of grazing, as shown in Fig. 7.32 (an exclosure protecting the above-mentioned *Dracaena* nursery on the Dixam plateau), make it abundantly clear why *Dracaena* is unable to regenerate in such degraded areas: vegetation cover has been almost completely denuded outside of exclosures.

Fig. 7.32 Livestock exclosure on the Dixam plateau. December 2008

In some areas, remaining older *Dracaena* trees are one of the few conspicuous signs of plant life in an otherwise virtually barren landscape (Fig. 7.33). This example also underlines Le Houérou's (1996) general statement regarding the probable minor impacts of climate change compared with the more immediate and ongoing threat of massive land degradation (see above). It should also be borne in mind that not only has the protective perennial vegetation layer been destroyed in such massively overgrazed areas, but that as a consequence, soils have become seriously degraded, in particular due to the loss of the thin organic topsoil.

Germination *per se* does not appear to be problematic in *Dracaena*. Adolt and Pavlis (2004) reported that germination rates as high as 77% could be achieved under greenhouse conditions, and that the mortality of seedlings amounted to only 10%. Even the comparably low rates of 35% germinability mentioned by Beyhl (1996b) would appear to be more than adequate to maintain populations, especially given the large numbers of seeds produced by each tree. Under harsh natural conditions, germination and, in particular, successful establishment are going to be substantially lower, even in best-case scenarios. Adolt and Pavlis (2004) note a "certain uniformity" of the age structure of *D. cinnabari* stands, and this is likely to be consistent with the rare recruitment success in many areas. It can be inferred from this observation that seedling mortality will be 100% in years of poor rainfall, especially in the more unfavourable sites. Furthermore, for such a long-lived species as *Dracaena*, it can be assumed from ecological theory that successful recruitment will be, and need only be, a rare event for the population to survive. It is unclear as to how often such

Fig. 7.33 *Dracaena cinnabari* in an otherwise virtually barren landscape. Dixam plateau, December 2008

"windows of recruitment opportunity" occur on Socotra, but in any case, assuming equal rainfall, recruitment events are likely to vary substantially with the nature of the substrate, specifically its water-holding capacity, the degree of shading and not least the amount of trampling by livestock.

Apart from a positive correlation with moisture index and annual temperature, Attorre et al. (2007) found that the current distribution of *Dracaena* is also linked to slope. It has been frequently observed (e.g. Beyhl 1996b; Beyhl and Mies 1996) that young plants of the species grow mainly on cliff ledges and other sites inaccessible to goats (Figs. 7.34 and 7.35), but are absent from the surrounding area. This was noted long ago for the Macaronesian *Dracaena draco* (Kunkel 1980). Apart from the grazing factor, which undoubtedly plays a major role, some of these sites, which represent in the literal sense "safe sites" for recruitment, are probably also characterised by better water storage capacity, allowing seeds to germinate and become established even in years of less than ideal rainfall.

For most monocots, including *D. cinnabari*, it is virtually impossible to assess the age of trees precisely using morphological or anatomical features. This includes methods based on the counting of the sausage-shaped branch units, because in the case of the morphologically similar *D. draco*, estimates for each of these vary from about 11 to 30 years (Adolt and Pavlis 2004). For practical purposes in the field, four morphological age classes of *Dracaena* can be distinguished: (I) single rosette, no trunk (Fig. 7.36); (II) plant with trunk and a single terminal rosette (Fig. 7.37); (III) plant with trunk, initial branching and several rosettes, crown diameter usually

Fig. 7.34 Younger dragon's blood trees are still fairly frequent in sites that are inaccessible for grazing livestock. Dixam plateau, December 2008

Fig. 7.35 Older dragon's blood trees with numerous younger individuals on inaccessible rock faces high in the Hagghier mountains. Fieri peaks, February 1999

Fig. 7.36 *Dracaena cinnabari*: age class I. Dixam plateau, December 2008

Fig. 7.37 *Dracaena cinnabari*: age class II. Muqadrihon pass, south of Hadibuh, December 2008

Fig. 7.38 *Dracaena cinnabari*: age class III. Dixam plateau, December 2008

less than 2.5 m (Fig. 7.38); (IV) plant with crown diameter in excess of 2.5 m, dense dracoid ramification (Fig. 7.39). As a rough estimate, Adolt and Pavlis (2004) assume the mean age of three separate mature populations of *D. cinnabari* at Firmihin and another one at Hamaderoh, belonging to age class IV, to range from 200 to 300 years.

In order to protect natural self-reproducing populations of *Dracaena*, Attorre et al. (2007) recommended that the boundaries of the Skent Nature Sanctuary should be extended to encompass Deidho in the eastern part of the Hagghier mountains. Although this section has focused on *D. cinnabari*, other tree species on Socotra, including frankincense (*Boswellia* spp.), are also suffering from the excessive grazing pressures, and appropriate conservation measures are required at least to maintain current population sizes, as discussed in the case of *Boswellia elongata* below.

7.5.2 The Conservation of **Boswellia elongata**

Protection of the *Boswellia elongata* stands at Homhil (see Sect. 7.3.2) should be regarded as a conservation priority, and indeed, the area has been designated as an Area of Special Botanical Interest. According to the zoning plan (see Sect. 7.6.1), Homhil therefore theoretically enjoys enhanced protection. Many activities have been banned there (Fig. 7.40). However, until the factors that have caused this massive degradation have been removed – first and foremost, serious overgrazing – it can be conclusively ruled out that natural regeneration of this woodland will proceed.

Fig. 7.39 *Dracaena cinnabari*: age class IV. December 2008

Fig. 7.40 Homhil Protected Area sign. December 2008

The most obvious form of protection would be to create temporary exclosures for several years, which would allow the partial recovery of the vegetation. It is highly probable that for the successful establishment of *Boswellia* seedlings, dense vegetation cover is required to protect them from temperature extremes and the rapid drying out of the substrate. However, as the vegetation is highly depleted, a more promising strategy would be to plant saplings, grown from seed collected from local populations of the trees, in temporary exclosures, protecting them from the elements by netting. Whether such exclosures would be acceptable to the local grazers is, however, open to question. As a final solution, it may be feasible simply to plant saplings in the grazed areas, protecting each one by netting or fencing. This would obviously be a case of the conservation of a single species rather than re-establishing natural ecosystem functioning, and would therefore be somewhat unsatisfactory, especially as the whole purpose of the protected area would be undermined.

7.6 Coordination of Conservation Efforts

The Socotra Archipelago Conservation and Development Programme (SCDP) was initiated by the Republic of Yemen within the Ministry of Planning and Development in Sana'a in 2002 to conserve and develop the Socotra Archipelago in a sustainable manner. The SCDP has received substantial international support from the United Nations Development Programme (UNDP), several European governments and a variety of international donors and non-governmental organisations. The main aim of the SCDP is to coordinate all government and donor efforts towards sustainable human development for the people of the Socotra Archipelago, while conserving the globally significant biodiversity of these unique islands. In this context, it should be noted that ecotourism is expected to play a major role in the economic development of the archipelago.

In 2003, Socotra was designated by UNESCO as a biosphere reserve, and in 2008, it was inscribed as a World Heritage Site (UNESCO 2008) on account of its important and significant natural habitats for *in situ* conservation of biological diversity, including those containing threatened species of outstanding universal value from the point of view of science or conservation. Biosphere reserves are areas of terrestrial and coastal ecosystems whose main role is to promote solutions that reconcile the conservation of biodiversity and its sustainable use. As a consequence, biosphere reserves are expected to fulfil three main functions, namely (1) conservation of landscapes, ecosystems, species and genetic variation; (2) promotion of economic and human development that is socioculturally and ecologically sustainable; (3) promotion of research, education and monitoring of the aspects listed in (1) and (2).

7.6.1 Conservation Zoning Plan

The entire Socotra Archipelago was recognised by the government of Yemen as a special conservation area of high global importance, and, with the involvement of Socotris and international experts, a conservation zoning plan was developed for the area, and ratified as per Presidential Decree 275 in 2000. This zoning plan focused

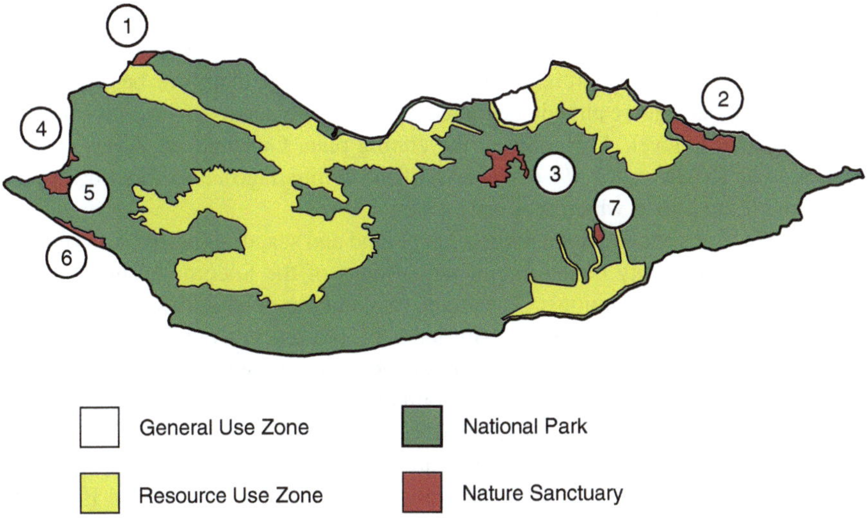

Fig. 7.41 Socotra Conservation Zoning Plan. Numbers are of designated terrestrial Nature Sanctuaries/Areas of Special Botanical Interest: 1 = Ditwah; 2 = Homhil; 3 = Skent; 4 = Shu'ab mangroves; 5 = Wadi Shu'ab; 6 = Niyt mangroves; 7 = Sarahin

on the preservation of traditional uses and practices of the local Socotri communities to ensure future sustainable use of natural resources. In most cases, traditional practices were deemed to be highly compatible with the objectives of biodiversity conservation. After an extensive consultation process with local community leaders, four key zones were designated in the zoning plan: (1) Resource Use Zone; (2) General Use Zone; (3) National Park; (4) Nature Sanctuary/Area of Special Botanical Interest. A map of the conservation zoning plan for Socotra itself (taken from Cheung and DeVantier 2006) is shown in Fig. 7.41. Detailed information on the Socotra Conservation Zoning Plan is available at http://www.socotraisland.org/plan/plan.html, and outlines of the four zones follow.

Resource Use Zones comprise areas that are managed to ensure long-term protection of the unique biological diversity of the archipelago, while at the same time providing a sustainable flow of natural products and services to meet community needs. Use of resources is in accordance with traditional practices to ensure their sustainable use, and these practices are enforced by the local communities themselves. A key objective of this zone is to protect the natural resource base from being alienated for other purposes that would be detrimental to the unique biological diversity, and to the subsistence of the local community that relies on it. The total terrestrial area included in the Resource Use Zone is 890 km^2, or 23% of the area of the archipelago.

General Use Zones are embedded within the Resource Use Reserve, and include sites that have already been significantly modified or degraded, or that will be earmarked to support vital economic interests of the local community. They will be used for the development of essential infrastructure to improve the standard of living of the local community (e.g. roads, power lines, transportation and health facilities). Infrastructure and commercial development efforts will therefore be focused in designated areas that are already degraded, thus limiting damage to the environment to an acceptable level. According to the zoning plan, General Use Zones cover a land area of 55 km^2 (ca. 1.4% of the area of the archipelago), and about 6% of the Resource Use Zone in which they are located.

National Parks encompass natural areas of land and sea, designated to: (1) protect the ecological integrity of the unique ecosystems of the Socotra Archipelago for present and future generations; (2) exclude exploitation or activities inimical to the objectives of conservation of biodiversity and of the preservation of natural landscapes; and (3) provide the foundation for scientific, educational, recreational and physical opportunities, all of which must be environmentally and culturally compatible. The terrestrial component of the National Parks covers an impressive area of 2,748 km^2, corresponding to nearly 73% of the land area of the archipelago. Two categories of enhanced protection fall within the National Parks. (1) Areas of Special Botanical Interest contain the highest concentration of rare and endemic plants in the archipelago. On account of their outstanding conservation significance, they will receive absolute priority with respect to management and protection. (2) Nature Sanctuaries are areas of unmodified or slightly modified land (and sea) that have retained their natural character without permanent or significant human habitation. They will be strictly protected and managed so as to preserve natural or near-natural conditions.

Compliance with the spirit of this zoning plan will be crucial to maintaining the long-term future of intact ecosystems on Socotra. Unfortunately, as illustrated by the example of Homhil (see Sect. 7.3.2), designated as a Nature Sanctuary and Area of Special Botanical Interest in the zoning plan, it is to be feared that despite the valiant efforts and best intentions of numerous individuals and organisations, Socotra is destined to suffer the same demise as another biodiversity hotspot in the region, namely Dhofar in southern Oman: green and pleasant to the casual tourist, but ecologically seriously degraded, as was already recognised in 1982 (see Fisher et al. 1998).

Appendices

Appendix 1: List of Vascular Plant Species of the Socotra Archipelago (Based primarily on Miller and Morris 2004)

Family/species	Additional to Miller and Morris (2004)
Pteridophyta	
Actiniopteridaceae	
Actiniopteris semiflabellata Pichi Serm.	
Adiantaceae	
Adiantum balfourii Baker	
Adiantum capillus-veneris L.	
Adiantum poirettii Wikstr.	
Consentinia vellea (Aiton) Tod.	Kilian and Hein (2006)
Doryopteris concolor (Langsd. & Fisch.) Kuhn	
Dryopteris schimperiana (A. Br.) C. Chr.	Kilian and Hein (2006)
Negripteris sciona Pichi Serm.	
Onychium divaricatum (Poir.) Alston	
Pellaea involuta (Sw.) Baker	
Pellaea viridis (Forssk.) Prantl	
Aspleniaceae	
Asplenium aethiopicum (Burm.f.) Bech.	
Asplenium schweinfurthii Baker[a]	
Asplenium trichomanes L.	
Ceterach phillipsianum Kümmerle	
Dennstaedtiaceae	
Pteridium aquilinum (L.) Kuhn	
Dryopteridaceae	
Hypodematium crenatum (Forssk.) Kuhn	
Polystichum fuscopaleaceum Alston	Kilian and Hein (2006)
Marsiliaceae	
Marsilea coromandeliana Willd.	

(continued)

G. Brown and B.A. Mies, *Vegetation Ecology of Socotra*, Plant and Vegetation 7,
DOI 10.1007/978-94-007-4141-6, © Springer Science+Business Media Dordrecht 2012

(continued)

Family/species	Additional to Miller and Morris (2004)
Oleandraceae	
Arthropteris orientalis (J.F. Gmel.) Posthumus	Kilian and Hein (2006)
Nephrolepis undulata (Afzel ex Sw.) J. Sm.	
Ophioglossaceae	
Ophioglossum polyphyllum A. Braun	
Ophioglossum reticulatum L.	
Parkeriaceae	
Ceratopteris cornuta (P. Beauv.) Lepr.	
Polypodiaceae	
Loxogramme abyssinica (Swartz) C. Presl	
Pleopeltis macrocarpa (Broy ex Willd.) Kaulf.	
Pteridaceae	
Pteris cretica L.	
Pteris quadriaurita Retz.	
Pteris vittata L.	
Thelypteridaceae	
Christella dentata (Forssk.) Brownsey & Jermy	
Gymnospermae	
Ephedraceae	
Ephedra foliata Boiss. ex C.A. Mey.	
Angiospermae: Monocotyledonae	
Amaryllidaceae	
Crinum balfourii Baker[a]	
Anthericaceae	
Chlorophytum filifolium Nordal & Thulin	
Chlorophytum graptophyllum (Baker) A.G. Mill.[a]	
Chlorophytum sp. *A*[a]	
Araceae	
Colocasia esculenta (L.) Schott	
Remusatia vivipara Schott	
Arecaceae	
Borassus aethiopicum Mart.	
Phoenix dactylifera L.	
Asparagaceae	
Asparagus africanus Lam.	
Asparagus sp. *A*[a]	
Asphodelaceae	
Aloe forbesii Balf.f.[a]	Mies (2001), Carter et al. (2011)
Aloe jawiyon Christie, Hannon & Oakham[a]	
Aloe perryi Baker[a]	
Aloe squarrosa Baker[a]	
Asphodelus fistulosus L.	
Colchicaceae	
Iphigenia socotrana Thulin[a]	
Commelinaceae	
Commelina albescens Hassk.	

(continued)

(continued)

Family/species	Additional to Miller and Morris (2004)
Commelina benghalensis L.	
Commelina forskalaei Vahl	
Cyanotis cristata Roem. & Schult.	
Cymodoceaceae	
Cymodocea rotundata Ehrenb. & Hempr. ex Aschers	
Cymodocea serrulata (R. Br.) Aschers & Magnus	
Halodule uninervis (Forssk.) Aschers	
Thalassodendron ciliatum (Forssk.) den Hartog	
Cyperaceae	
Bulbostylis barbata (Rottb.) C.B. Clarke	
Bulbostylis hispidula (Vahl) R. Haines	
Cladium mariscus R. Br.	
Cyperus amabilis Vahl	
Cyperus bulbosus Schrank	
Cyperus compressus L.	
Cyperus conglomeratus Rottb.	
Cyperus cyperoides (L.) Kuntze	
Cyperus difformis L.	
Cyperus dilatatus Schumach. & Thonn.	
Cyperus laevigatus L.	
Cyperus longus L.	
Cyperus rotundus L.	
Cyperus rubicundus Vahl	
Cyperus squarrosus L.	
Cyperus stramineo-ferrugineus Kük.	
Cyperus tenuiculmis Boeck.	
Eleocharis caduca (Del.) Schultes	
Eleocharis geniculata (L.) Roem. & Schult.	
Fimbristylis complanata (Retz) Link	
Fimbristylis cymosa R. Br.	
Fimbristylis dichotoma (L.) Vahl	
Fimbristylis ferruginea Vahl	
Fuirena ciliaris (L.) Roxb.	
Kyllinga brevifolia Rottb.	
Kyllinga microstyla C.B. Clarke	
Pycreus pumilus (L.) Nees	
Dioscoreaceae	
Dioscorea lanata Balf.f.[a]	
Dracaenaceae	
Dracaena cinnabari Balf.f.[a]	
Hyacinthaceae	
Dipcadi balfourii Baker[a]	
Dipcadi biflorum Ghazanfar	
Dipcadi guichardii Radcl.-Sm.[a]	
Dipcadi kuriensis A.G. Mill.[a]	
Drimia porphyrostachys (Baker) A.G. Mill.[a]	

(continued)

(continued)

Family/species	Additional to Miller and Morris (2004)
Ledebouria aff. *revoluta* (L.f.) Jessop	
Ledebouria grandifolia (Balf.f.) A.G. Mill.[a]	
Ledebouria insularis A.G. Mill.[a]	
Hydrocharitaceae	
Halophila decipiens Ostenfeld	
Halophila ovalis (R. Br.) Hooker	
Thalassia hemprichii (Ehr.) Ascherson	
Hypoxidaceae	
Hypoxis angustifolia Lam.	
Iridaceae	
Cyanixia socotrana (Hook.f.) Goldblatt & Manning[b]	
Romulea fischeri Pax	
Juncacaeae	
Juncus bufonius L.	
Juncus rigidus Desf.	
Juncus socotranus (Buchenau) Snogerup	
Najadaceae	
Najas graminea Delile	
Najas marina L.	
Orchidaceae	
Angraecum dives Rolfe	
Eulophia petersii (Rchb.f.) Rchb.f.	
Habenaria socotrana Balf.f.	
Holothrix socotrana Rolfe[a]	
Poaceae	
Acrachne racemosa Wight & Arn. ex Chiov.	
Aeluropus lagopoides (L.) Trin. ex Thwaites	
Andropogon bentii Stapf[a]	
Apluda mutica L.	
Aristida adscensionis L.	
Aristida anaclasta Cope[a]	
Aristida funiculata Trin. & Rupr.	
Arthraxon lancifolius (Trin.) Hochst.	
Arthraxon micans (Nees) Hochst.	
Bothriochloa insculpta (Hochst. ex A. Rich.) A. Camus	
Brachiaria arida (Metz) Stapf	
Brachiaria deflexa (Schumach.) C.E. Hubb. ex Robyns	
Brachiaria eruciformis (Sm.) Griseb. in Ledeb.	
Brachiaria lata (Schumach.) C.E. Hubb.	
Brachiaria ovalis Stapf	
Brachiaria ramosa (L.) Stapf	
Cenchrus ciliaris L.	
Cenchrus pennisetiformis Hochst. & Steud.	
Cenchrus setigerus Vahl	
Chloris pycnothrix Trin.	
Chloris virgata Sw.	

(continued)

(continued)

Family/species	Additional to Miller and Morris (2004)
Chrysopogon aucheri Stapf	
Chrysopogon plumulosus Hochst.	
Chrysopogon serrulatus Trin.	
Cymbopogon jwarancusa (Jones) Schult. ssp. *olivieri* (Boiss.) Soenarko	
Cynodon dactylon (L.) Pers.	
Dactyloctenium aegyptium (L.) Willd.	
Dactyloctenium aristatum Link	
Dactyloctenium hackelii Wagner & Vierh.[a]	
Dactyloctenium robecchii (Chiov.) Chiov.	
Desmostachya bipinnata (L.) Stapf	
Dichanthium annulatum (Forssk.) Stapf	
Dichanthium foveolatum (Delile) Roberty	
Digitaria ciliaris (Retz.) Koeler	
Digitaria rivae (Stapf) Chiov.	
Digitaria velutina (Forssk.) P. Beauv.	
Echinochloa colona (L.) Link	
Eleusine africana Kenn.-O'Byrne	Cope (2007)
Eleusine coracana (L.) Gaertn.	
Enneapogon desvauxii P. Beauv.	
Enneapogon persicus Boiss.	
Eragrostis aspera (Jacq.) Nees	
Eragrostis barrelieri Daveau	
Eragrostis cilianensis (All.) Vignolo ex Janch.	
Eragrostis ciliaris (L.) R. Br.	
Eragrostis lepida (A. Rich.) Hochst. ex Steud.	
Eragrostis minor Host	
Eragrostis papposa (Roem. & Schult.) Steud.	
Eragrostis tenuifolia (A. Rich.) Steud.	Kilian and Hein (2006)
Eragrostis viscosa (Retz.) Trin.	
Gastridium phleoides (Nees & Meyen) C.E. Hubb.	
Hackelochloa granularis Kuntze	
Halopyrum mucronatum (L.) Stapf	
Heteropogon contortus (L.) P. Beauv. ex Roem. & Schult.	
Hyparrhenia hirta (L.) Stapf	
Imperata cylindrica (L.) Rauesch.	
Ischaemum impressum Hack in DC.	Cope (2007)
Leptothrium senagalense (Kunth) Clayton	
Lepturus calcareus Cope[a]	
Lepturus nesiotes Cope[a]	Cope (2007)
Lepturus pulchellus (Balf.f.) Cope[a]	
Lepturus tenuis Balf.f.[a]	
Lolium persicum Boiss.	
Lolium temulentum L.	
Melanocenchris jacquemontii Jaub. & Spach	

(continued)

(continued)

Family/species	Additional to Miller and Morris (2004)
Melinis repens (Willd.) Zizka ssp. *grandiflora* Hochst.	
Microchloa kunthii Desv.	
Oplismenus burmannii (Retz.) P. Beauv.	
Oplismenus compositus (L.) P. Beauv.	
Panicum atrosanguineum Hochst. ex A. Rich.	
Panicum maximum Jacq.	
Panicum rigidum Balf.f.[a]	
Panicum socotranum Cope[a]	
Panicum turgidum Forssk.	
Paspalidium geminatum (Forssk.) Stapf	
Paspalum scrobiculatum L.	
Paspalum vaginatum Sw.	
Pennisetum divisum (J.F. Gmel.) Henrard	Cope (2007)
Pennisetum orientale Rich.	
Pennisetum setaceum (Forssk.) Chiov.	
Polypogon monspeliensis (L.) Desf.	
Schmidtia pappophoroides Steud.	
Setaria intermedia Roem. & Schult.	
Setaria pumila (Poir.) Roem. & Schult.	
Setaria verticillata (L.) P. Beauv.	
Sporobolus diander (Retz.) P. Beauv.	
Sporobolus ioclados (Nees ex Trin.) Nees	
Sporobolus minutus Link	
Sporobolus ruspolianus Chiov.	
Sporobolus spicatus (Vahl) Kunth	
Stipagrostis sokotrana (Vierh.) De Winter	
Tetrapogon tenellus (K.D.Koenig ex Roxb.) Chiov.	
Tetrapogon villosus Desf.	
Themeda quadrivalvis (L.) Kuntze	
Themeda triandra Forssk.	
Tragus berteronianus Schult.	
Tricholaena teneriffae (L.f.) Link	
Tricholaena vestita (Balf.f.) Stapf & C.E. Hubb. in Prain.[a]	
Trichoneura mollis (Kunth) Ekman	
Urochondra setulosa (Trin.) C.E. Hubb.	
Potamogetonaceae	
Potamogeton filiformis Pers.	
Potamogeton nodosus Poir.	
Potamogeton pectinatus L.	
Ruppia maritima L.	
Angiospermae: Dicotyledonae	
Acanthaceae	
Angkalanthus oligophylla Balf.f.[b]	
Anisotes diversifolius Balf.f.[a]	
Asystasia gangetica (L.) T. Anderson	

(continued)

(continued)

Family/species	Additional to Miller and Morris (2004)
Ballochia amoena Balf.f[b]	
Ballochia atro-virgata Balf.f.[b]	
Ballochia rotundifolia Balf.f.[b]	
Barleria aculeata Balf.f.[a]	
Barleria argentea Balf.f.[a]	
Barleria popovii Verdc.[a]	
Barleria tetracantha Balf.f.[a]	
Blepharis kuriensis Vierh.	
Blepharis maderaspatensis (L.) Roth	
Blepharis spiculifolia Balf.f.[a]	
Chorisochora minor (Balf.f.) Vollesen[a]	
Chorisochora striata (Balf.f.) Vollesen[a]	
Dicliptera effusa Balf.f.[a]	
Hypoestes pubescens Balf.f.[a]	
Justicia alexandrina R. Atkinson[a]	
Justicia heterocarpa T. Anderson	
Justicia rigida Balf.f.[a]	
Justicia takhinensis R. Atkinson[a]	
Neuracanthus aculeatus Balf.f.[a]	
Neuracanthus argyrophyllus Chiov.	
Neuracanthus capitatus Balf.f.	
Peristrophe paniculata (Forssk.) Brummitt	
Rhinacanthus scoparius Balf.f[a]	
Ruellia carnea Balf.f.	
Ruellia dioscoridis Napper[a]	
Ruellia insignis Balf.f.[a]	
Ruellia kuriensis Vierh[a]	
Ruellia patula Jacq.	
Ruellia paulayana Vierh.[a]	
Trichocalyx obovatus Balf.f.[b]	
Trichocalyx orbiculatus Balf.f.[b]	
Aizoaceae	
Aizoon canariense L.	
Corbichonia decumbens (Forssk.) Exell	
Gisekia pharnaceoides L.	
Glinus lotoides L.	
Mollugo cerviana Ser.	
Mollugo nudicaulis Lam.	
Tetragonia pentandra Balf.f.	
Zaleya pentandra (L.) C. Jeffrey	
Amaranthaceae	
Achyranthes aspera L.	
Aerva javanica (Burm.f.) Schult.	
Aerva lanata (L.) Juss.	
Aerva microphylla Moq.[a]	

(continued)

(continued)

Family/species	Additional to Miller and Morris (2004)
Aerva revoluta Balf.f.[a]	
Amaranthus dubius Thell.	
Amaranthus graecizans L.	
Amaranthus sparganiocephalus Thell.	
Digera muricata (L.) Mart.	
Gomphrena celosioides Mart.	
Psilotrichum aphyllum C.C. Towns.[a]	
Psilotrichum sericeum (Roxb.) Dalzell	
Pupalia lappacea (L.) Juss. v. *velutina* (Moq.) Hook.f.	
Anacardiaceae	
Lannea transulta (Balf.f.) Radcl.-Sm.[a]	
Rhus natalensis Krauss	Kilian and Hein (2006)
Rhus sp. *A*[a]	
Rhus thyrsiflora Balf.f.[a]	
Apiaceae	
Apium graveolens L.	
Centella asiatica (L.) Urb.	
Nirarathamnos asarifolius Balf.f.[b]	
Oreofraga morrisiana M.F. Watson & E.L. Barclay[b]	
Rughidia cordata (Balf.f.) M.F. Watson & E.L. Barclay[b]	
Rughidia milleri M.F. Watson & E.L. Barclay[b]	
Trachyspermum pimpinelloides (Balf.f.) H. Wolff	
Apocynaceae	
Acokanthera schimperi (A. DC.) Schweinf.	
Adenium obesum Roem.	
Calotropis procera (Ait.) W.T. Aiton	
Caralluma socotrana (Balf.f.) N.E. Br.	
Carissa spinarum L.	
Ceropegia affinis Vatke	
Ceropegia tihamana Lavranos	
Cryptolepis arbuscula (Radcl.-Sm.) Ventner[a]	
Cryptolepis intricata (Balf.f.) Ventner[a]	
Cryptolepis macrophylla (Radcl.-Sm.) Ventner[a]	
Cryptolepis socotrana (Balf.f.) Ventner[a]	
Cryptolepis volubilis (Balf.f.) Schwartz	
Cynanchum gerardii (Harv.) Liede	
Duvaliandra dioscoridis (Lavranos) M.G. Gilbert[b]	
Echidnopsis bentii N.E. Br. ex Hook.f.[a]	
Echidnopsis inconspicua Bruyns[a]	
Echidnopsis insularis Lavranos[a]	
Echidnopsis milleri Lavranos[a]	
Echidnopsis socotrana Lavranos[a]	
Edithcolea grandis N.E. Br.	

(continued)

(continued)

Family/species	Additional to Miller and Morris (2004)
Glossonema revoili Franch.	
Marsdenia robusta Balf.f.[a]	
Pergularia tomentosa L.	
Periploca visciformis (Vatke) K. Schum.	
Sarcostemma socotranum Lavranos[a]	
Sarcostemma viminale R. Br.	
Secamone cuneifolia Bruyns[a]	
Secamone socotrana Balf.f.[a]	
Socotrella dolichocnema Bruyns & A.G. Mill.[b]	
Vincetoxicum linifolium Balf.f.[a]	
Asteraceae	
Acanthospermum hispidum DC.	
Achyrocline luzuloides Vatke	
Achyrocline schimperi Schultz	
Ageratum conyzoides L.	
Bidens biternata (Lour.) Merr. & Sherff	
Bidens pilosa L.	
Blainvillea acmella (L.) Philipson.	Kilian and Hein (2006)
Conyza hochstetteri Sch. Bip.	
Cyanthillium cinereum (L.) H. Rob.	
Dichrocephala chrysanthemifolia (Blume) DC.	
Dicoma tomentosa Cass.	
Distephanus qazmi N. Kilian & A.G. Mill.[a]	
Eclipta prostrata (L.) L.	
Euryops arabicus Steud.	
Galinsoga parviflora Cav.	Kilian and Hein (2006)
Helichrysum aciculare Balf.f.[a]	
Helichrysum arachnoides Balf.f.[a]	
Helichrysum balfourii Vierh.[a]	
Helichrysum gracilipes Oliv. & Hiern	
Helichrysum nimmoanum Oliv. & Hiern[a]	
Helichrysum paulayanum Vierh.[a]	
Helichrysum rosulatum Oliv. & Hiern[a]	
Helichrysum sp. *A*[a]	
Helichrysum sp. *B*[a]	
Helichrysum sp. *C*[a]	
Helichrysum sp. *D*[a]	
Helichrysum sphaerocephalum Balf.f.[a]	
Helichrysum suffruticosum Balf.f.[a]	
Kleinia scottii (Balf.f.) P. Halliday[a]	
Launaea crassifolia (Balf.f.) C. Jeffrey	
Launaea crepoides Balf.f.[a]	
Launaea hafunensis Chiov.	
Launaea massauensis (Fresen.) Kuntze	
Launaea rhynchocarpa (Balf.f.) B.A. Mies[a]	
Launaea socotrana N. Kilian[a]	

(continued)

(continued)

Family/species	Additional to Miller and Morris (2004)
Launaea sp. *A*[a]	
Macledium canum (Balf.f.) S. Ortiz[a]	
Orbivestus cinerascens (Sch. Bip.) H. Rob.	
Pluchea glutinosa Balf.f.[a]	
Pluchea obovata Balf.f.[a]	
Prenanthes amabilis Balf.f.[a]	
Psiadia schweinfurthii Balf.f.[a]	
Pulicaria aromatica (Balf.f.) King-Jones & N. Kilian[a]	
Pulicaria dioscorides R. Atkinson[a]	
Pulicaria diversifolia Balf.f.[a]	
Pulicaria elegans E. Gamal-Eldin[a]	
Pulicaria lanata E. Gamal-Eldin[a]	
Pulicaria stephanocarpa Balf.f.[a]	
Pulicaria vieraeoides Balf.f.[a]	
Reichardia tingitana Roth	
Sigesbeckia orientalis L.	
Sonchus oleraceus L.	
Tripteris vaillantii Decne.	
Vernonia cockburniana Balf.f.[a]	
Vernonia unicata C. Jeffrey[a]	
Volutaria lippii Cass.	
Volutaria socotrensis Wagenitz[a]	Kilian and Hein (2006)
Avicenniaceae	
Avicennia marina (Forssk.) Vierh.	
Begoniaceae	
Begonia samhaensis M. Hughes & A.G. Mill.[a]	
Begonia socotrana Hook.f.[a]	
Boraginaceae	
Arnebia hispidissima (Lehm.) DC.	
Cordia crenata Del.	Kilian and Hein (2006)
Cordia obovata Balf.f.	
Cordia obtusa Balf.f.	
Cordia quercifolia Klotzsch	
Cystostemon socotranus Balf.f.[a]	
Echiochilon persicum (Burm.f.) I.M. Johnst.	
Echiochilon pulvinata A.G. Mill. & L. Urb.[a]	
Ehretia obtusifolia Hochst.	
Heliotropium aegyptiacum Lehm.	
Heliotropium balfourii Gürke[a]	
Heliotropium dentatum Balf.f.[a]	
Heliotropium derafontense Vierh.[a]	
Heliotropium kuriense Vierh.[a]	
Heliotropium nigricans Balf.f.[a]	
Heliotropium ophioglossum Stocks	
Heliotropium ovalifolium Forssk.	

(continued)

(continued)

Family/species	Additional to Miller and Morris (2004)
Heliotropium paulayanum Vierh.[a]	
Heliotropium rariflorum Stocks	
Heliotropium riebeckii Schweinf. & Vierh.[a]	
Heliotropium shoabense Vierh.[a]	
Heliotropium socotranum Vierh.[a]	
Heliotropium strigosum Willd.	
Heliotropium wagneri Vierh.[a]	
Heliotropium zeylanicum Lam.	
Trichodesma laxiflorum Balf.f.[a]	
Trichodesma microcalyx Balf.f.[a]	
Trichodesma scottii Balf.f.[a]	
Wellstedia socotrana Balf.f.[a]	
Brassicaceae	
Capsella bursa-pastoris L.	
Diceratella incana Balf.f.	
Erucastrum rostratum (Balf.f.) Gomez-Campo[a]	
Farsetia inconspicua A.G. Mill.[a]	
Farsetia longisiliqua Decaisne	
Farsetia socotrana B.L. Burtt[a]	
Farsetia stylosa R. Br.	
Hemicrambe fruticosa (C.C. Towns.) Gomez-Campo[a]	
Lachnocapsa spathulata Balf.f.[b]	
Nesocrambe socotrana A.G. Mill.[b]	
Sisymbrium erysimoides Desf.	
Burseraceae	
Boswellia ameero Balf.f.[a]	
Boswellia bullata Thulin & Gifri[a]	
Boswellia dioscorides Thulin & Gifri[a]	
Boswellia elongata Balf.f.[a]	
Boswellia nana Hepper[a]	
Boswellia popoviana Hepper[a]	
Boswellia socotrana Balf.f.[a]	
Boswellia sp. *A*[a]	
Commiphora kua (Royle) Vollesen	
Commiphora ornifolia (Balf.f.) Gillet[a]	
Commiphora parvifolia (Balf.f.) Engl.[a]	
Commiphora planifrons (Balf.f.) Engl.[a]	
Commiphora socotrana (Balf.f.) Engl.[a]	
Buxaceae	
Buxanthus pedicellatus Tiegh.	
Campanulaceae	
Campanula balfourii Wagner & Vierh.[a]	
Wahlenbergia lobelioides DC.	
Capparaceae	
Cadaba farinosa Forssk.	

(continued)

(continued)

Family/species	Additional to Miller and Morris (2004)
Cadaba insularis A.G. Mill.[a]	
Cadaba longifolia DC.	
Cadaba rotundifolia Forssk.	
Capparis cartilaginea Decne.	
Capparis decidua (Forssk.) Edgew.	
Maerua angolensis DC.	
Caryophyllaceae	
Arenaria foliacea Turrill	
Cometes abyssinica (R. Br.) Wall.	
Gymnocarpos bracteatus (Balf.f.) Thulin[a]	
Gymnocarpos kuriensis (Radcl.-Sm.) Thulin[a]	
Gypsophila montana Balf.f.	
Haya obovata Balf.f.[b]	
Polycarpaea balfourii Briq.[a]	
Polycarpaea caespitosa Balf.f.[a]	
Polycarpaea corymbosa (L.) Lam.	
Polycarpaea hassalensis D.F. Chamb.[a]	
Polycarpaea hayoides D.F. Chamb.[a]	
Polycarpaea kuriensis Wagner[a]	
Polycarpaea paulayana Wagner[a]	
Polycarpaea spicata Wight ex Arn.	
Silene apetala Willd.	
Spergula fallax (Lowe) Krause	
Sphaerocoma hookeri T. Anderson	
Celastraceae	
Maytenus sp. *A*[a]	
Chenopodiaceae	
Anabasis ehrenbergii Schweinf. ex Boiss.	
Arthrocnemum macrostachyum (Moric.) Moris & Delponte	
Atriplex farinosa Forssk.	
Atriplex griffithii Moq. ssp. *stocksii* (Boiss.) Boulos	
Chenopodium murale L.	
Cornulaca ehrenbergii Asch.	
Lagenantha cycloptera (Stapf) M.G. Gilbert & Friis	
Salsola spinescens Moq.	
Seidlitzia rosmarinus Bunge ex Boiss.	
Sevada schimperi Moq.	
Suaeda monoica Forssk. ex J.F. Gmel.	
Suaeda vermiculata Forssk. ex J.F. Gmel.	
Cleomaceae	
Cleome austroarabica D.F. Chamb. & Lamond	
Cleome brachycarpa Vahl ex DC.	
Cleome gynandra L.	
Cleome scaposa DC.	

(continued)

(continued)

Family/species	Additional to Miller and Morris (2004)
Cleome socotrana Balf.f.[a]	
Cleome tenella L.f.	
Cleome viscosa L.	
Clusiaceae	
Hypericum balfourii N. Robson[a]	
Hypericum fieriense N. Robson[a]	
Hypericum scopulorum Balf.f.[a]	
Hypericum socotranum Good[a]	
Hypericum tortuosum Balf.f.[a]	
Convolvulaceae	
Convolvulus arabicus Hochst.	
Convolvulus grantii Balf.f.[a]	
Convolvulus hildebrandtii Vatke[a]	
Convolvulus rhyniospermus Choisy	
Convolvulus sarmentosus Balf.f.[a]	
Convolvulus siculus L.	
Cressa cretica L.	
Cuscuta chinensis Lam.	
Cuscuta planiflora Ten.	
Dichondra repens J.R. Forst. & G. Forst.	
Evolvulus alsinoides L.	
Ipomoea cairica (L.) Sweet	
Ipomoea kotschyana Hochst.ex Choisy	
Ipomoea nil (L.) Roth	
Ipomoea obscura Ker Gawl.	
Ipomoea pes-caprae (L.) Roth	
Ipomoea sinensis (Desr.) Choisy	
Metaporana obtusa (Balf.f.) Staples[a]	
Seddera fastigiata (Balf.f.) Verdc.[a]	
Seddera glomerata (Balf.f.) Schwartz	
Seddera latifolia Hochst. & Steud.	
Seddera pedunculata (Balf.f.) Verdc.[a]	
Seddera semhaensis R.R. Mill[a]	Demissew and Mill (2009)
Seddera spinosa (Vierh.) Verdc.[a]	
Crassulaceae	
Crassula alata (Viv.) Berger ssp. *pharnaceoides*	
Crassula schimperi Fisch. & C.A. Mey. ssp. *phyturus*	
Kalanchoe farinacea Balf.f.[a]	
Kalanchoe robusta Balf.f.[a]	
Kalanchoe rotundifolia Haw.	
Umbilicus horizontalis (Guss.) DC.	
Cucurbitaceae	
Citrullus colocynthis Schrad.	
Citrullus lanatus (Thunb.) Matsum. & Nakai	
Cucumis dipsaceus Ehrenb.	

(continued)

(continued)

Family/species	Additional to Miller and Morris (2004)
Cucumis ficifolius A. Rich.	
Cucumis prophetarum L.	
Dendrosicyos socotrana Balf.f.[b]	
Eureiandra balfourii Cogn.[a]	
Momordica balsamina L.	
Momordica charantia L.	
Zehneria scabra (L.f.) Sond.	
Dirachmaceae	
Dirachma socotrana Schweinf.[a]	
Ebenaceae	
Euclea divinorum Hiern	
Erythroxylaceae	
Erythroxylon socotranum Thulin[a]	
Euphorbiaceae	
Acalypha indica L.	
Andrachne schweinfurthii (Balf.f.) Radcl.-Sm.[a]	
Andrachne telephioides L.	
Cephalocroton socotranus Balf.f.[a]	
Chrozophora oblongifolia (Delile) Juss.	
Croton pachyclados Radcl.-Sm.[a]	Mies (2001)
Croton sarcocarpus Balf.f.[a]	
Croton socotranus Balf.f.[a]	
Croton sulcifructus Balf.f.[a]	
Euphorbia abdelkuri Balf.f.[a]	
Euphorbia arbuscula Balf.f.[a]	
Euphorbia balsamifera Aiton	
Euphorbia granulata Forssk.	
Euphorbia hadramautica Baker	
Euphorbia hajhirensis Radcl.-Sm.[a]	
Euphorbia hamaderoensis A.G. Mill.[a]	
Euphorbia inaequilatera Sond.	
Euphorbia indica Lam.	
Euphorbia kischenensis Vierh.[a]	
Euphorbia kuriensis Vierh.[a]	
Euphorbia leptoclada Balf.f.[a]	
Euphorbia obcordata Balf.f.[a]	
Euphorbia schimperi Presley	
Euphorbia schweinfurthii Balf.f.[a]	
Euphorbia socotrana Balf.f.[a]	
Euphorbia spiralis Balf.f.[a]	
Flueggea leucopyrus Willd.	
Flueggea virosa (Roxb. ex Willd.) Voigt	
Jatropha unicostata Balf.f.[a]	
Meineckia filipes (Balf.f.) G.L. Webster[a]	
Phyllanthus maderaspatensis L.	

(continued)

(continued)

Family/species	Additional to Miller and Morris (2004)
Phyllanthus rotundifolius Willd.	
Ricinus communis L.	
Tragia balfouriana J.B. Gillet[a]	
Fabaceae	
Acacia edgeworthii T. Anderson	
Acacia nilotica (L.) Willd. ex Delile	
Acacia pennivenia Schweinf.[a]	
Acacia sarcophylla Chiov.	
Acacia aff. *tortilis* (Forssk.) Hayne	
Acacia sp. *A*[a]	
Alysicarpus monilifer (L.) DC.	
Alysicarpus vaginalis DC.	
Astragalus hauarensis Boiss.	
Canavalia africana Dunn	
Chamaecrista abus (L.) Irwin & Barneby	
Chamaecrista pumila (Lam.) V. Singh	Kilian and Hein (2006)
Chapmannia gracilis (Balf.f.) Thulin[a]	
Chapmannia reghidensis Thulin & McKean[a]	
Chapmannia sericea Thulin & McKean[a]	
Chapmannia tinireana Thulin[a]	
Crotalaria leptocarpa Balf.f.	
Crotalaria pteropoda Balf.f.	
Crotalaria retusa L.	
Crotalaria socotrana (Balf.f.) Thulin[a]	
Crotalaria spinosa Hochst.	
Crotalaria strigulosa Balf.f.[a]	
Cullen corylifolia (L.) Medik.	
Desmodium triflorum (L.) DC.	
Dichrostachys dehiscens Balf.f.[a]	
Erythrina melanacantha Taub. ex Harms	
Indigofera argentea Burm.f.	
Indigofera articulata Gouan	
Indigofera coerulea Roxb.	
Indigofera colutea (Burm.f.) Merr.	
Indigofera cordifolia Roth	
Indigofera hochstetteri Baker	
Indigofera marmorata Balf.f.[a]	
Indigofera nephrocarpa Balf.f.	
Indigofera nephrocarpoides J.B. Gillet[a]	
Indigofera nugalensis Thulin	
Indigofera oblongifolia Forssk.	
Indigofera pseudointricata J.B. Gillett	
Indigofera sokotrana Vierh.[a]	
Indigofera spiniflora Boiss.	
Indigofera tinctoria L.	

(continued)

(continued)

Family/species	Additional to Miller and Morris (2004)
Lotus garcinii DC.	
Lotus mollis Balf.f.[a]	
Lotus ononopsis Balf.f.[a]	
Lotus schimperi Steud.	
Medicago laciniata All.	
Medicago minima Lam.	
Medicago polymorpha L.	
Melilotus indica (L.) All.	
Microcharis disjuncta (J.B. Gillett) Schrire	
Paracalyx balfourii (Vierh.) Ali[a]	
Prosopis juliflora DC.	
Rhynchosia minima (L.) DC.	
Rhynchosia pulverulenta Stocks	
Senna holosericea (Fresen.) Greuter	
Senna italica Mill.	
Senna obtusifolia (L.) Irwin & Barneby	
Senna socotrana (Serrato) Lock[a]	
Tamarindus indica L.	
Taverniera sericophylla Balf.f.[a]	
Tephrosia apollinea DC.	
Tephrosia odorata Balf.f.[a]	
Tephrosia purpurea sensu Balf.f. non (L.) Pers.	
Tephrosia socotrana Thulin[a]	
Tephrosia strigosa (Dalzell) Santapau & Maheshwari	
Tephrosia subtriflora Hochst.	
Tephrosia uniflora Pers.	
Teramnus repens (Taub.) Baker f.	
Trigonella falcata Balf.f.[a]	
Vigna membranacea A. Rich.	
Zornia glochidiata DC.	
Zygocarpum coeruleum (Balf.f.) Thulin & Lavin[a]	
Flacourtiaceae	
Dovyalis abyssinica (Rich.) Warb.	
Frankeniaceae	
Frankenia pulverulenta L.	
Gentianaceae	
Centaurium tenuiflorum Fritsch	
Exacum affine Balf.f. ex Regel[a]	
Exacum caeruleum Balf.f.[a]	
Exacum socotranum Balf.f.[a]	
Sebaea microphylla Knobl.	
Geraniaceae	
Erodium cicutarium (L.) L'Hèr. ex Aiton	
Geranium biuncinatum Kokwaro	
Geranium mascatense Boiss.	

(continued)

(continued)

Family/species	Additional to Miller and Morris (2004)
Geranium trilophum Boiss.	
Pelargonium insularis Gibby & A.G. Mill.[a]	
Globulariaceae	
Poskea socotrana (Balf.f.) G. Taylor	
Goodeniaceae	
Scaevola socotraensis St. John[a]	
Lamiaceae	
Endostemon tenuiflorus (Benth.) Ashby	
Lavandula nimmoi Benth.[a]	
Leucas flagellifolia (Balf.f.) Guerke[a]	
Leucas hagghierensis Gifri & Cortés-Burns[a]	
Leucas kishenensis (Radcl.-Sm) Sebald[a]	
Leucas martinicensis (Jacq.) R. Br.	
Leucas neuflizeana Courbai	
Leucas penduliflora Gifri & Cortés-Burns[a]	
Leucas samhaensis Cortés-Burns & A.G. Mill.[a]	
Leucas spiculifolia (Balf.f.) Guerke[a]	
Leucas urticaefolia R. Br.	
Leucas virgata Balf.f.[a]	
Micromeria remota Balf.f.[a]	
Ocimum forskolei Benth.	
Orthosiphon ferrugineus Balf.f.[a]	
Orthosiphon pallidus Royle	
Plectranthus socotranus Radcl.-Sm.[a]	
Teucrium balfourii Vierh.[a]	
Teucrium sokotranum Vierh.[a]	
Lentibulariaceae	
Utricularia striatula Sm.	
Linaceae	
Linum trigynum L.	
Lythraceae	
Ammannia baccifera L.	
Ammannia multiflora Roxb.	
Lythrum hyssopifolium L.	
Punica protopunica Balf.f.[a]	
Malpighiaceae	
Acridocarpus socotranus Oliv.[a]	
Malvaceae	
Abutilon fruticosum Guill. & Perr.	
Abutilon pannosum Webb	
Gossypium barbadense L.	
Hibiscus dioscorides A.G. Mill.[a]	
Hibiscus diriffan A.G. Mill.[a]	
Hibiscus lobatus Kuntze	
Hibiscus macropodus Wagner & Vierh.[a]	

(continued)

(continued)

Family/species	Additional to Miller and Morris (2004)
Hibiscus malacophyllus Balf.f.[a]	
Hibiscus micranthus L.f.	
Hibiscus noli-tangere A.G. Mill.[a]	
Hibiscus palmatus Forssk.	
Hibiscus quattenensis A.G. Mill. & Thulin[a]	
Hibiscus scindicus Stocks	
Hibiscus scottii Balf.f.[a]	
Hibiscus sidaeformis Baill.	
Hibiscus socotranus G.Ll. Lucas[a]	
Hibiscus stenanthus Balf.f.[a]	
Hibiscus vitifolius L.	
Malva parviflora L.	
Pavonia arabica Hochst. & Steud.	
Senra incana Cav.	
Sida acuta Burm.f.	
Sida cordata (Burm.f.) Borss. Waalk.	
Sida cordifolia L.	
Sida ovata Forssk.	
Meliaceae	
Turraea socotrana Styles & F. White[a]	
Menispermaceae	
Cocculus balfourii Schweinf. ex Balf.f.	
Cocculus pendulus (J.R. Forst.) Diels	
Moraceae	
Dorstenia gigas Schweinf. ex Balf.f.[a]	
Dorstenia socotrana A.G. Mill.[a]	
Ficus cordata Thunb.	
Ficus vasta Forssk.	
Myrsinaceae	
Myrsine africana L.	
Nyctaginaceae	
Boerhavia diffusa L.	
Commicarpus boissieri (Heimerl) Cufod.	
Commicarpus heimerlii (Vierh.) Meikle[a]	
Commicarpus helenae (Roem. & Schult.) Meikle	
Commicarpus simonyi (Heimeri & Vierh.) Meikle[a]	
Oleaceae	
Jasminum fluminense Vell.	
Olea capensis L. ssp. *macrocarpa* (C.H. Wright) I. Verd.	Kilian and Hein (2006)
Onagraceae	
Ludwigia palustris Elliot	
Oxalidaceae	
Oxalis corniculata L.	
Papaveraceae	
Argemone mexicana L.	

(continued)

(continued)

Family/species	Additional to Miller and Morris (2004)
Pedaliaceae	
Pedalium murex L.	
Piperaceae	
Peperomia blanda (Jacq.) Kunth	
Peperomia tetraphylla Hook. & Arn.	
Pittosporaceae	
Pittosporum viridiflorum Sims	
Plantaginaceae	
Plantago aff. *afra* L.	Mies (photographic record from Hamaderoh)
Plantago amplexicaulis Cav.	
Plantago ovata Forssk.	
Plumbaginaceae	
Dyerophytum pendulum (Balf.f.) Kuntze[a]	
Dyerophytum socotranum J.R. Edm.[a]	
Limonium paulayanum (Vierh.) J.R. Edm.[a]	
Limonium sokotranum (Vierh.) Radcl.-Sm.[a]	
Polygalaceae	
Polygala abyssinica Fresen.	
Polygala erioptera DC.	
Polygala kuriense A.G. Mill.[a]	
Polygonaceae	
Calligonum comosum L'Hèr.	
Emex spinosa (L.) Campd.	
Oxygonum sinuatum (Hochst. & Steud.) Dammer	
Persicaria barbata (L.) Hara	
Persicaria glabra (Willd.) M. G mez	
Portulacaceae	
Portulaca kuriensis M.G. Gilbert[a]	
Portulaca oleracea L.	
Portulaca quadrifida L.	
Portulaca samhaensis A.G. Mill.[a]	
Portulaca sedifolia A.G. Mill.[a]	
Talinum portulacifolium (Forssk.) Asch. & Schweinf.	
Primulaceae	
Anagallis arvensis L.	
Resedaceae	
Caylusea hexagyna (Forssk.) M.L. Green	
Ochradenus socotranus A.G. Mill.[a]	
Reseda viridis Balf.f.[a]	
Rhamnaceae	
Helinus integrifolius Lam.	
Ziziphus spina-christi (L.) Willd.	
Rubiaceae	
Carphalea obovata (Balf.f.) Verdc.[a]	

(continued)

(continued)

Family/species	Additional to Miller and Morris (2004)
Diodia aulacosperma K. Schum.	
Gaillonia puberula Balf.f.[a]	
Gaillonia putorioides (A.R. Sm.) Petruss. & Thulin[a]	
Gaillonia thymoides Balf.f.[a]	
Gaillonia tinctoria Balf.f.[a]	
Galium setaceum Lam.	
Galium spurium L.	
Kohautia socotrana Bremek.[a]	
Kohautia subverticellata (S. Schum.) Mantell	
Kraussia socotrana Bridson[a]	
Oldenlandia aretioides Vierh.[a]	
Oldenlandia balfourii Bremek.[a]	
Oldenlandia bicornuta (Balf.f.) Bremek.[a]	
Oldenlandia corymbosa var. *caespitosa* (Benth.) Verdc.	
Oldenlandia herbacea (L.) Roxb.	
Oldenlandia ocellata Bremek.[a]	
Oldenlandia pulvinata (Balf.f.) Bremek.[a]	
Placopoda virgata Balf.f.[b]	
Pyrostria socotrana (Radcl.-Sm) Bridson[a]	
Tamridaea capsulifera (Balf.f.) Thulin & B. Bremer[b]	
Valantia hispida L.	
Rutaceae	
Citrus aurantium Willd.	
Ruta chalapensis L.	
Thamnosma hirschii Schweinf.	
Thamnosma socotrana Balf.f.[a]	
Salvadoraceae	
Salvadora persica L.	
Santalaceae	
Osyris quadripartita Decne.	
Sapindaceae	
Allophylus rubifolius (A. Rich.) Engl.	
Dodonaea viscosa Jacq.	
Sapotaceae	
Sideroxylon fimbriatum Balf.f.[a]	
Spiniluma discolor (Radcl.-Sm.) Friis[a]	
Scrophulariaceae	
Anticharis arabica Endl.	
Aptosimum pumilum Benth.	
Bacopa monnieri (L.) Pennell	
Camptoloma lyperiiflorum (Vatke) Hilliard	
Campylanthus spinosus Balf.f.	
Cistanche phelypaea (L.) Cout.	
Craterostigma pumilum Hochst.	
Graderia fruticosa Balf.f.[a]	

(continued)

(continued)

Family/species	Additional to Miller and Morris (2004)
Lindenbergia sokotrana Vierh.[a]	
Misopates orontium (L.) Raf.	
Nanorrhinum hastatum (Benth.) Ghebrehewit	
Nanorrhinum kuriensis (Radcl.-Sm.) Ghebrehewit[a]	
Orobanche cernua Loefl.	
Orobanche minor Sm.	
Orobanche ramosa L.	
Schweinfurthia pedicellata Benth. & Hook.	
Scrophularia arguta Aiton	
Striga asiatica Kuntze	
Striga densiflora Benth.	
Striga gesnerioides (Willd.) Vatke	
Veronica polita Fr.	
Xylocalyx aculeolatus S. Carter[a]	
Xylocalyx asper Balf.f.[a]	
Solanaceae	
Datura fastuosa L.	
Lycium sokotranum Wagner & Vierh.[a]	
Nicandra physaloides Gaertn.	
Nicotiana tabacum L.	
Physalis minima L.	
Solanum cordatum Forssk.	
Solanum incanum L.	
Solanum villosum Willd.	
Withania adunensis Vierh.[a]	
Withania riebeckii Schweinf.[a]	
Withania somnifera (L.) Dunal	
Sterculiaceae	
Melhania milleri Abedin[a]	
Melhania muricata Balf.f.	
Sterculia africana (Lour.) Fiori	
Waltheria indica L.	
Tamaricaceae	
Tamarix aphylla (L.) Karst.	
Tamarix nilotica (Ehrenb.) Bunge	
Thymelaeaceae	
Gnidia socotrana (Balf.f.) Gilg[a]	
Tiliaceae	
Corchorus aestuans L.	
Corchorus depressus (L.) Stocks	
Corchorus erodiodes Balf.f.[a]	
Grewia bilocularis Balf.f.[a]	
Grewia erythraea Schweinf.	
Grewia mansouriana Abedin	
Grewia milleri Abedin[a]	

(continued)

(continued)

Family/species	Additional to Miller and Morris (2004)
Grewia turbinata Balf.f.[a]	
Triumfetta sp. *A*[a]	
Urticaceae	
Forsskaolea viridis Ehrenb.	
Parietaria debilis G. Forst.	
Pilea tetraphylla (Steud.) Blume	
Pouzolzia auriculata Wight	
Urtica urens L.	
Valerianaceae	
Valerianella affinis Balf.f.[a]	
Verbenaceae	
Clerodendrum galeatum Balf.f.[a]	
Clerodendrum leucophloeum Balf.f.[a]	
Coelocarpum haggierensis A.G. Mill.[a]	
Coelocarpum socotranum Balf.f.[a]	
Phyla nodiflora (L.) Greene	
Priva socotrana Moldenke[a]	
Violaceae	
Hybanthus enneaspermus (L.) F. Muell.	
Viola cinerea Boiss.	
Viscaceae	
Korthalsella taenioides (Comm. ex DC.) Engel	
Vitaceae	
Cissus hamaderohensis Radcl.-Sm.[a]	
Cissus paniculata (Balf.f.) Planch.[a]	
Cissus subaphylla (Balf.f.) Planch.[a]	
Zygophyllaceae	
Fagonia luntii Baker	
Fagonia paulayana Wagner & Vierh.	
Tribulus terrestris L.	
Zygophyllum album L.f.	
Zygophyllum decumbens Del.	
Zygophyllum qatarense Hadidi	
Zygophyllum simplex L.	

[a]endemic species
[b]endemic genus

Appendix 2: List of Bryophytes of Socotra (Based on Kürschner 2006a)

Anthocerotopsida

Anthoceros punctatus L.

Hepaticopsida

Asterella pappii (Gola) Grolle

Athalamia spathysii (Lindenb.) S. Hatt.

Exormotheca pustulosa Mitt.

Frullania ericoides (Nees) Mont.

Frullania schimperi Nees

Frullania socotrana Mitt.

Frullania trinervis (Lehm. & Lindenb.) Gottsche et al.

Lejeunea cavifolia (Ehrh.) Lind.

Lejeunea rhodesiae (Sim) R.M. Schust.

Mannia androgyna (L.) A. Evans

Marchesinia excavata (Mitt.) Schiffn.

Mastigolejeunea auriculata Steph.

Microlejeunea africana Steph.

Oxymitra incrassata (Brot.) Sérgio & Sim-Sim

Pellia epiphylla (L.) Corda

Plagiochasma appendiculatum Lehm. & Lindenb.

Plagiochasma beccarianum Steph.

Plagiochasma rupestre (Forst.) Steph.

Plagiochila fusifera Taylor

Radula appressa Mitt.

Radula comorensis Steph.

Radula quadrata Gottsche

Radula voluta Taylor ex Gottsche et al.

Riccia argenteolimbata O.H. Volk & Perold

Riccia atromarginata Levier var. *jovet-astiae* Rauh & Buchloh

Riccia crinita Taylor

Riccia macrocarpa Levier

Riccia trabutiania Steph.

Southbya nigrella (De Not.) Henriq.

Targionia hypophylla L.

Bryopsida

Anoectangium aestivum (Hedw.) Mitt.

Anomobryum julaceum (P. Gaertn. et al.) Schimp.

Barbula schweinfurthiana Müll. Hal.[a]

Brachymenium nepalense Hook.

Brachymenium sp.

Braunia secunda (Hook.) Bruch & Schimp.

Bryum arachnoideum Müll. Hal.

Bryum caespiticium Hedw.

Bryum capillare Hedw.

Bryum dichotomum Hedw.

(continued)

(continued)
Campylopus pilifer Brid.
Chionoloma bombayense (Müll. Hal.) Sollm.
Didymodon cordatus Jur.
Entosthodon fascicularis (Hedw.) Müll. Hal.
Eucladium verticillatum (Brid.) Bruch & Schimp.
Fabronia socotrana Mitt.
Fissidens flaccidus Mitt.
Fissidens megalotis W.P. Schimp. ex C. Müll.
Fissidens pellucidus Hornsch.
Fissidens serratus Müll. Hal.
Grimmia laevigata (Brid.) Brid.
Grimmia longirostris Hook.
Grimmia trichophylla Grev.
Gyroweisia reflexa (Brid.) Schimp.
Hyophila involuta (Hook.) A. Jaeger
Hyophila punctulata (Mitt.) Kindb.
Macrocoma abyssinica (Müll. Hal.) Vitt
Macrocoma tenuis (Hook. & Grev.) Vitt
Palamocladium leskeoides (Hook.) E. Britton
Papillaria crocea (Hampe) A. Jaeger
Philonotis pungens (Mitt.) Mitt.
Pleurochaete squarrosa (Brid.) Lindb.
Pterogoniadelphus assimilis (Müll. Hal.) Ochyra & Zijlstra
Racopilum capense Müll. Hal.
Schlotheimia balfourii Mitt.[a]
Scorpiurium circinatum (Brid.) M. Fleisch. & Loeske
Sematophyllum socotrense Buck[a]
Semibarbula indica (F. Weber) Wijk.& Margad.
Syntrichia fragilis (Taylor) Ochyra
Tortella humilis (Hedw.) Jenn.
Tortella inclinata (R. Hedw.) Limpr.
Tortella nitida (Lindb.) Broth.
Tortella smithii Townsend[a]
Trichostomum brachydontium Bruch
Trichostomum crispulum Bruch
Weissia artocosana (Mitt.) Zander[a]
Weissia condensa (Voit) Lindb.
Weissia socotrana Mitt.[a]

[a]endemic taxon

Appendix 3: List of Lichens of Socotra

Arthoniaceae	
Arthonia applanata (Müll. Arg.) Stiz.[a]	
Arthonia calospora Müll. Arg.[a]	
Arthonia cinnabarina Wallr.	
Arthonia perpallens Nyl.	
Arthonia polymorpha Ach.	
Arthonia speciosa Nyl.	
Arthonia stictaria Nyl.	
Arthothelium abnorme Müll. Arg.	
Arthothelium crozalsianum (Bouly de Lesd.) Bouly de Lesd.	(5)
Arthothelium emersum Müll. Arg.[a]	
Arthothelium variabile J. Steiner[a]	
Caliciaceae	
Calicium leucinum Müll. Arg.	
Catillariaceae	
Catillaria obfuscata (Müll. Arg.) Zahlbr.[a]	
Catillaria sigmoidea (Müll. Arg.) Zahlbr.[a]	(5)
Chiodectonaceae	
Chiodecton circumscissum Müll. Arg.[a]	
Chiodecton farinaceum Fée	
Chiodecton nanum Müll. Arg.[a]	
Chiodecton perplexum Nyl.	
Chrysotrichaceae	
Chrysothrix candelaris (L.) Laundon	(5)
Cladoniaceae	
Cladonia verticillata (Hoffm.) Schaerer	
Collemataceae	
Collema coccophorum Tuck.	(4)
Collema crispum (Huds.) Wigg.	(4)
Collema fragile Taylor	(4)
Collema fuscovirens (With.) J.R. Laundon	(4)
Collema multipartitum Sm. in Sm. & Sowerby	
Collema nigrescens (Huds.) DC.	
Collema pulcellum Ach. var. *subnigrescens* (Müll. Arg.) Degel.	(5)
Collema rugosum Krempelh.	(5)
Collema subflaccidum Degel.	(4)
Collema tenax (Sw.) Ach. em. Degel.	(4)
Leptogium cyanescens (Rabenh.) Körber	(5)
Leptogium diaphanum Nyl.	
Leptogium menziesii Mont.	
Ectolechiaceae	
Calopadia puiggarii (Müll. Arg.) Vezda	(5)
Tapellaria epiphylla (Müll. Arg.) R. Sant.	(5)

(continued)

(continued)

Gloeoheppiaceae	
Gloeoheppia erosa (J. Steiner) Marton	(4)
Gloeoheppia turgida (Ach.) Gyelnik	(4)
Graphidaceae	
Diploschistes actinostomus (Pers. ex Ach.) Zahlbr.	
Graphina balfourii Müll. Arg.[a]	
Graphina lactea Müll. Arg.[a]	
Graphina socotranum (Müll. Arg.) Müll. Arg.[a]	
Graphina varians Müll. Arg.[a]	
Graphis brachycarpa Müll. Arg.[a]	
Graphis comma Nyl.	
Graphis tenella Ach.	
Helminthocarpon euphorbicolum J. Steiner[a]	
Helminthocarpon leucocarpum (Müll. Arg.) Müll. Arg.	
Helminthocarpon scriptellum J. Steiner[a]	
Phaeographina balfourii (Müll. Arg.) Müll. Arg. in Balf.f.[a]	
Phaeographis inusta Müll. Arg. var. *radians* Müll. Arg. in Balf.f.[a]	
Phaeographis inusta (Ach.) Müll. Arg. var. *simpliciuscula* (Leighton) Müll. Arg. in Balf.f.	
Heppiaceae	
Heppia adglutinata (Krempelh.) Mass.	(4)
Heppia arenacea M. Schultz	(6)
Heppia despreauxii (Mont.) Tuck.	(4)
Heppia exigua (Müll. Arg.) Müll. Arg.[a]	(4)
Heppia solorinoides (Nyl.) Nyl.	(4)
Heppsoraceae	
Heppsora bullata (Müll. Arg.) Lumbsch & Mies	(5)
Lecanoraceae	
Lecanora carpinea Vain.	
Lecanora cretacea (Müll. Arg.) Stiz.[a]	
Lecanora impressa (Kremp.) Zahlbr.	
Lecanora leprosa Fée	
Lecanora notha Fée[a]	
Lecanora subfusca Ach.	
Lecanora sulphurella Hepp	(5)
Lecanora sulphurescens Fée[a]	
Lecanora variata (Müll. Arg.) Stiz.[a]	
Lecideaceae	
Lecidea contractula Müll. Arg.[a]	
Lichinaceae	
Digitothyrea divergens (Henss.) Moreno & Egea	
Lempholemma intricatum (Arn.) Zahlbr.	(4)
Lempholemma socotranum M. Schultz	(3)
Lempholemma polycarpum M. Schultz	(4), (6)
Lichinella algerica (J. Steiner) Moreno & Egea	(4)
Lichinella cribellifera (Nyl.) Moreno & Egea	(4)

(continued)

(continued)

Lichinella iodopulchra (Crozals) Moreno & Egea	(4)
Lichinella nigritella (Lettau) Moreno & Egea	(4)
Lichinella robusta Henss.	(4)
Lichinella stipatula Nyl.	(4)
Paulia aldabrensis Henss.	(1)
Paulia gibbosa Henss.	
Paulia myriocarpa (Zahlbr.) Henss.	(1)
Paulia nitidula (Müll. Arg.) M. Schultz (Syn. *Synalissa nitidula*)	
Paulia perforata (Pers.) Asah.	(1)
Paulia schroederi (Zahlbr.) Henss.	(1)
Peccania arizonica Tuck. ex Herre	(4)
Peccania cerebriformis Henss. & Büdel	(4)
Peccania coralloides (Mass.) Mass.	(4)
Peccania fontqueriana Moreno & Egea	(4)
Phloeopeccania hispanica Willems & Henssen	(4)
Phloeopeccania pulvinulina J. Steiner	(4)
Phylliscum tenue Henss.	(4)
Pseudopaulia tessellata M. Schultz	(2), (4)
Psorotichia diffracta (Nyl.) Forss.	(4)
Psorotichia hassei Fink in Hedrick	(4)
Psorotichia murorum Mass.	(4)
Psorotichia obtenebrans (Nyl.) Forss.	(4)
Psorotichia schaerei (Mass.) Arn.	(4)
Pterygiopsis sp.	(4)
Pyrenopsis cf. *conferta* (Born.) Nyl.	(4)
Pyrenopsis sp. 1	(4)
Pyrenopsis sp. 2	(4)
Synalissa symphorea (Ach.) Nyl.	(4)
Thyrea confusa Henss. in Henss. & P.M. Jorg.	(4)
Thyrea pachyphylla (Müll. Arg.) Henss.	(4)
Thyrea plectopsora Mass.	(4)
Lobariaceae	
Pseudocyphellaria aurata (Ach.) Vainio	
Pseudocyphellaria crocata (L.) Vainio	
Melaspileaceae	
Melaspilea stigmatea Müll. Arg.[a]	
Mycoblastaceae	
Tephromela atra (Huds.) Hafellner in Kalb	
Ochrolechiaceae	
Ochrolechia sp.	(5)
Opegraphaceae	
Cresponea chloroconia (Tuck.) Egea & Torr.	
Cresponea flava (Vain.) Egea & Torr.	(5)
Enterographa affinis Müll. Arg.[a]	
Enterographa fraterculans Müll. Arg.[a]	
Enterographa lactea Müll. Arg.[a]	

(continued)

(continued)

Helminthocarpon scriptellum J. Steiner	(5)
Lecanographa cretacea (Müll. Arg.) Egea & Torr.[a]	
Lecanographa elegans (Müll. Arg.) Mies & Schultz[a] (Syn. *Opegrapha elegans* Müll.Arg.)	(5)
Lecanographa grumulosa (Duf.) Egea & Torr.	(5)
Lecanographa lyncea (Sm.) Egea & Torr.	
Lecanographa subcalcarea (Müll. Arg.) Egea & Torr.[a]	
Lecanographa vestita (Müll. Arg.) Mies[a]	
Opegrapha bonplandi Fée	
Opegrapha caesio-atra J. Steiner[a]	
Opegrapha dracaenarum Müll. Arg.[a]	
Opegrapha microspora Müll. Arg.[a]	
Opegrapha sororiella Müll. Arg.	
Parmeliaceae	
Parmelia latissima Fée f. *sorediata* Nyl.	
Parmelia urceolata Eschw. var. *nuda* Müll. Arg.	
Parmotrema chinense (Osbeck) Hale & Ahti	
Parmotrema crinita (Ach.) Hale	
Parmotrema nilgherrense (Nyl.) Hale	
Parmotrema reticulatum (Taylor in Mackay) M. Choisy	
Parmotrema soyauxii (Müll. Arg.) Hale	
Parmotrema tinctorum (Nyl.) Hale	
Pseudoparmelia carneopruinata (Zahlbr.) Hale	
Rimelia cetrata (Ach.) Hale & Fletcher	
Usnea articulata (L.) Hoff.	
Usnea dasypoga Rohling	
Usnea florida Wigg.	
Xanthoparmelia convexula (Müll. Arg.) Hale[a]	
Xanthoparmelia sublaevis (P. Cout.) Hale	
Peltulaceae	
Peltula bolanderi (Tuck.) Wetmore	(4)
Peltula euploca (Ach.) Poelt	(4)
Peltula imbricata R. Filson	(4)
Peltula impressa (Vain.) Swinscow & Krog	(4)
"Peltula leptophylla"	see (4)
Peltula michoacanensis (Bouly de Lesd.) Wetmore	(4)
Peltula obscurans (Nyl.) Gyelnik	(4)
Peltula patellata (Bagl.) Swinscow & Krog	(4)
Peltula placodizans (Zahlbr.) Wetmore	(4)
Peltula radicata Nyl.	(4)
Pertusariaceae	
Pertusaria flavens Nyl.	
Pertusaria lutescens (Eschw. in Martius) Kremp.	
Pertusaria pertusa Tuck. var. *neocaledonica* Nyl.	
Pertusaria schizostoma Müll. Arg.[a]	
Pertusaria socotrana Müll. Arg.[a]	

(continued)

(continued)

Pertusaria subflavens Müll. Arg.[a]	
Pertusaria xantholeuca Müll. Arg.[a]	
Placolecidaceae	
Placolecis opaca (Fr.) Hafellner	(5)
Placynthiaceae	
Placynthium nigrum (Huds.) S. Gray	(4)
Pyrenulaceae	
Pyrenula immissa (Stirton) Zahlbr.	(5)
Pyrenula obscurata Müll. Arg.[a]	
Physciaceae	
Buellia afra Vain. in Welw.	
Buellia albinea Müll. Arg.[a]	
Buellia areolata Fée	
Buellia brachyspora Müll. Arg.[a]	
Buellia disciformis (Fr.) Mudd. (Syn. *Buellia parasema* Körb.)	
Buellia glaucotheca Malme	
Buellia granularis (Müll. Arg.) Mayrh.[a]	
Buellia innata Müll. Arg.[a]	
Buellia leucina Müll. Arg.[a]	
Buellia protothallina Vain.	
Buellia substigmatea Müll. Arg.[a]	
Buellia substigmatea Müll. Arg. var. *obfuscata* Müll. Arg.[a]	
Diploicia subcanescens (R.G. Werner) Hafellner & Poelt	(5)
Dirinaria aegialita (Afz. ex Ach.) Moore	(5)
Dirinaria applanata (Fée) Awasthi	
Dirinaria confluens (Fr.) Awasthi	(5)
Dirinaria leopoldii (J. Steiner) Awasthi	(5)
Dirinaria picta (Sw.) Clem.& Shear in Clem.	
Heterodermia comosa (Eschw.) Follm. & Redon	(5)
Heterodermia lepidota Swinscow & Krog	(5)
Heterodermia leucomelas (L.) Poelt	
Heterodermia obscurata (Nyl.) Trev.	(5)
Heterodermia speciosa (Wulfen) Trev.	
Physcia atrostriata Moberg	(5)
Physcia caesia (Hoffm.) Fürnr.	(4)
Physcia crispa Nyl.	
Physcia endopyxinea Müll. Arg.[a]	
Physcia integrata Nyl. var. *obsessa* (Nyl.) Vainio	
Physcia obscurella Müll. Arg.[a]	
Physcia obscurella Müll. Arg. var. *fusca* Müll. Arg.[a]	
Physcia tribacioides Nyl.	(5)
Pyxine cocoes Nyl.	
Pyxine convexa Müll. Arg.[a]	
Pyxine endoleuca (Müll. Arg.) Vainio	
Rinodina substellulata Müll. Arg.	
Rinodina teichophila Arn. var. *corticola* Arn.	

(continued)

(continued)

Ramalinaceae	
Bacidia decussata (Müll. Arg.) Zahlbr.[a]	
Bacidia socotrana (Müll. Arg.) Zahlbr.[a]	
Mycobilimbia melanospila (Müll. Arg.) Vouaux	
Phyllopsora furfuracea (Pers.) Zahlbr.	(5)
Ramalina curnowii Crombie in Nyl.	
Ramalina dendriscoides Nyl. var. *minor* Müll. Arg.[a]	
Ramalina dendriscoides Nyl. var. *nodulosa* Müll. Arg.[a]	
Ramalina farinacea Ach.	
Ramalina fastigiata (Pers.) Ach.	
Ramalina furcellata Mont.	
Ramalina nervulosa (Müll. Arg.) Des Abb.	(5)
Ramalina testudinaria Nyl.	
Toninia massata (Tuck.) Herre	(5)
Roccellaceae	
Dirina approximata Zahlbr.	(4)
Dirina ceratoniae (Ach.) Fr.	
Dirina immersa Müll. Arg.	
Dirina massiliensis Dur. & Mont.	
Dirina massiliensis Dur. & Mont. f. *sorediata* (Müll. Arg.) Tehler	
Dirina paradoxa (Fée) Tehler ssp. *africana* (Fée) Tehler	(5)
Feigeana socotrana Mies, Lumbsch & Tehler[a]	
Minksia caesiella Müll. Arg.[a]	
Minksia candida Müll. Arg.[a]	
Roccella balfourii Müll. Arg.	
Roccella belangeriana Fée	(5)
Roccella fuciformis (L.) Ach.	
Roccella hertelii Mies & Schultz[a]	(5) see footnote (b)
Roccella montagnei Bél. var. *peruensis* Krempelh.	
Roccella tinctoria DC. in Lamarck & DC	
Roccellina cinerea (Müll. Arg.) Tehler[a]	
Roccellina cinerea (Müll. Arg.) Tehler f. *sorediosa*[a]	
Roccellographa cretacea J. Steiner	
Simonyella laevigata J. Steiner	
Scoliciosporaceae	
Scoliciosporum umbrinum (Ach.) Arnold	(5)
Sphinctrinaceae	
Sphinctrina microcephala Nyl.	
Sphinctrina tubaeformis Mass.	(5)
Stictidaceae	
Thelopsis paucispora Breuss & M. Schultz[a]	(7)
Teloschistaceae	
Caloplaca albido-caerulescens (Müll. Arg.) J. Steiner[a]	
Caloplaca aurantiaca Th. Fr.	
Caloplaca aurantiaca Th. Fr. var. *isidiosellum* (Crombie) Zahlbr.	
Caloplaca balfourii (Müll. Arg.) Zahlbr.[a]	

(continued)

(continued)

Caloplaca citrina Th. Fr.	
Caloplaca deplanata (Müll. Arg.) Zahlbr.[a]	
Caloplaca granulifera (Müll. Arg.) Jatta[a]	
Caloplaca granulifera (Müll. Arg.) Jatta var. *subvitellina* Müll. Arg.[a]	
Caloplaca poliotera (Müll. Arg.) J. Steiner	
Caloplaca pyracea Th. Fr.	
Caloplaca steropea Poetsch in Poetsch & Schiderm.	
Teloschistes flavicans (Sw.) Norm.	
Teloschistes flavicans (Sw.) Norm. var. *maximus* Zahlbr.	
Tornabea scutellifera (With.) Laundon	(5)
Thelenellaceae	
Thelenella luridella (Nyl.) Mayrh.	
Trypetheliaceae	
Laurera pauperrima (Müll. Arg.) Zahlbr.[a]	
Laurera velata (Müll. Arg.) Zahlbr.[a]	
Verrucariaceae	
Catapyrenium lacinulatum (Ach.) Breuss	(4)
Catapyrenium semaforonensis Breuss	(4)
Catapyrenium squamulosum (Ach.) Breuss	(4)
Heteroplacidium podolepis (Breuss) Breuss	(5)
Normandina pulchella Nyl.	
Placidium lacinulatum (Ach.) Breuss	(5)
Placidium semaforonense (Breuss) Breuss	(5)
Placidium squamulosum (Ach.) Breuss	(5)
Polyblastiopsis tropica Müll. Arg.[a]	
Verrucaria libricola (Fée) Nyl.	
Verrucaria prominens Müll. Arg.[a]	
Verrucaria rupestris Schrader var. *alocizoides* Müll. Arg.[a]	

Based on Mies (2001), with additional records denoted in the second column as follows: (1) Schultz et al. (1999); (2) Schultz (2002); (3) Schultz (2003); (4) Schultz and Mies (2003); (5) Mies and Schultz (2004); (6) Schultz (2005); (7) Breuss and Schultz 2007

[a] endemic taxon

[b] In a recent paper, Tehler et al. (2010) concluded that *R. hertelii* can probably be regarded as a form of *R. balfourii*. We do not share this opinion, based on extensive field experience of the variability of this species. It should also be noted that Tehler et al. (2010) did not include *R. hertelii* in their phylogenetic analyses.

References

Abdel-Rahman A-FM, Doig R (1987) The Rb-Sr geochronological evolution of the Ras Gharib segment of the northern Nubian Shield. J Geol Soc Lond 144:577–586

Acocks JPH (1953) Veld types of South Africa. Mem Bot Surv S Afr 28:1–192

Adams JA, Endo AS, Stolzy LH, Rowlands PG, Johnson HB (1982) Controlled experiments on soil compaction produced by off-road vehicles in the Mojave Desert, California. J Appl Ecol 19:167–175

Adolt R, Pavlis J (2004) Age structure and growth of *Dracaena cinnabari* populations on Socotra. Trees 18:43–53

Agnew C, Warren A (1996) A framework for tackling drought and land degradation. J Arid Environ 33:309–320

Akhani H, Trimborn P, Ziegler H (1997) Photosynthetic pathways in Chenopodiaceae from Africa, Asia and Europe with their ecological, phytogeographical and taxonomical importance. Plant Syst Evol 206:187–221

Al-Gifri AN, Kürschner H, Mies B (1995) New records, additions and a new species, *Sematophyllum socotrense* Buck (Sematophyllaceae, Musci) to the bryophyte flora of Socotra (Yemen). Studies in Arabian bryophytes. 19. Nova Hedwigia 61:467–480

Alexander D, Miller A (1995) Socotra's misty future. New Sci 147:32–35

Alexander D, Miller A (1996) Saving the spectacular flora of Socotra. Plant Talk 7:19–22

Allen AP, Pockman WT, Restrepo C, Milne BT (2008) Allometry, growth and population regulation of the desert shrub *Larrea tridentata*. Funct Ecol 22:197–204

Althawadi AM, Grace J (1986) Water use by the desert cucurbit *Citrullus colocynthis* (L.) Schrad. Oecologia 70:475–480

Andrus N, Trusty J, Santos-Guerra A, Jansen RK, Francisco-Ortega J (2004) Using molecular phylogenies to test phytogeographical links between East/South Africa–Southern Arabia and the Macaronesian islands – a review, and the case of *Vierea* and *Pulicaria* section Vieraeopsis (Asteraceae). Taxon 53:333–346

Anonymous (1924) Obituary notices of fellows deceased: Sir Isaac Bayley Balfour, 1853–1922. Philos R Soc Lond B, 96:i–xxxiv

Antonovics J, Bradshaw AD, Turner RG (1971) Heavy metal tolerance in plants. Adv Ecol Res 7:1–85

Arvidsson L (1982) A monograph of the lichen genus *Coccocarpia*. Opera Bot 67:1–96

Attorre F, Francesconi F, Taleb N, Scholte P, Saed A, Alfo M, Bruno F (2007) Will dragonblood survive the next period of climate change? Current and future potential distribution of *Dracaena cinnabari* (Socotra, Yemen). Biol Conserv 138:430–439

Attorre F, Taleb N, de Sanctis M, Farcomeni A, Guillet A, Vitale M (2011) Developing conservation strategies for endemic tree species when faced with time and data contraints: *Boswellia* spp. on Socotra (Yemen). Biodivers Conserv 20:1483–1499

DOI 10.1007/978-94-007-4141-6,

Axelrod DI, Raven PH (1978) Late Cretaceous and tertiary vegetation history of Africa. In: Werger MJA (ed) Biogeography and ecology of Southern Africa. Dr W Junk, The Hague

Baas P, Jansen S, Smets E (2001) Vegetative anatomy and affinities of *Dirachma socotrana* (Dirachmaceae). Syst Bot 26:231–241

Bainbridge DA (2007) A guide for desert and dryland restoration: new hope for arid lands. Society for Ecological Restoration International. Island Press, Washington, DC

Baker HG (1948) Dimorphism and monomorphism in the Plumbaginaceae: I. A survey of the family. Ann Bot 12:207–219

Balfour IB (1881) On the island of Socotra. Rep Annu Meet Br Assoc Adv Sci 1881:482–494

Balfour IB (1882a) Diagnoses plantarum novarum et imperfecte descriptorum phanerogamarum Socotrensium. Proc R Soc Edinb B 11:498–514, 834–843

Balfour IB (1882b) The dragon's blood tree of Socotra (*Dracaena cinnabari* Balf.fil.). Trans R Soc Edinb-Earth 30:619–623

Balfour IB (1883) Diagnoses plantarum novarum et imperfecte descriptorum phanerogamarum Socotrensium. Pars tertia. Proc R Soc Edinb B 12:76–98, 402–411

Balfour IB (1884) Diagnoses plantarum novarum et imperfecte descriptorum phanerogamarum Socotrensium. Pars quarta (supplementum). Proc R Soc Edinb B 13:402–411

Balfour IB (1888) Botany of Sokotra. Trans R Soc Edinb 31:1–446

Balfour IB (1903) Botany of Sokotra and Abd-el-Kuri: Angiospermae. In: Forbes HO (ed) The natural history of Sokotra and Abd-el-Kuri. Bulletin of the Liverpool Museums (Special Bulletin). Henry Young, London

Balkwill M-J, Balkwill K (1997) Delimitation and infra-generic classification of *Baleria* (Acanthaceae). Kew Bull 52:535–573

Balkwill M-J, Balkwill K (1998) A preliminary analysis of distribution patterns in a large, pantropical genus, *Barleria* L. (Acanthaceae). J Biogeogr 25:95–110

Ballard HE Jr, Sytsma KJ (2000) Evolution and biogeography of the woody Hawaiian violets (*Viola*, Violaceae): Arctic origins, herbaceous ancestry and bird dispersal. Evolution 54:1521–1532

Barbour M (1973) Desert dogma reexamined: root/shoot productivity and plant spacing. Am Midl Nat 89:41–57

Barbour MG, Burk JH, Pitts WD, Gilliam FS, Schwartz MW (1998) Terrestrial plant ecology, 3rd edn. Benjamin/Cummings, California

Barth H-J (1999) Desertification in the Eastern Province of Saudi Arabia. J Arid Environ 43:399–410

Baskin CC, Baskin JM (1998) Seeds: ecology, biogeography, and evolution of dormancy and germination. Academic, London

Basson PW, Burchard JE, Hardy JT, Price ARG (1977) Biotopes of the western Arabian Gulf. Marine life and environments of Saudi Arabia. ARAMCO, Dhahran

Bazara'a MAR, Guarino L, Miller AG, Morris M, Obadi N (1991) Finger millet on Socotra: a disappearing crop. FAO/IBPGR Plant Genet Resour Newsl 88/89:67–68

Beals EW (1968) Spatial patterns of shrubs on a desert plain in Ethiopia. Ecology 49:744–746

Beier B-A, Nylander JAA, Chase MW, Thulin M (2004) Phylogenetic relationships and biogeography of the desert plant genus *Fagonia* (Zygophyllaceae), inferred by parsimony and Bayesian model averaging. Mol Phylogenet Evol 33:91–108

Bellstedt DU, van Zyl L, Marais EM, Bytebier B, de Villiers CA, Makwarela AM, Dreyer LL (2008) Phylogenetic relationships, character evolution and biogeography of southern African members of *Zygophyllum* (Zygophyllaceae) based on three plastid regions. Mol Phylogenet Evol 47:932–949

Benabid A, Cuzin F (1997) Populations de dragonnier (*Dracaena draco* L. subsp. *ajgal* Benabid et Cuzin) au Maroc: valeurs taxinomique, biogéographique et phytosociologique. C R Acad Sci Paris, Science de la Vie 320:267–277

Benz BW, Martin CE (2006) Foliar trichomes, boundary layers, and gas exchange in 12 species of epiphytic *Tillandisia* (Bromeliaceae). J Plant Physiol 163:648–656

Beydoun ZR, Bichan HR (1970) The geology of Socotra Island, Gulf of Aden. Q J Geol Soc Lond 125:413–446

Beyhl FE (1996a) The Emta tree (*Euphorbia arbuscula* Balf. fil.), a succulent tree of dracoid habitus from the island of Soqotra (Dicotyledones: Tricococcae: Euphorbiaceae). In: Dumont HJ (ed) Proceedings of the first international symposium on Soqotra Island: present and future. United Nations, New York

Beyhl FE (1996b) Attempts at raising the Soqotran dragon tree, *Dracaena cinnabari* Balf. fil., outside the island (Monocotyledones: Lililes: Agavaceae). In: Dumont HJ (ed) Proceedings of the first international symposium on Soqotra Island: present and future. United Nations, New York

Beyhl FE (2000) Die Drachenbaum-Wolfsmilch (*Euphorbia arbuscula*), ein Baum von drachenbaumartigem Wuchs. Mitt Deut Dendrol Ges 85:91–101

Beyhl FE, Mies BA (1996) Remarks on the vegetation ecology of Socotra. In: University of Aden (ed) First international scientific symposium on Socotra Island: present and future, vol 4. Aden, 24–28 Mar 1996,pp 25–62

Beyhl FE, Mies BA (1999) Die Soqotrinische Wüstenrose *Adenium socotranum* Vierh. und die "Insel der Seligen". Acta Biol Benrodis 10:13–20

Birks HJB (1973) Past and present vegetation of the Isle of Skye. Cambridge University Press, Cambridge

Birse ACR, Bott WF, Morrison J, Samuel MA (1997) The Mesozoic and Early Tertiary tectonic evolution of the Socotra area, eastern Gulf of Aden, Yemen. Mar Petrol Geol 14:675–684

Biswell HH (1956) Ecology of California grasslands. J Range Manage 9:19–24

Bobe R (2006) The evolution of arid ecosystems in eastern Africa. J Arid Environ 66:564–584

Botting D (1957) Report on the Oxford University expedition to Socotra. Oxford Univ Explor Club Bull 8:45–57

Botting D (1958a) The Oxford University expedition to Socotra. Geogr J 124:200–209

Botting D (1958b) The island of the dragon's blood. Hodder and Staughton, London

Bramwell D (1985) Contribucion a la biogeografia de las Islas Canarias. Botanica Macaronesica 14:3–34

Bremer B (2009) A review of molecular phylogenetic studies of Rubiaceae. Ann Mo Bot Gard 96:4–26

Bremer B, Thulin M (1998) Collapse of Isertieae, re-establishment of Mussaendeae, and a new genus of Sabiceeae (Rubiaceae); phylogenetic relationships based on rbcL data. Plant Syst Evol 211:71–92

Breuss O, Schultz M (2007) *Thelopsis paucispora*, a new lichen species from Socotra (Yemen). Lichenologist 39:35–40

Bridson DM (1995) A new *Kraussia* (Rubiaceae–Octotropideae) from Socotra. Kew Bull 50:773–776

Briggs JC (2003) The biogeographic and tectonic history of India. J Biogeogr 30:381–388

Brockmann-Jerosch H (1919) Baumgrenze und Klimacharakter. Beiträge zur geobotanischen Landesaufnahme 6. Rascher, Zürich

Brown G (2001) Vegetation ecology and biodiversity of degraded desert areas in north-eastern Arabia. Habilitation thesis, University of Rostock

Brown G (2003) Factors maintaining plant diversity in degraded areas of northern Kuwait. J Arid Environ 54:183–194

Brown G (2006) The sabkha vegetation of the United Arab Emirates. In: Khan MA, Böer B, Kust GS, Barth H-J (eds) Sabkha ecosystems, vol II. Springer, Berlin

Brown G (2009) Observations on the cooling behaviour, and associated habitat, of four desert lark species (Alaudidae) in two areas of Kuwait. Sandgrouse 31:6–14

Brown G, Al-Mazrooei S (2003) Rapid vegetation regeneration in a seriously degraded *Rhanterium epapposum* community in northern Kuwait after four years of protection. J Environ Manage 68:387–395

Brown G, Brinkmann K (1992) Heavy metal tolerance in *Festuca ovina* L. from contaminated sites in the Eifel Mountains, Germany. Plant Soil 143:239–247

Brown G, Porembski S (1998) Flora and vegetational aspects of miniature dunes in a sand-depleted *Haloxylon salicornicum* community in the Kuwait desert. Flora 193:133–140

Brown G, Peacock J, Loughland R, Alhadrami GA (2006) Coastal and terrestrial ecosystem management requirements in the Arabian Peninsula. In: Amer KM, Böer B, Brook MC, Adeel Z, Clüsener-Godt M, Saleh W (eds) Policy perspectives for ecosystem and water management in the Arabian Peninsula. United Nations University, Hamilton

Brown G, Aspinall S, Sakkir S (2007) The vegetation of the coastal white sands at Taweela (Abu Dhabi Emirate). Tribulus 17:5–15

Brown G, Böer B, Sakkir S (2008) The coastal vegetation of the western and southern Gulf – characterisation and conservation aspects. In: Abuzinada AH, Barth H-J, Krupp F, Böer B, Abdessalaam TZ (eds) Protecting the Gulf's marine ecosystems from pollution. Birkhäuser Verlag, Basel

Brown JH, Valone TJ, Curtin CG (1997) Reorganization of an arid ecosystem in response to recent climate change. Proc Natl Acad Sci USA 94:9729–9733

Bruijnzeel LA (2005) Tropical montane cloud forest: a unique hydrological case. In: Bonell M, Bruijnzeel LA (eds) Forests, water and people in the humid tropics. Cambridge University Press, Cambridge

Bruyns PV (1988) A revision of the genus *Echidnopsis* Hook.f. (Asclepiadaceae). Bradleya 6:1–48

Bruyns PV (2000) Phylogeny and biogeography of the stapeliads. Plant Syst Evol 221:227–244

Bruyns PV (2004) New species in *Echidnopis* and *Secamone* and new records of *Ceropegia* (Apocynaceae) from Socotra. Edinb J Bot 61:7–19

Bruyns PV, Miller AG (2002) *Socotrella*, a new genus of stapeliad (Apocynaceae–Asclepiadoideae) from the island of Socotra. Novon 12:330–333

Burgoyne PM, van Wyk AE, Anderson JM, Schrire BD (2005) Phanerozoic evolution of plants on the African plate. J Afr Earth Sci 43:13–52

Carter RWG (1988) Coastal environments. An introduction to the physical, ecological and cultural systems of coastlines. Academic, London

Carter S (1962) Revision of *Xylocalyx* Balf. f. Kew Bull 16:147–152

Carter S (1992) New pair-spined species of *Euphorbia* (Euphorbiaceae) from Somalia. Nord J Bot 12:403–421

Carter S, Lavranos JJ, Newton LE, Walker CC (2011) Aloes: the definitive guide. Royal Botanic Gardens, Kew

Casper SJ (2000) Die Geschichte des Kanarischen Drachenbaumes in Wissenschaft und Kunst. Vom Arbor Gadensis des Posidonius zur *Dracaena draco* (L.) L. Haussknechtia Beih 10:1–128

Caujapé-Castells J, Tye A, Crawford DJ, Santos-Guerra A, Sakai A, Beaver K, Lobin W, Florens FBV, Moura M, Jardim R, Gómes I, Kueffer C (2010) Conservation of oceanic island floras: present and future global challenges. Perspect Plant Ecol Evol Syst 12:107–129

Cerny V, Pereira L, Kujanova M, Vasikova A, Hajek M, Morris M, Mulligan CJ (2009) Out of Arabia – the settlement of island Soqotra as revealed by mitochondrial and Y chromosome genetic diversity. Am J Phys Anthropol 138:439–447

Chapman VJ (1974) Salt marshes and salt deserts of the world, 2nd edn. Cramer, Lehre

Charley JL, West NE (1975) Plant-induced soil chemical patterns in some shrub-dominated semidesert ecosystems of Utah. J Ecol 63:945–963

Cheung C, DeVantier L (2006) Socotra – A natural history of the islands and their people. Odyssey Book and Guides, Hong Kong

Cook LG, Crisp MD (2005) Directional asymmetry of long-distance dispersal and colonization could mislead reconstructions of biogeography. J Biogeogr 32:741–754

Cope TA (2007) Flora of the Arabian Peninsula and Socotra, vol 5, part 1. Edinburgh University Press, Edinburgh

Cortés-Burns H, Schrire BD, Pennington RT, Miller AG (2004) A taxonomic revision of Socotran Indigofereae (Leguminosae – Papilionoideae) with insights into the phytogeographical links of the Socotran Archipelago. Nord J Bot 22:693–711

Cowan PJ, Brown GM (2001) Prostrate desert gourd plants as apparent cooling sites for larks in heat of day. Sandgrouse 23:59–60

Cronk QCB (1986) Botanical survey of Socotra for conservation and plant genetic resources. WWF conservation yearbook 1985/1986. WWF, Gland

Cronk QCB (1989) The past and present vegetation of St Helena. J Biogeogr 16:47–64
Cronk QCB (1997) Islands: stability, diversity, conservation. Biodivers Conserv 6:477–493
Crosby MR, Schultze-Motel U, Schultze-Motel W (1983) Katalog der Laubmoose von Madagaskar und den umliegenden Inseln. Willdenowia 13:187–255
Cunningham GL, Strain BR (1969) An ecological significance of seasonal leaf variability in a desert shrub. Ecology 50:400–408
Danin A (1978) Plant species diversity and plant succession in a sandy area in the northern Negev. Flora 167:409–422
Danin A (1996) Plants of desert dunes. Springer, Heidelberg
Danin A, Orshan G (1990) The distribution of Raunkiaer life forms in Israel in relation to the environment. J Veg Sci 1:41–48
Darbishire OV (1903) Botany of Sokotra and Abd-el-Kuri: Thallophyta: Lichenes. In: Forbes HO (ed) A natural history of the islands of Sokotra and Abd-el-Kuri. Bulletin of the Liverpool Museums (Special Bulletin). Henry Young, London
Davis CC, Bell CD, Fritsch PW, Mathews S (2002) Phylogeny of *Acridocarpus-Brachylophon* (Malpighiaceae): implications for tertiary tropical floras and Afroasian biogeography. Evolution 56:2395–2405
Davis MA, Grime JP, Thompson K (2000) Fluctuating resources in plant communities: a general theory of invasibility. J Ecol 88:528–534
de Nicolás JP, Fernandez-Palacios JM, Ferrer FJ, Nieto E (1989) Inter-island floristic similarities in the Macaronesian region. Vegetatio 84:117–125
de Quieroz A (2005) The resurrection of oceanic dispersal in historical biogeography. Trends Ecol Evol 20:68–73
de Winter B (1971) Floristic relationships between the northern and southern arid areas in Africa. Mitt Bot Staatssamml Muenchen 10:424–437
Deil U (1991) Rock communities in tropical Arabia. Flora et Vegetatio Mundi 9:175–187
Deil U (1996) Zur Kenntnis der Adiantetea-Gesellschaften des Mittelmeerraumes und umliegender Gebiete. Phytocoenologia 26:481–536
Deil U (1998) Coastal and sabkha vegetation. In: Ghazanfar SA, Fisher M (eds) Vegetation of the Arabian Peninsula. Kluwer, Dordrecht
Deil U (2000) Halophytic vegetation along the Arabian coast – azonal or linked to climatic zones? Phytocoenologia 30:591–611
Deil U, Al-Gifri A-N (1998) Montane and wadi vegetation. In: Ghazanfar SA, Fisher M (eds) Vegetation of the Arabian Peninsula. Kluwer, Dordrecht
Deil U, Müller-Hohenstein K (1996) An outline of the vegetation of Dubai (UAE). Verh Ges Ökologie 35:77–95
Demissew S, Mill R (2009) Revision of the genus *Seddera* (Convolvulaceae). Kew Bull 64: 179–233
Doe B (1992) Socotra. Island of tranquillity. Immel Publishing, London
Dregne HE (1986) Desertification of arid lands. In: El-Baz F, Hassam MHA (eds) Physics of desertification. Martinus Nijhoff Publishers, Dordrecht
Durn G (2003) Terra rossa in the Mediterranean region: parent materials, composition and origin. Geol Croat 56:83–100
Ebert TA, McMaster GS (1981) Regular pattern of desert shrubs: a sampling artefact? J Ecol 69:559–564
Edgell HS (2006) Arabian deserts: nature, origin and evolution. Springer, Berlin
Eggli U (2002) The illustrated handbook of succulent plants. Dicotyledons. Springer, Berlin
Ehleringer JR (1985) Annuals and perennials of warm deserts. In: Chabot BF, Mooney HA (eds) Physiological ecology of North American plant communities. Chapman and Hall, New York
Ehleringer JR, Björkman O (1976) A comparison of photosynthetic characteristics of *Encelia* species possessing glabrous and pubescent leaves. Plant Physiol 62:185–190
Ehleringer JR, Mooney HA (1978) Leaf hairs: effects on physiological activity and adaptive value to a desert shrub. Oecologia 37:183–200
Ehleringer JR, Björkman O, Mooney HA (1976) Leaf pubescence: effects on absorptance and photosynthesis in a desert shrub. Science 192:376–377

El-Bana MI, Li ZQ, Nijs I (2007) Role of host identity in effects of phytogenic mounds on plant assemblages and species richness on coastal arid dunes. J Veg Sci 18:635–644

Ellison AM (2000) Mangrove restoration: do we know enough? Restor Ecol 8:219–229

Ellner S, Shmida A (1981) Why are adaptations for long-range seed dispersal rare in desert plants? Oecologia 51:133–144

Engler A (1910) Socotra. Die Pflanzenwelt Afrikas, insbesondere seiner tropischen Gebiete. Grundzüge der Pflanzenverbreitung in Afrika und die Charakterpflanzen Afrikas. I. Allgemeiner Überblick über die Pflanzenwelt Afrikas und ihre Existenzbedingungen. W Engelmann, Leipzig

Engler A (1913) Burseraceae. Bot Jahrb 48:447

Engler A, Drude O (1915) Die Pflanzenwelt Afrikas, insbesondere seiner tropischen Gebiete. Grundzüge der Pflanzenverbreitung in Afrika und die Charakterpflanzen Afrikas. Heft III-1: Die dicotyledonen Angiospermen, Casuarinaceae bis Dichapetalaceae. W Engelmann, Leipzig

Eren M, Kadir S, Hatipoglu Z, Gül M (2008) Quaternary calcrete development in the Mersin area, southern Turkey. Turk J Earth Sci 17:763–784

Evenari M (1985) Adaptations of plant and animals to the desert environment. In: Evenari M, Noy-Meir I, Goodall DW (eds) Ecosystems of the world 12A. Hot deserts and arid shrublands, A. Elsevier, Amsterdam

Evenari M, Shanan L, Tadmor N (1982) The Negev: the challenge of a desert. Harvard University Press, Cambridge

Fisher M, Ghazanfar SA, Chaudhary SA, Seddon PJ, Robertson EF, Omar S, Abbas JA, Böer B (1998) Diversity and conservation. In: Ghazanfar SA, Fisher M (eds) Vegetation of the Arabian Peninsula. Kluwer, Dordrecht

Fleitmann D, Burns SJ, Neff U, Mangini A, Matter A (2003) Changing moisture sources over the last 330,000 years in Northern Oman from fluid-inclusion evidence in speleothems. Quat Res 60:223–232

Fleitmann D, Matter A, Burns SJ, Al-Subbary A, Al-Aowah MA (2004) Geology and Quaternary climate history of Socotra. In: Krupp F (ed) Fauna of Arabia. Socotra Archipelago, vol 20. Libri, Basel

Forbes HO (1899) The English expedition to Socotra. Geogr J 13:633–637

Forbes HO (1903) The natural history of Sokotra and Abd-el-Kuri. Bulletin of the Liverpool Museums (Special Bulletin). Henry Young, London

Forman LL (1980) *Cocculus balfourii* (Menispermaceae) in Oman (Dhofar). Kew Bull 35:379–381

Francisco-Ortega J, Jansen RK, Santos-Guerra A (1996) Chloroplast DNA evidence of colonization, adaptive radiation, and hybridization in the evolution of the Macaronesian flora. Proc Natl Acad Sci USA 93:4085–4090

Frey W, Kürschner H (1988) Bryophytes of the Arabian Peninsula and Socotra. Floristics, phytogeography and definition of the xerothermic Pangaean element. Studies in Arabian bryophytes 12. Nova Hedwigia 46:37–120

Friedel MH (1991) Range condition assessment and the concept of thresholds: a viewpoint. J Range Manage 44:422–426

Gadgil S (2003) The Indian monsoon and its variability. Annu Rev Earth Planet Sci 31:429–467

Galley C, Bytebier B, Bellstedt DU, Linder HP (2007) The Cape element in the Afrotemperate flora: from Cape to Cairo? Proc R Soc Lond B 274:535–543

Galloway DJ (1988) Plate tectonics and the distribution of cool temperate Southern Hemisphere macrolichens. Bot J Linn Soc Lond 96:45–55

Gams H (1964) Die Bedeutung der afromontanen und afroalpinen Floren für die Geschichte der mediterran-montanen und alpinen Floren. Phyton 11:1–17

Gates DM, Aldrefer R, Taylor E (1968) Leaf temperatures of desert plants. Science 159:994–995

Gibson AC (1996) Structure-function relations of warm desert plants. Springer, Berlin

Gilbert MG (1980) *Duvaliandra*, a new genus of stapeliad. Cact Succ J Gr Brit 42:99–101

Gillett JB (in Popov GB) (1957) Species referred to in G.B. Popov's paper, "The vegetation of Socotra". J Linn Soc Lond 55:717–720

Gillett JB (1966) The species of *Ormocarpum* Beauv. and *Arthrocarpum* Balf.f. (Leguminosae) in south-western Asia and Africa (excluding Madagascar). Kew Bull 20:323–355

Gillett JB (1991) Burseraceae. In: Polhill RM (ed) Flora of tropical East Africa. Balkema, Rotterdam

Gleason HA (1926) The individualistic concept of the plant association. Bull Torrey Bot Club 53:7–26

Glendening GE (1952) Some quantitative data on the increase of mesquite and cactus on a desert grassland range in southern Arizona. Ecology 33:319–328

Goldblatt P, Manning JC, Davies J, Savolainen V, Rezai S (2003) *Cyanixia*, a new genus for the Socotran endemic *Babiana socotrana* (Iridaceae-Crocoideae). Edinb J Bot 60:517–532

Goudie AS (1973) Duricrusts in tropical landscapes. Clarendon Press, Oxford

Grainger A (1992) Characterization and assessment of desertification processes. In: Chapman GP (ed) Desertified grasslands: their biology and management. Academic, London

Grime JP (1977) Evidence for the existence of three primary strategies in plants and its relevance to ecological and evolutionary theory. Am Nat 111:1169–1194

Grime JP (2001) Plant strategies, vegetation processes and ecosystem properties. Wiley, Chichester

Guebourg J-L (1998) Socotra, une île hors du temps. Presses Universitaires Bourdeaux, Bourdeaux

Gutterman Y (1993) Seed germination in desert plants. Springer, Berlin

Gutterman Y (1996) Environmental influences during seed maturation and storage affecting germinability in *Spergularia diandra* genotypes inhabiting the Negev Desert, Israel. J Arid Environ 34:313–323

Guzmán B, Vargas P (2009) Long-distance colonization of the Western Mediterranean by *Cistus ladanifer* (Cistaceae) despite the absence of special dispersal mechanisms. J Biogeogr 36:954–968

Gwynne MD (1968) Socotra. Acta Phytogeogr Suec 54:179–185

Haake B, Rixen T, Ittekkot V (1993) Variability of moonsonal upwelling signals in the deep western Arabian Sea. Mitt Geol-Paläont Inst Univ Hamburg (SCOPE/UNEP special volume) 76:85–96

Habrova H, Cermak Z, Pavlis J (2009) Dragon's blood tree – Threatened by overmaturity, not by extinction: dynamics of a *Dracaena cinnabari* woodland in the mountains of Soqotra. Biol Conserv 142:772–778

Halliday P (1988) *Kleinia scottii*. Noteworthy species of *Kleinia* (Halliday, P.) Hooker's Icon Pl 39(4): t3897

Hardin G (1968) The tragedy of the commons. Science 162:1243–1248

Heady HF (1958) Vegetational changes in the California annual type. Ecology 39:402–416

Hedberg O (1970) Evolution of the afroalpine flora. Biotropica 2:16–23

Heinrichs J, Lindner M, Groth H, Hentschel J, Feldberg K, Renker C, Engel JJ, von Konrat M, Long DG, Schneider H (2006) Goodbye or welcome Gondwana? – insights into the phylogenetic biogeography of the leafy liverwort *Plagiochila* with a description of *Proskauera*, gen. nov. (Plagiochilaceae, Jungermanniales). Plant Syst Evol 258:227–250

Henderson S, Dawson TP, Whittaker RJ (2006) Progress in invasive plants research. Prog Phys Geogr 30:25–46

Heywood VH, Brummitt RK, Culham A, Seberg O (2007) Flowering plant families of the world. Royal Botanic Gardens, Kew

Higgins SI, Nathan R, Cain ML (2003) Are long-dispersal events in plants usually caused by non-standard means of dispersal? Ecology 84:1945–1956

Hildebrandt A, Al Aufi M, Amerjeed M, Shammas M, Eltahir EAB (2007) Ecohydrology of a seasonal cloud forest in Dhofar: 1. Field experiment. Water Resour Res 43:W10411

Hjertson M (2003) Revision of the disjunct genus *Campylanthus* (Scrophulariaceae). Edinb J Bot 60:131–174

Hjertson M, Henrot J, Thulin M (2008) *Campylanthus hajarensis* sp. nov. and a new record of *Campylanthus* (Scrophulariaceae) from Oman. Nord J Bot 26:35–37

Hobohm C (2000) Plant species diversity and endemism on islands and archipelagos, with special reference to the Macaronesian Islands. Flora 195:9–24

Holling CS (1973) Resilience and stability of ecological systems. Annu Rev Ecol Syst 4:1–23

Howarth DG, Gustafsson MHG, Baum DA, Motley TJ (2003) Phylogenetics of the genus *Scaevola* (Goodeniaceae): implication for dispersal patterns across the Pacific Basin and colonization of the Hawaiian Islands. Am J Bot 90:915–923

Hufford KM, Mazer SJ (2003) Plant ecotypes: genetic differentiation in the age of ecological restoration. Trends Ecol Evol 18:147–155

Hughes M, Miller AG (2002) A new endemic species of *Begonia* (Begoniaceae) from the Socotra Archipelago. Edinb J Bot 59:273–281

Hunt DR (1971) *Wellstedia socotrana* Balf.f. Hooker's Icon Pl 38 (7):5 ser, t3665

Hunt DR, Lebrun JP (1975) A new *Wellstedia* (Boraginaceae) from Ethiopia. Kew Bull 30:222

Hunter FM (1876) Notes on Socotra. J Anthropol Inst 7:364–372

IUCN (2010) IUCN Red list of threatened species. Version 2010.4. http://www.iucnredlist.org. Accessed 27 Oct 2010

IUSS Working Group WRB (2006) World reference base for soil resources 2006, 2nd edn, World Soil Resources Reports No. 103. FAO, Rome

Joubert L, Venter HJT, Verhoeven RL (2008) A new species of *Cryptolepis* (Apocynaceae: Periplocoideae) from Kenya and Tanzania, Africa. Bot J Linn Soc Lond 157:343–346

Juan C, Emerson BC, Oromí P, Hewitt GM (2000) Colonization and diversification: towards a phylogeographic synthesis for the Canary Islands. Trends Ecol Evol 15:104–109

Kämmer F (1974) Klima und Vegetation von Teneriffa besonders im Hinblick auf den Nebelniederschlag. Scripta Geobotanica 7:1–78

Kappen L, Straka H (1988) Pollen and spores transport into the Antarctic. Polar Biol 8:173–180

Kärnefelt I (1979) The brown fruticose species of *Cetraria*. Opera Bot 46:1–150

Kassas M (1957) On the ecology of the Red Sea coastal land. J Ecol 45:187–203

Kers LE (1993) New taxa in *Maerua* (Capparaceae) proposed for the Flora of Ethiopia. Novon 3:50–54

Khalaf FI (1989) Desertification and aeolian processes in the Kuwait desert. J Arid Environ 16:125–145

Khan MA (2003) Light, salinity, and temperature effects on the seed germination of perennial grasses. Am J Bot 90:131–134

Kier G, Kreft H, Lee TM, Jetz W, Ibisch PL, Nowicki C, Mutke J, Barthlott W (2009) A global assessment of endemism and species richness across island and mainland regions. Proc Natl Acad Sci USA 106:9322–9327

Kilian N (1997) Revision of *Launaea* Cass. (Compositae, Lactuceae, Sonchinae). Englera 17:1–478

Kilian N, Hein P (2006) New and noteworthy records for the vascular plant flora of Socotra Island, Yemen. Englera 28:57–77

Kilian N, Hein P, Bahah SO (2002) A new species of *Campylanthus* (Scrophulariaceae) from Ras Fartak, Al-Mahra, and notes on other species of the genus in Yemen. Willdenowia 32:271–279

King TL, Woodell SRJ (1973) The causes of regular pattern in desert perennials. J Ecol 61:761–765

King TL, Woodell SRJ (1984) Are regular patterns in desert shrubs artefacts of sampling? J Ecol 72:295–298

Klackenberg J (1985) The genus *Exacum* (Gentianaceae). Opera Bot 84:1–144

Klackenberg J (1992) Taxonomy of *Secamone* (Asclepiadaceae) in Asia and Australia. Kew Bull 47:595–612

Klackenberg J (2002) Tribe Exaceae. In: Struwe L, Albert VA (eds) Gentianaceae, systematics and natural history. Cambridge University Press, Cambridge

Klaus R, Turner JR (2004) The marine biotopes of the Socotra Archipelago. In: Krupp F (ed) Fauna of Arabia. Socotra Archipelago, vol 20. Libri, Basel

Knapp R (1968) Höhere Vegetationseinheiten von Äthiopien, Somalia, Natal, Transvaal, Kapland und einigen Nachbargebieten. Geobot Mitt 56:1–36

Knapp PA (1991) The response of semi-arid vegetation assemblages following the abandonment of mining towns in south-western Montana. J Arid Environ 20:205–222

Kopp H (1999) Abiotische Geofaktoren. In: Wranik W (ed) Sokotra. Mensch und Natur. Jemen-Studien Band 14. Dr Ludwig Reichert Verlag, Wiesbaden

Kossmat F (1907) Geologie der Inseln Sokotra, Semha und Abd El Kuri. Denkschr Kais Akad Wiss Wien, Mat-Nat Kl 71:1–62

Král K, Pavliš J (2006) The first detailed land-cover map of Socotra Island by Landsat/ETM+ data. Int J Remote Sens 27:3239–3250

Kreft H, Jetz W, Mutke J, Kier G, Barthlott W (2008) Global diversity of island floras from a macroecological perspective. Ecol Lett 11:116–127

Krueger WC (1990) Animal range relations on arid and semi-arid rangelands. In: Halwagy R, Taha FK, Omar SA (eds) Advances in range management in arid lands. Kegan Paul International, London, pp 201–205

Krupp F, Apel M, Zajonz U (2002) Meeresbiologische Untersuchungen im Sokotra-Archipel (Jemen). Natur und Museum 132:273–294

Kunkel G (1980) Die Kanarischen Inseln und ihre Pflanzenwelt. Fischer, New York

Kürschner H (1986) Omanisch-makranische Disjunktionen. Ein Beitrag zur pflanzengeographischen Stellung und zu den florengenetischen Beziehungen Omans. Bot Jahrb Syst 106:541–562

Kürschner H (1998) Biogeography and introduction to vegetation. In: Ghazanfar SA, Fisher M (eds) Vegetation of the Arabian Peninsula. Kluwer, Dordrecht

Kürschner H (2000a) Bryophyte flora of the Arabian Peninsula and Socotra. Bryophytorum Bibliotheca 55:1–131

Kürschner H (2000b) *Marchesinia mackaii* (Hook.) S.F. Gray (Lejeuneaceae, Hepaticopsida), and further leafy epiphytic liverworts new to Socotra Island. Additions to the bryophyte flora of the Arabian Peninsula and Socotra 1. Nova Hedwigia 71:503–508

Kürschner H (2003a) Nineteen new records to the bryophyte flora of Socotra Island. Additions to the bryophyte flora of the Arabian Peninsula and Socotra 5. Willdenowia 33:445–458

Kürschner H (2003b) Further new bryophytes to Socotra and the Yemen mainland, including *Papillaria crocea* (Hampe) A. Jaeger (Meteoriaceae, Bryopsida), a palaeotropical Asian species. Additions to the bryophyte flora of the Arabian Peninsula and Socotra 3. Nova Hedwigia 76:257–267

Kürschner H (2004a) Phytosociology, ecology and life strategies of a remarkable epiphytic bryophyte community of Socotra Island, Yemen. Bot Jahrb Syst 125:377–395

Kürschner H (2004b) Life strategies and adaptations in bryophytes from the Near and Middle East. Turk J Bot 28:73–84

Kürschner H (2006a) A bryophyte flora of Socotra Island, Yemen. Englera 28:97–162

Kürschner H (2006b) Syntaxonomy, synecology and life strategies of a saxicolous bryophyte community of Socotra Island, Yemen. Englera 28:163–174

Kürschner H, Ghazanfar SA (1998) Bryophytes and lichens. In: Ghazanfar SA, Fisher M (eds) Vegetation of the Arabian Peninsula. Kluwer, Dordrecht

Kürschner H, Ochyra R (2004) Remarkable new record to the bryophyte flora of Yemen (al-Mahra and Socotra Island). Additions to the bryophyte flora of the Arabian Peninsula and Socotra 6. Cryptog Bryol 25:69–81

Kürschner H, Sollman P (2004) New and noteworthy bryophyte records for the flora of Yemen. Additions to the bryophyte flora of the Arabian Peninsula and Socotra 8. Willdenowia 34:301–308

Kürschner H, Al-Gifri A-N, Al-Subai MY, Rowaished AK (1998) Vegetational patterns within coastal salines in southern Yemen. Feddes Repert 109:147–159

Kürschner H, Hein P, Kilian N, Hubaishan MA (2006a) Diversity and zonation of the forests and woodlands of the mountains of northern Socotra, Yemen. Englera 28:11–55

Kürschner H, Kilian N, Hein P, Mukram A (2006b) The Adenio obesi-Sterculietum africanae, a relic Arabian mainland community vicarious to the Socotran *Adenium-Sterculia* woodland. Englera 28:79–96

Kutiel P, Danin A (1987) Annual-species diversity and aboveground phytomass in relation to some soil properties in the sand dunes of the northern Sharon Plains, Israel. Vegetatio 70: 45–49

Lange OL (1959) Untersuchungen über Wärmehaushalt und Hitzeresistenz mauretanischer Wüsten- und Savannenpflanzen. Flora 147:595–651

Larcher E (2003) Physiological plant ecology. Springer, Berlin

Larson DW, Matthes U, Kelly PE (2000) Cliff ecology: pattern and process in cliff ecosystems. Cambridge University Press, Cambridge

Lathrop EW (1983) Recovery of perennial vegetation in military maneuver areas. In: Webb RH, Wilshire HG (eds) Environmental effects of off-road vehicles. Springer, New York

Lavin M, Beyra Matos A (2008) The impact of ecology and biogeography on legume diversity, endemism, and phylogeny in the Caribbean Region: a new direction in historical biogeography. Bot Rev 74:178–196

Lavin M, Thulin M, Labat J-N, Pennington RT (2000) Africa, the odd man out: molecular biogeography of dalbergioid legumes (Fabaceae) suggests otherwise. Syst Bot 25:449–467

Lavin M, Schrire BP, Lewis G, Pennington RT, Delgado-Salinas A, Thulin M, Hughes CE, Beyra Matos A, Wojciechowski MF (2004) Metacommunity process rather than continental tectonic history better explains geographically structured phylogenies in legumes. Philos Trans R Soc B 359:1509–1522

Lavranos JJ (1966) On the occurence and taxonomy of *Adenium* in tropical Arabia and on Soqotra. Cact Succ J (USA) 38:19–23

Lavranos JJ (1969) The genus *Aloe* L. in the Socotran Archipelago. Indian Ocean. A revision. Cact Succ J (USA) 41:202–207

Lavranos JJ (1970) The Stapeliae of Socotra. Cact Succ J (USA) 42:134–136

Lavranos JJ (1971a) *Caralluma dioscoridis*. Hooker's Icon Pl (37) 5 ser7(4): t3687

Lavranos JJ (1971b) The island of Abd-el-Kuri and its peculiar *Euphorbia*. Cact Succ J (USA) 43:109–111

Lavranos JJ (1972a) A new *Sarcostemma* from Socotra. Nat Cact Succ J 27:37

Lavranos JJ (1972b) Notes on the succulent flora of North-east Africa and southern Arabia. Cact Succ J (USA) 44:257–261

Lavranos JJ (1974) Plant hunting on Socotra. Excelsa 4:25–28

Lavranos JJ (1979) *Caralluma dioscorides* Lavranos. Asclepiadaceae 18:10–12

Lavranos JJ (1993) Two new species of *Echidnopsis* (Asclepiadaceae-Stapeliae) from Socotra. Cact Succ J (USA) 65:293–295

Lavranos JJ (1994) Impressions from Socotra & Abd Al-Kuri. Euphorbia J 9:190–209

Laycock WA (1991) Stable states and thresholds of range condition on North American rangelands: a viewpoint. J Range Manage 44:427–433

Le Houérou HN (1986) The desert and arid zones of Northern Africa. In: Evenari M, Noy-Meir I, Goodall DW (eds) Ecosystems of the world 12B. Hot deserts and arid shrublands, B. Elsevier, Amsterdam

Le Houérou HN (1996) Climate change, drought and desertification. J Arid Environ 34:133–185

Le Houérou HN (1997) Climate, flora and fauna changes in the Sahara over the past 500 million years. J Arid Environ 37:619–647

Le Houérou HN (2000) Restoration and rehabilitation of arid and semiarid Mediterranean ecosystems in North Africa and West Asia: a review. Arid Soil Res Rehab 14:3–14

Le Houérou HN (2002) Man-made deserts: desertization process and threats. Arid Land Res Manage 16:1–36

Le Houérou HN (2005) Diffuse vs. contracted vegetation patterns: an objective demarcation between arid and desert vegetations. Isr J Plant Sci 53:177–182

Lems K (1960) Botanical notes on the Canary Islands. II. The evolution of plant forms in the islands: *Aeonium*. Ecology 41:1–17

Lems K (1961) Botanical notes on the Canary Islands. III. The life-form spectrum and its interpretation. Ecology 42:569–572

Leslau W (1938) Lexique Socotri (sudarabique moderne) avec comparaisons et explications étymologiques. Klincksieck, Paris

Levey DL, Tewksbury JJ, Bolker BM (2008) Modelling long-distance seed dispersal in heterogeneous landscapes. J Ecol 96:599–608

Levin SA, Muller-Landau HC, Nathan R, Chave J (2003) The ecology and evolution of seed dispersal: a theoretical perspective. Annu Rev Ecol Syst 34:575–604

Link DA (1991) *Dirachma somalensis* D.A. Link sp. nov. A new species of a remarkable and highly endangered monogeneric family. B Jard Bot Nat Belg 61:3–13

Liston A, Rieseberg LH, Ellias TS (1989) Genetic similarity is high between intercontinental disjunct species of *Senecio* (Asteraceae). Am J Bot 76:383–388

Long DG (1986) Some Socotran mosses collected by Schweinfurth. Studies in Arabian bryophytes. 5. Nova Hedwigia 43:443–445

Lösch R, Beyhl FE, Mies BA, Schweihofen B (1990) Relative Standortkonstanz der Federbuschvegetation auf den mittelatlantischen Inseln und das Fehlen klimatisch-orographischer Voraussetzungen für eine Waldklimax auf den Kapverden. Cour Forschungsinst Senckenb 129:75–82

Ludwig JA, Tongway DJ (1996) Rehabilitation of semiarid landscapes in Australia. II. Restoring vegetation patches. Restor Ecol 4:398–406

MacArthur RH, Wilson EO (1967) The theory of island biogeography. Princeton University Press, Princeton

Macey JR, Kuehl JV, Larson A, Robinson MD, Ugurtas IH, Ananjeva NB, Rahman H, Javed HI, Osman RM, Doumma A, Papenfuss TJ (2008) Socotra Island the forgotten fragment of Gondwana: unmasking chameleon lizard history with complete mitochondrial genomic data. Mol Phylogenet Evol 49:1015–1018

Mandaville JP (1990) Flora of Eastern Saudi Arabia. Kegan Paul International, London

Marrero A, Almeida RS, González-Martin M (1998) A new species of the wild dragon tree, *Dracaena* (Dracaenaceae) from Gran Canaria and its taxonomic and biogeographic implications. Bot J Linn Soc Lond 128:291–314

McGinnies WG, Goldman BJ, Paylore P (1968) Deserts of the world. University of Arizona Press, Tucson

Middleton N, Thomas D (1997) World atlas of desertification. UNEP/Arnold, London

Mies BA (1994) Checkliste der Gefäßpflanzen, Moose und Flechten und botanische Bibliographie der Insel Sokotra und des Sokotrinischen Archipels (Jemen, Indischer Ozean). Senckenb Biol 74:213–258

Mies BA (1995a) Die Sukkulenten des Sokotrinischen Archipels (Jemen, Indischer Ozean). (Teil 1) Die Vegetationszonen der Sukkulenten. Kakt Sukk 46:92–96

Mies BA (1995b) On the comparison of the flora and vegetation of the island groups of Socotra and Macaronesia. Bol Mus Munic Funchal Supl 4:455–471

Mies BA (1998a) The phytogeography of Soqotra. Evidence for disjunctive taxa, especially with Macaronesia. In: Dumont HJ (ed) Proceedings of the first international symposium on Soqotra Island: present and future. United Nations, New York

Mies BA (1998b) Biogeographie der Insel Sokotra (Jemen). Geogr Rundsch 50:722–729

Mies BA (1998c) Leben auf großem Fuß. Untersuchungen in Caudexpflanzen der Insel Sokotra (Jemen). Kakt Sukk 49:49–52, 85–90

Mies BA (1999a) Flora und Vegetation der Insel Sokotra. In: Wranik W (ed) Sokotra. Mensch und Natur. Jemen-Studien Band 14. Dr Ludwig Reichert Verlag, Wiesbaden

Mies BA (1999b) Bottle-trees and heat on Socotra (Yemen): a big-foot story. A temperature study of *Adenium socotranum* (Apocynaceae) and *Dendrosicyos socotrana* (Cucurbitaceae). Cact Succ J (USA) 71:60–73

Mies BA (1999c) Neue Pflanzenassoziationen des Crotonion socotrani Mi. 1999 von der Insel Sokotra (Jemen, Indischer Ozean). Acta Biol Benrodis 10:27–48

Mies BA (2000) Vegetation ecology of North African shrubland. Studies from the Canaries, Arabia, Horn of Africa and Soqotra. In: Breckle S-W, Schweitzer B, Arndt U (eds) Results of worldwide ecological studies. Proceedings of the 1st symposium of the A.F.W. Schimper-Foundation established by H. and E. Walter, Hohenheim, October 1998. Verlag Günter Weidenbach, Stuttgart

Mies BA (2001) Flora und Vegetationsölogie der Insel Soqotra. Essener Ökologische Schriften 15. Westarp Wissenschaften, Hohenwarsleben

Mies BA, Aschan G (1995) Radiation regime and temperature conditions in the canopy of the succulent shrub *Euphorbia balsamifera*. Vieraea 24:115–125

Mies BA, Beyhl FE (1998) The vegetation ecology of Socotra. In: Dumont HJ (ed) Proceedings of the first international symposium on Soqotra Island: present and future. United Nations, New York

Mies BA, Beyhl FE (1999) Ein Szenario von Entwaldung und Desertifikation im Tal von Qalansiyah (Insel Soqotra, Jemen, Indischer Ozean). Acta Biol Benrodis 10:49–55

Mies BA, Printzen C (1997) Notes on the lichens of Socotra. Bibl Lichenol 68:223–239

Mies BA, Schultz M (2004) New and interesting lichen records from Soqotra Island (Yemen, Indian Ocean). Bibl Lichenol 88:433–452

Mies BA, Zimmer H (1993) Die Vegetation der Insel Sokotra im Indischen Ozean. Natur und Museum 123:253–264

Mies BA, Zimmer H (1994) Die Populationen von *Adenium socotranum* (Balfour) Vierhapper und *Dendrosicyos socotrana* Balfour bei Ras Hebak (Insel Sokotra, Jemen) und ihre Gefährdung. Kakt Sukk 45:1–5

Mies BA, Printzen C, Beyhl FE (1994) Zweite Forschungsreise nach Sokotra (Jemen, Indischer Ozean) im Januar 1994. Akt Lich Mitt Bryol-Lichenol Arbeitsgem Mitteleur 6:1–6

Mies BA, Lumbsch HT, Tehler A (1995) *Feigeana socotrana*, a new genus and species from Socotra, Yemen (Roccellaceae; Euascomycetidae). Mycotaxon 54:155–162

Mies BA, Jimenez MS, Morales D (1996) Ecophysiology and distribution of the endemic leafless spurge *Euphorbia aphylla* and the introduced *E. tirucalli* (Euphorbiaceae, *Euphorbia* sect. Tirucalli) in the Canary Islands. Plant Syst Evol 202:27–36

Mies BA, Lavranos JJ, James GJ (2000) Frankincense on Socotra Island (*Boswellia*, Burseraceae; Yemen). Cact Succ J (USA) 72:265–278

Millennium Ecosystem Assessment (2005) Ecosystems and human well-being: desertification synthesis. World Resources Institute, Washington, DC

Miller AG (1984) A revision of *Ochradenus*. Notes R Bot Gard Edinb 41:491–504

Miller AG (1996) Studies in the flora of Arabia. XXIX. Two new species from Yemen. Edinb J Bot 53:145–147

Miller AG, Cope TA (1996) Flora of the Arabian Peninsula and Socotra, vol 1. Edinburgh University Press, Edinburgh

Miller JR, Hobbs RJ (2007) Habitat restoration – do we know what we're doing? Restor Ecol 15:382–390

Miller AG, Morris M (1988) Plants of Dhofar. The southern region of Oman. Traditional, economic and medicinal uses. Diwan of the Royal Court, Sultanate of Oman

Miller AG, Morris M (2001) Conservation and sustainable use of the biodiversity of Socotra Archipelago. Final Report (Botany). YEM/96/G32. Royal Botanic Garden, Edinburgh

Miller AG, Morris M (2004) Ethnoflora of the Socotra Archipelago. Royal Botanic Garden, Edinburgh

Miller AG, Short M, Sutton DA (1982) A revision of *Schweinfurthia*. Notes R Bot Gard Edinb 40:23–40

Miller AG, Atkinson R, Al Khulaidi AW, Taleeb N (2002) *Nesocrambe*, a new genus of Cruciferae (Brassiceae) from Soqotra, Yemen. Willdenowia 32:61–67

Milne RI (2006) Northern hemisphere plant disjunctions: a window on Tertiary land bridges and climate change? Ann Bot 98:465–472

Mitten W (1888) Muscineae. In: Balfour IB (ed) Botany of Socotra. Trans R Soc Edinb 31:330–336

Monod T (1954) Modes contractés et diffus de la végétation saharienne. In: Cloudsley-Thompson JL (ed) Biology of deserts. University of London, London

Monod T (1971) Remarques sur les symmétries floristiques des zone sèches nord et sud en Afrique. Mitt Bot Staatssamml Muenchen 10:375–423
Morley RJ (2007) Cretaceous and Tertiary climate change and the past distribution of megathermal rainforests. In: Bush MB, Flenley JR (eds) Tropical rainforest responses to climate change. Springer, Berlin
Morucchio GB (1969) Adumbratio florae aethiopicae. 20. Globulariaceae. Webbia 24:619–635
Mott JJ, McComb AJ (1974) Patterns in annual vegetation and soil microrelief in an arid region of western Australia. J Ecol 62:115–126
Müller C (1901) Genera muscorum frondosorum. Kummer, Leipzig
Müller DH (1907) Zur Geschichte der Südarabischen Expedition. Denkschr Kais Akad Wiss Wien, Mat-Nat Kl 71:321–490
Muller J (1981) Fossil pollen records of extant angiosperms. Bot Rev 47:1–142
Müller-Argau J (1882a) Diagnoses Lichenum Socotrensium novorum a particibus expeditionum Prof. Bayley Balfour et Dr. Schweinfurth lectorum. Philos R Soc Edinb 1881–1882:457–472
Müller-Argau J (1882b) Lichenes Arabici. Bull Herb Boissier 1:130–131
Müller-Argau J (1887) *Enterographa affine*, *E. fraterculans*. Mém Soc Phys Hist Nat Genève 29:69
Müller-Argau J (1888) Lichenes. In: Balfour IB (ed) Botany of Socotra. Trans R Soc Edinb 31:342–390
Müller-Dombois D, Ellenberg H (1974) Aims and methods of vegetation ecology. Wiley, New York
Nagy ZT, Joger U, Wink M, Glaw F, Vences M (2003) Multiple colonization of Madagascar and Socotra by colubrid snakes: evidence from nuclear and mitochondrial gene phylogenies. Proc R Soc Lond B 270:2613–2621
Nathan R (2006) Long-distance dispersal of plants. Science 313:786–788
Nathan R, Schurr FM, Spiegel O, Steinitz O, Trakhtenbrot A, Tsoar A (2008) Mechanisms of long-distance seed dispersal. Trends Ecol Evol 23:638–647
Naumkin VV (1993) Island of the Phoenix. An ethnographic study of the people of Socotra. Ithaca Press, Reading
Nobel PS (2009) Physicochemical and environmental plant physiology. Academic, Oxford
Nogales M, Medina FM, Quilis V, González-Rodíguez M (2001) Ecological and biogeographical implications of Yellow-legged Gulls (*Larus cachinnans* Pallas) as seed dispersers of *Rubia fruticosa* Ait. (Rubiaceae) in the Canary Islands. J Biogeogr 28:1137–1145
Nordenstam B (1968a) The genus *Euryops*. Morphology, cytology and taxonomy. Thesis, University of Lund
Nordenstam B (1968b) The genus *Euryops*. Part I. Taxonomy. Opera Bot 20:1–409
Nordenstam B (1969) Phytogeography of the genus *Euryops* (Compositae). A contribution to the phytogeography of southern Africa. Opera Bot 23:1–77
Novikoff G (1983) Desertification by overgrazing. Ambio 12:102–105
Noy-Meir I (1973) Desert ecosystems: environment and producers. Annu Rev Ecol Syst 4:25–51
Olesen JM, Jordano P (2002) Geographic patterns in plant-pollinator mutualistic networks. Ecology 83:2416–2424
Olson ME (2003) Stem and leaf anatomy of the arborescent Cucurbitaceae *Dendrosicyos socotrana* with comments on the evolution of pachycauls from lianas. Plant Syst Evol 239:199–214
Olsvig-Whittaker L, Shachak M, Yair A (1983) Vegetation patterns related to environmental factors in a Negev Desert watershed. Vegetatio 54:153–165
Orshan G (1986) The deserts of the Middle East. In: Evenari M, Noy-Meir I, Goodall DW (eds) Ecosystems of the world 12B. Hot Deserts and Arid Shrublands, B. Elsevier, Amsterdam
Ortiz S (2001) Reinstatement of the genus *Macledium* Cass. (Asteraceae, Mutisieae): morphological and phylogenetic arguments. Taxon 50:733–744
Ortiz S, Netnou NC (2005) A new species of *Dicoma* (Asteraceae, Mutisieae) from South Africa. Bot J Linn Soc 147:509–513
Othman WA (1996) Hydrogeological conditions of Socotra Island. In: University of Aden (ed) First international scientific symposium on Socotra Island: present and future, vol 4. Aden, 24–28 Mar 1996. pp 51–76

Pandey CB, Singh JS (1991) Influence of grazing and soil conditions on secondary savanna vegetation in India. J Veg Sci 2:95–102

Pax F (1910) *Cephalocrotonopsis socotrana*. Pflanzenreich IV 147:11–15

Pax F, Hoffmann O (1931) Euphorbiaceae. In: Engler A, Harms H (eds) Die natürlichen Pflanzenfamilien, 2nd edn, Band 19C. W Engelmann, Leipzig

Peacock JM, Ferguson ME, Alhadrami GA, McCann IR, Al Hajoj A, Saleh A, Karnik R (2003) Conservation through utilization: a case study of the indigenous forage grasses of the Arabian Peninsula. J Arid Environ 54:15–28

Pennington RT, Cronk QCB, Richardson JA (2004) Introduction and synthesis: plant phylogeny and the origin of major biomes. Philos Trans R Soc B 359:1455–1464

Petrusson L, Thulin M (1996) Taxonomy and biogeography of *Gymnocarpus* (Caryophyllaceae). Edinb J Bot 53:1–26

Peucker K (1899) The Austrian expedition to Southern Arabia and Sokotra. Geogr J 13: 638–640

Pfanz H (2008) Bark photosynthesis. Trees 22:137–138

Philippi T (1993) Bet-hedging germination of desert annuals: beyond the first year. Am Nat 142:474–487

Phillips RC, Loughland RA, Youssef AM (2004) The seagrass resources of Abu Dhabi Emirate. In: Loughland RA, Al Muhairi FS, Fadel SS, Almehdi AM, Hellyer P (eds) Marine atlas of Abu Dhabi. Emirates Heritage Club, Abu Dhabi

Pichi-Sermolli REG (1955) Tropical East Africa (Ethiopia, Somaliland, Kenya, Tanganyika). Arid Zone Res 6:302–360

Pichi-Sermolli REG (1957) Una carta geobotanica dell'Africa Orientale. Webbia 13:15–132

Pichi-Sermolli REG (1976) Botanical collections in Africa from August 1970. Boissiera 24:597

Pierce S, Vianelli A, Cerabolini B (2005) From ancient genes to modern communities: the cellular stress response and the evolution of plant strategies. Funct Ecol 19:763–776

Pietsch D (2006) Böden der Trockentropen. Prozess- und Strukturindikatoren-gestützte Analyse geschichteter, polygenetischer und degradierter Böden der Insel Socotra (Jemen). Jemen-Studien Band 17. Dr Ludwig Reichert Verlag, Wiesbaden

Pietsch D, Kühn P (2009) Soil developmental stages of Cambisols and Calcisols on Socotra island. Soil Sci 174:292–302

Pietsch D, Morris M (2010) Modern and ancient knowledge of conserving soils in Socotra Island, Yemen. In: Zdruli P, Pagliai M, Kapur S, Cano AF (eds) Land degradation and desertification: assessment, mitigation and remediation. Springer, Berlin

Pole MS (2001) Can long-distance dispersal be inferred from the New Zealand plant fossil record? Aust J Bot 49:357–366

Popov G (1957) The vegetation of Socotra. J Linn Soc Bot Lond 55:706–720

Porembski S (1996) Notes on the vegetation of inselbergs in Malawi. Flora 191:1–8

Prentice IC, Werger MJA (1985) Clump spacing in a desert dwarf shrub community. Vegetatio 63:133–139

Price RS (1911) The roots of some North African desert grasses. New Phytol 10:328–339

Rabinowitz PD, Coffin MF, Falvey D (1983) The separation of Madagascar and Africa. Science 220:67–69

Radcliffe-Smith A (1973) An account of the genus *Cephalocroton* Hochst. (Euphorbiaceae). Kew Bull 28:123–132

Radcliffe-Smith A (1989) Notes on African Euphorbiaceae: XX. *Acalypha* (ii), etc. Kew Bull 44:439–454

Raunkiaer C (1904) Om biologiske typer, med hensyn til planternes tilpasning til at overleve ungustige aarstider. Botanisk Tidsskrift 26:XIV

Raunkiaer C (1907) Planterigets livsformer of deres betydning for geografien. Nordisk Forlag, Kobenhavn-Kristiania

Raunkiaer C (1934) Life forms of plants and statistical plant geography. Clarendon Press, Oxford

Raven PH, Axelrod DI (1974) Angiosperm biogeography and past continental movements. Ann Mo Bot Gard 61:539–673

Raxworthy CJ, Forstner MRJ, Nussbaum RA (2002) Chameleon radiation by oceanic dispersal. Nature 415:784–787

Řičánek M, Hanáček P (2006) In search of *Socotrella*. Cact Succ J (USA) 78:6–9

Richardson DM, Pysek P, Rejmanek M, Barbour MG, Panetta FD, West CJ (2000) Naturalization and invasion of alien plants: concepts and definitions. Divers Distrib 6:93–107

Robson NBK (1985) Studies in the genus *Hypericum* L. (Guttiferae. 3. Sections 1. Campylosporus to 6a Umbraculoides. Bull Br Mus Nat Hist Bot 12:163–325

Robson NKB (1993) Parallel evolution in tropical montane *Hypericum*. Opera Bot 121:263–274

Robson NKB (1996) Studies in the genus *Hypericum* L. (Guttiferae). 6. Sections 20. Myriandra to 28. Elodes. Bull Nat Hist Mus Bot 26:75–217

Rogers ZS (2009) A revision of Malagasy *Gnidia* (Thymelaeaceae, Thymelaeoideae). Ann Mo Bot Gard 96:324–368

Ronse De Craene LP, Miller AG (2004) Floral development and anatomy of *Dirachma socotrana* (Dirachmaceae): a controversial member of the Rosales. Plant Syst Evol 249:111–127

Rowley GD (1980) The pollination mechanism of *Adenium* (Apocynaceae). Natl Cact Succ J 35:2–5

Royal Geographical Society (1978) Socotra (Samhah, Darsa). Map 1:125000. Royal Geographical Society, London

Rundel PW (1982) Water uptake by organs other than roots. In: Lange OL, Nobel PS, Osmond CB, Ziegler H (eds) Physiological plant ecology. Water relations and carbon assimilation. Encyclopedia of Plant Physiology, N.S. 12B. Springer, Berlin

Rundel PW, Gibson AC (1996) Ecological communities and processes in a Mojave Desert ecosystem. Cambridge University Press, Cambridge

Russian Military Map (1978) Suqutra, Samhah, Darsa. Military Map 1:200 000, Moskwa

Samuel MA, Harbury N, Bott R, Thabet AM (1997) Field observations from the Socotran platform: their interpretation and correlation to Southern Oman. Mar Petrol Geol 14:661–673

Sax DF, Gaines SD (2008) Species invasions and extinction: the future of native biodiversity on islands. Proc Natl Acad Sci USA 105:11490–11497

Schatz GE (1996) Malagasy/Indo-Australo-Malesian phytogeographic connections. In: Lourenco WR (ed) Biogeography of Madagascar. Editions ORSTOM, Paris

Schils T, Coppejans E (2003) Phytogeography of upwelling areas in the Arabian Sea. J Biogeogr 30:1339–1356

Schlesinger WH, Reynolds JF, Cunningham GL, Huenneke LF, Jarrell WM, Virginia RA, Whitford WG (1990) Biological feedbacks in global desertification. Science 247:1043–1048

Scholte P, de Geest P (2010) The climate of Socotra Island (Yemen): a first-time assessment of the timing of the monsoon wind reversal and its influence on precipitation and vegetation patterns. J Arid Environ 74:1507–1515

Schulter D (2000) The ecology of adaptive radiation. Oxford University Press, Oxford

Schultz M (2002) *Pseudopaulia tessellata*, a new genus and species in the Lichinaceae (lichenized Ascomycotina) from Socotra, Yemen. Mycotaxon 82:443–449

Schultz M (2003) *Lempholemma socotranum*, a new species of the Lichinaceae from Socotra (Yemen, Indian Ocean). Bibl Lichenol 86:155–159

Schultz M (2005) *Heppia arenacea* and *Lempholemma polycarpum*, two new species from southern Yemen and Socotra. Lichenologist 37:227–235

Schultz M, Büdel B (2002) Key to the genera of the Lichinaceae. Lichenologist 34:39–62

Schultz M, Mies BA (2003) The saxicolous and terricolous, cyanobacterial lichens of Socotra (Yemen, Indian Ocean). Nova Hedwigia 77:73–97

Schultz M, Mies BA, Al-Gifri A-N (1999) The genus *Paulia* (Lichinaceae, lichenized Ascomycetes) on Socotra Island. Bryologist 102:61–66

Schultz M, Brown G, Büdel B (2000) Cyanophilous lichens from Kuwait. Nova Hedwigia 70:193–216

Schultz M, Arendholz W-R, Büdel B (2001) Origin and evolution of the lichenized ascomycete order Lichinales: monophyly and systematic relationships inferred from ascus, fruiting body and SSU rDNA evolution. Plant Biol 3:116–123

Schupp EW, Jordano P, Gómez JM (2010) Seed dispersal effectiveness revisited: a conceptual review. New Phytol 188:333–353

Schuster JL (1998) Restoring structure and function of degraded rangeland ecosystems. In: Omar SAS, Misak R, Al-Ajmi D, Al-Awadhi N (eds) Sustainable development in the arid zones, vol 2, Management and improvement of desert resources. Balkema, Rotterdam

Schweinfurth G (1884a) Allgemeine Betrachtungen über die Flora von Socotra. Englers Bot Jahrb 5:40–49

Schweinfurth G (1884b) Ein Besuch auf Socotra mit der Riebeckschen Expedition. Freiburg i. B

Schweinfurth G (1891a) Erinnerungen an eine Fahrt nach Socotra. Westermanns Illustr Deut Monatshefte 35/69:603–626

Schweinfurth G (1891b) Erinnerungen von einer Fahrt nach Sokotra. Westermanns Illustr Deut Monatshefte 35/70:29–53

Schweinfurth G (1894) Sammlung arabisch-äthiopischer Pflanzen. Ergebnisse von Reisen in den Jahren 1881, 88, 89, 91, 92 und 94. I. Bull Herb Boissier 2:1–113

Schweinfurth G (1896) Sammlung arabisch-äthiopischer Pflanzen. Ergebnisse von Reisen in den Jahren 1881, 88, 89, 91, 92 und 94. II. Bull Herb Boissier 4:114–266

Schweinfurth G (1899) Sammlung arabisch-äthiopischer Pflanzen. Ergebnisse von Reisen in den Jahren 1881, 88, 89, 91, 92 und 94. III. Bull Herb Boissier 7:267–340

Scott A (1881) Remarks on the natural history of Socotra. W Blackwood and Sons, Edinburgh

Sen DN (1973) Ecology of Indian desert III. Survival adaptations of vegetation in dry environment. Vegetatio 27:201–265

Sen DN, Chawan DD, Sharma KD (1972) Ecology of Indian desert. V. On the water relations of *Salvadora* species. Flora 161:463–471

Shachak M, Sachs M, Moshe I (1998) Ecosystem management of desertified shrublands in Israel. Ecosystems 1:475–483

Shakun JD, Burns SJ, Fleitmann D, Kramers J, Matter A, Al-Subary A (2007) A high-resolution, absolute-dated deglacial speleothem record of Indian Ocean climate from Socotra Island, Yemen. Earth Planet Sci Lett 259:442–456

Shantz HL (1927) Drought resistance and soil moisture. Ecology 8:145–157

Sheppard CRC, Price ARG, Roberts CM (1992) Marine ecology of the Arabian region: patterns and processes in extreme tropical environments. Academic, London

Shmida A (1985) Biogeography of the desert flora. In: Evenari M, Noy-Meir I, Goodall DW (eds) Ecosystems of the world 12A. Hot deserts and arid shrublands, A. Elsevier, Amsterdam

Shmida A, Werger MJA (1992) Growth form diversity on the Canary Islands. Vegetatio 102:183–199

Shobrak M (1998) Notes on the breeding and cooling behaviour of Hoopoe Lark *Alaemon alaudipes* in central Saudi Arabia. Sandgrouse 20:53–55

Silvertown J (2004) The ghost of competition past in the phylogeny of island endemic plants. J Ecol 92:168–173

Silvertown J, Francisco-Ortega J, Carine M (2005) The monophyly of island radiations: an evaluation of niche pre-emption and some alternative explanations. J Ecol 93:653–657

Simon E (1978) Heavy metals in soils, vegetation development and heavy metal tolerance in plant populations from metalliferous areas. New Phytol 81:175–188

Sipman HJM (1983) A monograph of the lichen family Megalosporaceae. Bibl Lichenol 18:1–241

Smith TM, Shugart HH, Woodward FI (1997) Plant functional types: their relevance to ecosystem properties and global change. Cambridge University Press, Cambridge

Smith SE, Riley E, Tiss JL, Fendenheim DM (2000) Geographical variation in predictive seedling emergence in a perennial desert grass. J Ecol 88:139–149

Sprugel DG (1991) Disturbance, equilibrium, and environmental variability: what is 'natural' vegetation in a changing environment? Biol Conserv 58:1–18

St John H (1962) A new *Scaevola* (Goodeniaceae) from Socotra Island. Webbia 17:45–48

Stefanovi S, Austin DF, Olmstead RG (2003) Classification of Convolvulaceae: a phylogenetic approach. Syst Bot 28:791–806

Steiner J (1907) Bearbeitung der von O. Simony 1898 und 1899 in Südarabien, auf Sokotra und den benachbarten Inseln gesammelten Flechten. Denkschr Kais Akad Wiss Wien, Mat-Nat Kl 71:93–102

Steinmann VW, Porter JM (2002) Phylogenetic relationships in Euphorbieae (Euphorbiaceae) based on ITS and ndhF sequence data. Ann Mo Bot Gard 89:453–490

Stock WD, Dlamini TS, Cowling RM (1999) Plant induced fertile islands as possible indicators of desertification in a succulent desert ecosystem in northern Namaqualand, South Africa. Plant Ecol 142:161–167

Stone CP, Smith CW, Tunison JT (1992) Alien plant invasions in native ecosystems in Hawaii: Management and research. Cooperative National Park Resources Studies Unit, University of Hawaii, Honolulu

Storey BC (1995) The role of mantle plumes in continental breakup: case histories from Gondwanaland. Nature 377:301–308

Storey M, Mahoney JJ, Saunders AD, Duncan RA, Kelley SP, Coffin MF (1995) Timing of hot spot-related volcanism and the breakup of Madagascar and India. Science 267:852–855

Tansley AG (1935) The use and abuse of vegetational concepts and terms. Ecology 16:284–307

Tehler A (1983) The genera *Dirina* and *Roccellina* (Roccellaceae). Opera Bot 70:1–86

Tehler A (1990) A new approach to the phylogeny of Euascomycetes with a cladistic outline of Arthoniales focussing on Roccellaceae. Can J Bot 68:2458–2492

Tehler A, Irestedt M, Wedin M, Ertz D (2010) The Old World *Roccella* species outside Europe and Macaronesia: taxonomy, evolution and phylogeny. Syst Biodivers 8:223–246

Thalen DCP (1979) Ecology and utilization of desert shrub rangelands in Iraq. Dr W Junk, The Hague

Thiv M, Meve U (2007) A phylogenetic study of *Echidnopsis* Hook. f. (Apocynaceae-Asclepiadoideae) - taxonomic implications and the colonization of the Socotran archipelago. Plant Syst Evol 265:71–86

Thiv M, Thulin M, Kilian N, Linder HP (2006) Eritreo-Arabian affinities of the Socotran flora as revealed from the molecular phylogeny of *Aerva* (Amaranthaceae). Syst Bot 31:560–570

Thiv M, van der Niet T, Rutschmann F, Thulin M, Brune T, Linder HP (2011) Old-New World and trans-African disjunctions of *Thamnosa* (Rutaceae): intercontinental long-distance dispersal and local differentiation in the succulent biome. Am J Bot 98:76–87

Thulin M (1987) A new species of *Xylocalyx* (Scrophulariaceae) from Somalia. Nord J Bot 7:267–269

Thulin M (1993) Flora of Somalia., vol 1, Pteridophyta; Gymnospermae; Angiospermae (Annonaceae-Fabaceae). Royal Botanic Gardens, Kew

Thulin M (1995) Flora of Somalia, vol 4, Angiospermae (Hydrocharitaceae-Pandanaceae). Royal Botanic Gardens, Kew

Thulin M (1996) *Erythroxylon* (Erythroxylaceae) on Socotra. Nord J Bot 16:301–302

Thulin M (1997) A new species and a new record of *Tephrosia* (Leguminosae) from Socotra. Nord J Bot 17:349–351

Thulin M (1998a) A new combination for a *Crotalaria* (Leguminosae) from Socotra. Kew Bull 53:460

Thulin M (1998b) *Gaillonia* (Rubiaceae-Paederieae) in Africa and Arabia. Nord J Bot 18:31–38

Thulin M (1999a) *Thamnosma* (Rutaceae) in Africa. Nord J Bot 19:5–11

Thulin M (1999b) *Chapmannia* (Leguminosae-Stylosanthinae) extended. Nord J Bot 19:597–607

Thulin M (2001) *Exacum* (Gentianaceae) on the Arabian Peninsula and Socotra. Nord J Bot 21:243–247

Thulin M (2006) Flora of Somalia, vol 3. Royal Botanic Gardens, Kew

Thulin M, Al-Gifri A-N (1998) The frankincense trees *(Boswellia* spp.) of Soqotra. In: Dumont HJ (ed) Proceedings of the first international symposium on Soqotra Island: present and future. United Nations, New York

Thulin M, Johansson ANB (1996) Taxonomy and biogeography of the anomalous genus *Wellstedia*. In: van der Maesen LJG, van der Burgt XM, de Rooy JM (eds) The biodiversity of African plants. Proceedings of the XIVth AETFAT Congress, Wageningen. Kluwer, Dordrecht

Thulin M, Lavin M (2001) Phylogeny and biogeography of the *Ormocarpum* group (Fabaceae): a new genus *Zygocarpum* from the Horn of Africa region. Syst Bot 26:299–317

Thulin M, Bremer B, Richardson J, Niklasson J, Fay MF, Chase MW (1998) Family relationships of the enigmatic rosid genera *Barbeya* and *Dirachma* from the Horn of Africa region. Plant Syst Evol 213:103–119

Tilman D, May RM, Lehman CL, Nowak MA (1994) Habitat destruction and the extinction debt. Nature 371:65–66

Tongway DJ, Ludwig JA (1996) Rehabilitation of semiarid landscapes in Australia. I. Restoring productive soil patches. Restor Ecol 4:388–397

Townsend CC (1969) Bryophytes from Socotra. Musci. Trans Br Bryol Soc 5:790–795

Triantis KA, Borges PAV, Ladle RJ, Hortal J, Cardoso P, Gaspar C, Dinis F, Mendonça E, Silveira LMA, Gabriel R, Melo C, Santos AMC, Amorim IR, Ribeiro SP, Serrano ARM, Quartau JA, Whittaker RJ (2010) Extinction debt on oceanic islands. Ecography 33:285–294

Troll C (1948) Der asymmetrische Aufbau der Vegetationszonen und Vegetationsstufen auf der Nord- und Südhalbkugel. Bericht über das Geobotanische Forschungsinstitut Rübel in Zürich für das Jahr 1947:46–83

Tüxen R (1956) Die heutige potentielle natürliche Vegetation als Gegenstand der Vegetationskartierung. Angew Pflanzensoz 13:5–42

UNESCO (2008) Socotra Archipelago. http://whc.unesco.org/en/list/1263/documents/. Accessed Apr 2011

Uvarov BP, Popov GB (1957) The saltatorial Orthoptera of Socotra. J Linn Soc Zool 43:359–389

Valenti GS (1965) Adumbratio florae aethiopicae. 12. Buxaceae. Webbia 20:771–778

Valido A, Dupont YL, Olesen JM (2004) Bird-flower interactions in the Macaronesian islands. J Biogeogr 31:1945–1953

Valladares F, Pearcy RW (1998) The functional ecology of shoot architecture in sun and shade plants of *Heteromeles arbutifolia* M. Roem., a Californian chaparral shrub. Oecologia 114:1–10

van Auken OW (2000) Shrub invasions of North American semiarid grasslands. Annu Rev Ecol Syst 31:197–215

van Damme K, Banfield L (2011) Past and present human impacts on the biodiversity of Socotra Island (Yemen): implications for future conservation. Zool Middle East Supplementum 3:31–88

Venable DL, Lawlor L (1980) Delayed germination and dispersal in desert annuals: escape in time and space. Oecologia 46:272–282

Venter HJT, Verhoeven RL (1997) A tribal classification of the Periplocoideae (Apocynaceae). Taxon 46:705–720

Verdcourt B (1969) The arid corridor between northeast and southwest areas of Africa. Palaeoecol Afr 4:140–144

Verstraete MM (1986) Defining desertification: a review. Clim Chang 9:5–18

Vesey-Fitzgerald DF (1957) The vegetation of central and eastern Arabia. J Ecol 45:779–798

Vetaas OR (1992) Micro-site effects of trees and shrubs in dry savannas. J Veg Sci 3:337–344

Vierhapper J (1903) Neue Pflanzen aus Sokótra, Abdal Kuri und Semhah. I., II. Österr Bot Z 53:433–436, 481–482

Vierhapper J (1904) Neue Pflanzen aus Sokótra, Abdal Kuri und Semhah. III., IV., V. Österr Bot Z 54:32–36, 61–64, 286–287

Vierhapper J (1905) Neue Pflanzen aus Sokótra, Abdal Kuri und Semhah. VI., VII., VIII. Österr Bot Z 55:88–91, 439–440, 462–466

Vierhapper J (1906) Neue Pflanzen aus Sokótra, Abdal Kuri und Semhah. IX., X. Österr Bot Z 56:256–262, 298–305

Vierhapper J (1907) Beiträge zur Kenntnis der Flora Südarabiens und der Inseln Sokotra, Semha und Abd El Kuri. Bearbeitung der von Dr. St. Paulay und Professor Dr. O. Simony auf der Expedition der Kaiserlichen Akademie der Wissenschaften nach Südarabien und den Inseln

Sokótra, Sémha und Abd el Kûri vom December 1898 bis Mitte März 1899 gesammelten Gefäßpflanzen. I. Die Gefäßpflanzen der Inseln Sokótra, Sémha und Abd el Kuri. Denkschr Kais Akad Wiss Wien, Mat-Nat Kl 71:321–490
Vierhapper J (1909) Plantae novae Arabiae meridionalis atque Sokótrae insulae. Repertorium novarum specierum regni vegetabilis 7(140–143):158–165
Volkens G (1887) Die Flora der ägyptisch-arabischen Wüste auf Grundlage anatomisch-physiologischer Forschungen. Gebrüder Bornträger, Berlin
von Heuglin T (1861) Skizze der Inselgruppe von Socotra. Mittheilungen aus Justus Perthes' Geographischer Anstalt über wichtige neue Erforschungen auf dem Gesammtgebiet der Geografie von Dr. A. Petermann. 149–151
von Wettstein R (1906) Socotra. In: Karsten G, Schenk H (eds) Vegetationsbilder. Gustav Fischer Verlag, Jena
von Willert DJ, Eller BM, Werger MJA, Brinckmann E, Ihlenfeldt H-D (1992) Life strategies of succulents in deserts with special reference to the Namib Desert. Cambridge University Press, Cambridge
Wallace AR (1911) Island life, 3rd edn. Macmillan, London
Walter H, Breckle S-W (1999) Vegetation und Klimazonen, 7th edn. Ulmer, Stuttgart
Walter H, Breckle S-W (2004) Ökologie der Erde. Band 2. Spezielle Ökologie der Tropischen und Subtropischen Zonen. Elsevier, Amsterdam
Ward D (2009) The biology of deserts. Oxford University Press, Oxford
Warren A, Sud YC, Rozanov B (1996) The future of deserts. J Arid Environ 32:75–89
Webb DA (1977) An Irish flora. Dundalgan Press, Dundalk
Webb RH, Stielstra SS (1979) Sheep grazing effects on Mojave Desert vegetation and soils. Environ Manage 3:517–529
Webb RH, Wilshire HG (1980) Recovery of soils and vegetation in a Mojave Desert ghost town, Nevada, USA. J Arid Environ 3:291–303
Wedin M (1995) The lichen family Sphaerophoraceae (Caliciales, Ascomycotina) in temperate areas of the Southern Hemisphere. Symb Bot Upsal 31:1–102
Wellsted JR (1835) Memoir on the Island of Socotra. J R Geogr Soc Lond 5:129–229
Werger MJA (1978) The Karoo-Namib region. In: Werger MJA (ed) Biogeography and ecology of Southern Africa, vol 1. Dr W Junk, The Hague
Werger MJA (1986) Karoo and southern Kalahari. In: Evenari M, Noy-Meir I, Goodall DW (eds) Ecosystems of the world 12B. Hot deserts and arid shrublands, B. Elsevier, Amsterdam
Western AR (1989) The flora of the United Arab Emirates. An introduction. UAE University, Al-Ain
Westoby M, Walker B, Noy-Meir I (1989) Opportunistic mangement for rangelands not at equilibrium. J Range Manage 42:266–274
White F (1983) The vegetation of Africa. UNESCO, Paris
Whittaker RH (1953) A consideration of climax theory: the climax as population and pattern. Ecol Monogr 23:41–78
Whittaker RJ, Fernández-Palacios JM (2007) Island biogeography: ecology, evolution, and conservation. Oxford University Press, Oxford
Whittaker RJ, Bush MB, Richards K (1989) Plant recolonization and vegetation succession on the Krakatau Islands, Indonesia. Ecol Monogr 59:59–123
Willson MF (1993) Dispersal mode, seed shadows, and colonization patterns. Vegetatio 107/108:261–280
Wilson JB, Lee WG (2000) C-S-R triangle theory: community-level predictions, tests, evaluation of criticisms, and relation to other theories. Oikos 91:77–96
Woldewahid G, van der Werf W, Sykora K, Abate T, Mostofa B, van Huis A (2007) Description of plant communities on the Red Sea coastal plain of Sudan. J Arid Environ 68:113–131
Wolfe AD, Randle CP, Liu L, Steiner KE (2005) Phylogeny and biogeography of Orobanchaceae. Folia Geobot 40:115–134
Wranik W (1999) Die Tierwelt der Insel Sokotra. In: Wranik W (ed) Sokotra. Mensch und Natur. Jemen-Studien Band 14. Dr Ludwig Reichert Verlag, Wiesbaden

Wranik W, Reddy KB, Raqeeb Thabet A (1986) Socotra Island. Contributions to the flora and fauna. Aden University, Aden, pp 1–150

Wullstein LH, Bruening ML, Bollen WB (1979) Nitrogen fixation associated with sand grain root sheaths (rhizosheaths) of certain xeric grasses. Physiol Plant 46:1–4

Yuan Y-M, Wohlhauser S, Möller M, Chassot P, Mansion G, Grant J, Küpfer P, Klackenberg J (2003) Monophyly and relationships of the tribe Exaceae (Gentianaceae) inferred from nuclear ribosomal and chloroplast DNA sequences. Mol Phylogenet Evol 28:500–517

Yuan Y-M, Wohlhauser S, Möller M, Klackenberg J, Callmander MW, Küpfer P (2005) Phylogeny and biogeography of *Exacum* (Gentianaceae): a disjunctive distribution in the Indian Ocean basin resulted from long distance dispersal and extensive radiation. Syst Biol 54:21–34

Zerbe S (1998) Potential natural vegetation: validity and applicability in landscape planning and nature conservation. Appl Veg Sci 1:165–172

Zohary M (1937) Die verbreitungsökologischen Verhältnisse der Pflanzen Palästinas. I. Die antitelechorischen Erscheinungen. Beih Bot Centralblatt 56:1–55

Zohary M (1973) Geobotanical foundations of the Middle East. Gustav Fischer Verlag, Stuttgart

Index of Scientific Plant Names

Scientific names of plant species, genera and families mentioned in the text. A full list of species occurring on Socotra is given in Appendix 1 (vascular plants), Appendix 2 (bryophytes) and Appendix 3 (lichens). * denotes image

DOI 10.1007/978-94-007-4141-6, © Springer Science+Business Media Dordrecht 2012

R

S

T

Subject Index

G. Brown and B.A. Mies, *Vegetation Ecology of Socotra*, Plant and Vegetation 7,
DOI 10.1007/978-94-007-4141-6,

D

T

MIX
Papier aus verantwortungsvollen Quellen
Paper from responsible sources
FSC® C105338

If you have any concerns about our products,
you can contact us on
ProductSafety@springernature.com

In case Publisher is established outside the EU,
the EU authorized representative is:
Springer Nature Customer Service Center GmbH
Europaplatz 3, 69115 Heidelberg, Germany

Printed by Libri Plureos GmbH
in Hamburg, Germany